Origin of Cosmic Rays

NATO ADVANCED STUDY INSTITUTES SERIES

*Proceedings of the Advanced Study Institute Programme, which aims
at the dissemination of advanced knowledge and
the formation of contacts among scientists from different countries*

The series is published by an international board of publishers in conjunction
with NATO Scientific Affairs Division

A	Life Sciences	Plenum Publishing Corporation
B	Physics	London and New York
C	Mathematical and Physical Sciences	D. Reidel Publishing Company Dordrecht and Boston
D	Behavioral and Social Sciences	Sijthoff International Publishing Company Leiden
E	Applied Sciences	Noordhoff International Publishing Leiden

Series C – Mathematical and Physical Sciences

Volume 14 – Origin of Cosmic Rays

Origin of Cosmic Rays

*Proceedings of the NATO Advanced Study Institute
held in Durham, England, August 26–September 6, 1974*

edited by

J. L. OSBORNE and A. W. WOLFENDALE

Physics Department, University of Durham, Durham, England

D. Reidel Publishing Company

Dordrecht-Holland / Boston-U.S.A.

Published in cooperation with NATO Scientific Affairs Division

Library of Congress Cataloging in Publication Data

NATO Advanced Study Institute, Durham, Eng., 1974.
    Origin of cosmic rays.

    Bibliography: p.
    1. Cosmic rays—Congresses. I. Osborne, J. L., ed.
Title. II. Wolfendale, A. W., ed. III.
QC484.8.N37   1974   539.7'223   75–2436

ISBN 90 277 0585 2

Published by D. Reidel Publishing Company
P.O. Box 17, Dordrecht, Holland

Sold and distributed in the U.S.A., Canada, and Mexico
by D. Reidel Publishing Company, Inc.
306 Dartmouth Street, Boston, Mass. 02116, U.S.A.

All Rights Reserved
Copyright © 1975 by D. Reidel Publishing Company, Dordrecht
No part of this book may be reproduced in any form, by print, photoprint, microfilm,
or any other means, without written permission from the publisher

Printed in the Netherlands by D. Reidel, Dordrecht

TABLE OF CONTENTS

Preface	VII
List of Participants	IX
Introductory Cosmic Rays A.W. Wolfendale	1
The Galaxy and the Interstellar Medium J.L. Osborne	13
Extragalactic Cosmic Rays G.R. Burbidge	25
Sidereal Daily Variations in Cosmic Ray Intensity and their Relationships to Solar Modulation and Galactic Anisotropies T. Thambyahpillai	37
Energy Spectrum and Mass Composition of Cosmic Ray Nuclei from 10^{12} to 10^{20} eV A.A. Watson	61
Nuclear Mass Composition at 'Low' Energies (i.e. $<10^{12}$ eV/nucleon) I.L. Rasmussen	97
Nucleosynthesis and Galactic Cosmic Rays H. Reeves	135

Galactic Structure, Magnetic Fields and Cosmic 165
Ray Containment

K.O. Thielheim

Galactic Propagation of Cosmic Rays Below 10^{14}eV 203

J.L. Osborne

Possible Explanations of the Spectral Shape 221

A.W. Wolfendale

The Cosmic Ray Electron Component 233

P. Meyer

Gamma Ray Astrophysics 267

F.W. Stecker

Observation of Celestial Gamma Rays 335

K. Pinkau

Collapsed Stars, Pulsars and the Origin 371
of Cosmic Rays

F. Pacini

Historical Searches for Supernovae 399

F.R. Stephenson

Supernovae and the Origin of Cosmic Rays (I) 425

S.A. Colgate

Supernovae and the Origin of Cosmic Rays (II), 447
a Model of Cosmic Ray Production in Supernovae

S.A. Colgate

PREFACE

The cosmic radiation was discovered over 60 years ago and since that time many exciting results have appeared, results of interest to a variety of branches of science, primarily Nuclear Physics, Astrophysics and Geophysics. Perhaps the most basic aspect is the intriguing one of the origin of the radiation and the present publication comprises a reproduction of the lectures on the topic given at the Advanced Study Institute held from August 26 to September 6, 1974 in Durham, England.

We endeavoured to organise the Institute as a School as distinct from a Conference and the lecturers cooperated to the full in making their contributions very clear and comprehensive. In producing the manuscripts for these Proceedings the lecturers were greatly assisted by the Scientific Secretaries (Research Students from our Department, listed overleaf) and we are very grateful to them for their efforts.

Many members of the Physics Department and staff of St. Mary's College (where participants were accommodated), helped with the organisation and we thank them most sincerely. Our particular thanks go to our Departmental Superintendent, Mr. C.F. Cleveland, for his many contributions to the successful outcome of the meeting. We are grateful to Mrs. S.M. Naylor for her help in preparing this printed version of the Proceedings from the manuscripts.

Durham, October 3, 1974

J.L. Osborne
A.W. Wolfendale

Institute Directors: J.L. Osborne and A.W. Wolfendale

Physics Department
University of Durham, Durham, England

Advisors: G.R. Burbidge
University of California
San Diego, U.S.A.

H. Elliot
Imperial College, London, England

P. Meyer
University of Chicago, U.S.A.

H. Reeves
Centre d'Etudes Nucleaires de Saclay
France

Scientific Secretaries: G.J. Dickinson A.W. Strong
D. Dodds M. White
D.K. French D.M. Worral
K. Pimley

We are very grateful to the Scientific Affairs Division of NATO for sponsoring this Advanced Study Institute.

LIST OF PARTICIPANTS

Lecturers

Burbidge, G.R., University of California, San Diego
Colgate, S.A., New Mexico Institute of Mining and Technology
Meyer, P., Chicago
Osborne, J.L., Durham
Pacini, F., Frascati
Pinkau, K., Garching
Rasmussen, Lyngby
Reeves, H., Saclay
Stecker, F.W., Goddard Space Flight Center
Stephenson, F.R., Newcastle upon Tyne
Thambyahpillai, T., Imperial College, London
Thielheim, K.O., Kiel
Watson, A.A., Leeds
Wolfendale, A.W., Durham

Participants

Arens, M., Amsterdam
Ashton, F., Durham
Barrett, M.L., Leeds
Bazer-Bachi, A.R., Toulouse
Bignami, G.F., Goddard Space Flight Center
Burn, B.J., I.O.A., Cambridge
Capelato, H.V., Saclay
Carter, P.D., North London Polytechnic
Cetincelik, M., T.N.E.K., Ankara
Cherki, G., Saclay
Cherry, M.L., Chicago
di Cocco, G., Bologna
Edelstein, W.A., Glasgow
Felten, J.E., Steward Observatory
Frangos, A.C., Athens
Fuchs, B., Kiel
Goned, A.M.S., Cairo
Gregory, J.C., Alabama

Grupen, C., Gesamthochschule Siegen
Hainebach, K.L., Rice University
Hermsen, W., Leiden
Hillas, A.M., Leeds
Jokisch, H., Kiel
Juliusson, E., Chicago
Karr, G.R., Alabama
Kidd, J.M., Naval Research Laboratory, Washington
Kinaci, S.R., Ege Universy
Lindstam, S., Lund
Linsley, J., University of New Mexico
Mandolesi, N., Bologna
Martin, J.W., University of Washington
McIvor, I., D.A.M.T.P., Cambridge
Meikle, W.P.S., Glasgow
Meneguzzi, M., Saclay
Moyano, C.E., Garching
Nelson, A.H., Cardiff
Nissen, D., Kiel
O'Sullivan, C.T.O., Cork
Paizis, C., Milan
Pizzichini, G., Bologna
Plieninger, T., Heidelberg
Robba, N., Palermo
Robson, E.I., Queen Mary College, London
Rothermel, H., Garching
Sacco, B., Palermo
Schlickeiser, R., Kiel
Schmidt, W., Garching
Schwartz, S.J., D.A.M.T.P., Cambridge
Simon, M., Garching
Skilling, J., D.A.M.T.P., Cambridge
Sun, M.P., Imperial College, London
Thompson, M.G., Durham
Tumer, T., M.E.T.U., Ankara
Turtelli, A., Jr., Unicamp
Turver, K.E., Durham
Valtaoja, E.J.J., Turku
Verma, S.D., Louisiana
Wdowczyk, J., Lodz
Westergaard, N.J., Lyngby
Wild, P., Leeds
Wilson, A.S, Sterrewacht te Leiden

INTRODUCTORY COSMIC RAYS

A.W. Wolfendale

Physics Department, University of Durham, U.K.

I. SCOPE OF THE PRESENT WORK

The history of cosmic ray studies is one of the romances of modern science. From the observations of a small residual ionization in carefully shielded ionization chambers at the turn of the Century sprang the development of a subject embracing many disciplines of physics and leading to fundamental advances in knowledge in many areas.

Despite the identification of the particles present in the cosmic ray beam and a host of measurements on the energy spectra of the components and of the manner in which the particles propagate through the atmosphere and below ground the origin of the bulk of the primary particles is still unclear. Only at 'low' energies, below about a GeV has it been possible to identify the sun as a source of some of the particles. The subject of solar cosmic rays, a topic of great interest, can be regarded as a subject in its own right and attention will not be given to it here; instead, the origin of those components, largely of higher energy and coming from more distant sources, is our prime concern.

It is necessary to state rather clearly what is known about the various primary components and also about Galactic and Extra-galactic space before endeavouring to suggest specific origin models. The first Chapter gives a very brief introduction to the question of the primary components; later Chapters give much more detail and deal with the Astronomical setting.

II. THE PRIMARY COMPONENTS

1. Definitions

The term 'primary' component is taken to mean the component present above the atmosphere, that is before any secondary interactions in the gas of the atmosphere have taken place. In view of the presence of the earth's magnetic field the intensity of the charged particles will depend on the latitude; when the 'primary spectrum' is quoted, corrections have usually been applied to allow for this field and the spectrum then refers to what would have been observed were the earth's field switched off.

An idea of which particles might be expected to be present in the primary beam comes from an analysis of the 'universal abundances' - data which come from studies of stellar spectra, meteorites, etc. These abundances include a wide variety of nuclear masses with hydrogen as the biggest component; the expectation is borne out with the cosmic ray beam being mainly populated by protons, at least at energies below about 10^{13}eV where direct measurements of primary masses have been made. There are notable differences between the concentrations of various elements in the primary beam and the Universal abundances, however. Some of these are probably due to the cosmic rays at their places of acceleration (the so-called primordial particles) not being representative and others are certainly due to propagation effects. The significant flux of Li, Be and B in the primary beam is virtually certain to be due to the fragmentation of heavier nuclei in their passage through interstellar matter.

In addition to the nuclei, primary electrons and positrons have been identified, as have γ-rays. Neutrinos and some neutrons will also be present but these have not, as yet, been identified.

2. Energy Density

Some idea of the astrophysical significance of the various components present can be gauged from their energy densities and values are given in Table I. For comparison, energy densities near the earth of other components (visible light, etc.) are also shown.

Component		Energy Density (eV cm^{-3})
Charged primaries (from summary by Wolfendale, 1973)	above 10^9eV 10^{12}eV 10^{15}eV 10^{18}eV	$\sim 5.10^{-1}$ $\sim 2.10^{-2}$ $\sim 10^{-4}$ $\sim 10^{-8}$
Electrons and positrons (from summary spectrum of Meyer, 1971)	above 10^9eV 10^{10}eV 10^{11}eV	$\sim 4.10^{-3}$ $\sim 1.10^{-3}$ $\sim 2.10^{-4}$
γ-rays, diffuse background (from summary by Strong et al., 1974)	above 10^7eV above 10^8eV	$\sim 1.10^{-5}$ $\approx 2.10^{-6}$
Starlight (from Allen, 1974)		$\sim 4.10^{-1}$
2.7K Black body radiation		$2 \cdot 4.10^{-1}$

Table I: Energy Densities of 'Cosmic' Components near the Earth calculated from the expression
$$\sigma = \frac{4\pi}{c} \int j(E) \, EdE,$$
where $j(E)$ is the differential energy spectrum of the appropriate component. By 'charged primaries' is meant protons and heavier nuclei.

Examination of the cosmic ray components alone, above the same energy, say 10^9eV, shows that there is a wide disparity between their energy densities. Thus, with respect to the charged primaries, (by which is meant protons and heavier nuclei) the electrons and positrons carry about 10^{-2} of the energy and the diffuse γ-ray background carries $\approx 10^{-7}$ of the energy. At this point it should be remarked that, for γ-rays, there is also a component carrying more energy which is of galactic origin; that this is galactic is shown by the fact that there is a very broad peak towards the galactic centre (with a width in longitude of very approximately $\pm 40°$; Fichtel et al., 1973). However, the increase in γ-ray energy density is probably less than an order of magnitude.

Although there is a disparity between the energy densities of the individual cosmic ray components, rather remarkable near-

coincidences occur between the value for charged primaries
($\sim 5 \times 10^{-1}$ eV cm^{-3}) and some energies relevant to the galaxy.
Thus, the energy density of starlight is $\sim 4 \times 10^{-1}$ eV cm^{-3} and
that associated with the galactic magnetic field is
$\simeq 6 \times 10^{-1}$ eV cm^{-3} for a mean field of 5µgauss (actually the mean
field is probably uncertain to a factor 2 and therefore the energy
density to a factor 4). Furthermore, the energy density associated
with the motions of gas clouds in the galaxy, averaged over the
nearby galactic region, is also in the range (1-10) $\times 10^{-1}$ eV cm^{-3}.
Such near agreement is suggestive of an equilibrium situation for
a system in which the bulk of the cosmic rays originate in
galactic sources but there can be no certainty about it. For
example, the black body radiation also has the same order of energy
density (2.4 $\times 10^{-1}$ eV cm^{-3}) and this radiation is not of galactic
origin but almost certainly pervades the whole Universe.

3. Energy Spectra

Although later Chapters will deal with the energy spectra in
much greater detail a brief summary is given here.

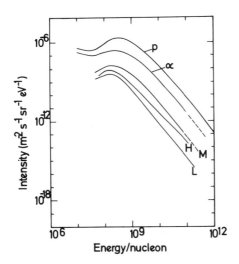

Fig. 1: Summary of measurements on the primary spectrum of
protons and nuclei corrected for geomagnetic effects. The summary
is that given by Wolfendale (1973) modified to allow for more
recent data which indicate that the iron spectrum probably has a
somewhat smaller exponent than the other components.

The groupings of nuclei are as follows: L: $3 \leqslant Z \leqslant 5$;
M: $6 \leqslant Z \leqslant 9$ and H: $10 \leqslant Z$.

Figure 1 gives a rather superficial summary of measurements on the primary spectrum of protons and nuclei; the ranges of nuclear charge corresponding to L (light), M (medium) and H (heavy) are given in the caption. Nuclei of charge much above 26 have also been observed but there are insufficient data as yet to give a good energy spectrum (what evidence there is does not suggest much of a difference from that of the other heavy nuclei). Related to the flux of the iron group, the heaviest nuclei, with $Z > 96$, have a flux of about 5×10^{-7} (Blandford et al., 1971).

As can be seen in Figure 1 there is some evidence for the spectrum of the H-group being somewhat flatter than that of the other components. In fact, there is a suggestion that a number of groups of nuclei have a smaller exponent than that of protons For example, Balasubrahmanyan and Ormes (1973) quote differential exponents of 2.56 ± 0.04 for carbon and oxygen, 2.44 ± 0.07 for $10 < Z < 14$ and 2.0 ± 0.14 for iron, the energy range in each case being about 1-100 GeV/nucleon. These exponents can be compared with the value of $2.6 - 2.7$ for protons and α-particles.

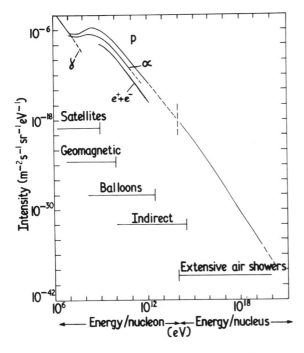

Fig. 2: Summary of measurements of some of the primary components. Protons and α's as in Fig. 1, $e^+ + e^-$ and γ are from the references given in Table I.

Extending the region of interest to higher energies, Figure 2 shows the whole energy range; the techniques used in the measurements are also indicated. The spectra of protons and α particles alone are indicated below about 10^{12}eV. At energies above this value the evidence as to the mass composition of the nuclei becomes increasingly vague. As indicated in the Figure, above about 10^{15}eV the spectrum is derived from measurements on extensive air showers; these are essentially calorimetric and the total energy brought in by the primary nucleus is determined.

Figure 2 shows that the primary spectrum of all nuclei summed together can be written in the form $j(E) \propto E^{-\gamma(E)}$ where $\gamma(E)$ is approximately constant (at $\sim$ 2.6) for $10^{10} < E < 3.10^{15}$eV and has a different value ($\sim$ 3.2) for $E > 3.10^{15}$eV. In fact, the actual situation may be somewhat more complicated than this, as will be seen in a later Chapter.

Also shown in Figure 2 are the spectra of electrons and positrons, and the diffuse γ-ray spectrum; it was these spectra which gave rise to the energy densities given in Table I.

III. INTERACTIONS OF PRIMARIES IN THE GALAXY AND BEYOND

1. General Remarks

The previous section has dealt with the form of the energy spectrum of the various primary components. This information, together with such data as are available on the directional properties of the components, provides the basis on which origin theories are erected.

Clearly, a distinction must be made between the interactions for γ-rays, which propagate in straight lines, and might be expected to give rather specific clues as to the origin and those for charged particles, which are subjected to the deflections caused by the various magnetic fields in space. In the following sections a brief analysis of the more important interactions will be given.

2. Interactions of Charged Particles

The interactions can be divided into those with magnetic fields, matter and with radiation. Figure 3 summarises the situation for protons and heavier nuclei

The relation in Figure 3 indicates that, in a typical galactic field of 5μgauss, a proton of momentum 10^{18}eV/c has a Larmor radius of $\sim$ 200 pc, i.e. about the half-thickness of the galactic disc in our vicinity. This means that at momenta much

INTRODUCTORY COSMIC RAYS

below this value (say $\lesssim 10^{16}$ eV/c) the galactic field and its irregularities are likely to smear out the arrival directions of particles which may have been produced in specific galactic sources unless these sources are comparatively local. Only at much higher momenta will specific galactic sources be likely to stand out when 'viewed' with charged particles.

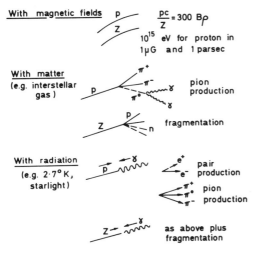

Fig. 3: Types of interaction of cosmic ray protons and nuclei.

Nucleons and nuclei interacting with interstellar gas atoms generate pions and the usual mixture of other secondaries (kaons, etc.) just as they do when they interact with the gas atoms in the earth's atmosphere. It is almost certain that the galactic γ-rays referred to earlier originate in this way although whether the increased γ-ray emissivity which appears to exist at $\sim$5 kpc from the galactic centre comes from an increased rate of cosmic ray proton production in this region (e.g. Stecker et al., 1973), or whether it is due to an excess of gas there (molecular hydrogen etc.) is still an open question (Dodds et al., 1974). It is likely that some, at least, of the primary electrons and positrons (particularly those below 100 MeV) originate from the secondary γ-rays.

The fragmentation process is an important one for understanding the propagation of the cosmic ray components in the galaxy and many detailed studies have been made. From a knowledge of the relative numbers of the various secondary nuclei produced when heavy nuclei fragment, Shapiro et al. (1971 and later publications) have used the measured primary composition to work back to the primordial composition, i.e. the relative numbers of nuclei. Figure 4 gives a brief summary of the situation, where the primordial abundances are given with respect to those in the solar photosphere. It should be remarked that this area is one in which there is much contemporary activity: later Chapters give a more up to date account.

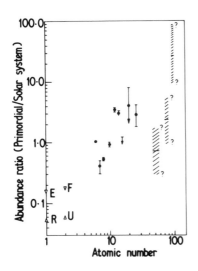

Fig. 4: The ratio of primordial cosmic ray abundances to those in the solar system (photosphere) plotted against atomic number, Z. The figure is taken from the work of Shapiro et al. (1971). Carbon is adopted as the datum. The ratios shown for hydrogen E and R assume that the primordial spectrum follows a power law in energy or rigidity respectively. U denotes the ratio for helium using measurements of solar spectra by Unsold (1969) and F denotes that from measurements by Biswas et al. (1966) of energetic solar particles. The shaded areas for heavy nuclei are based on the measurements of Fowler et al. (1970).

Turning to interactions with radiation fields, these are presumably of importance in the vicinities of sources; for example, near pulsars the intense radiations will quickly cause fragmentation of such heavy nuclei as may be initially accelerated to energies above threshold (threshold being such as to give photon energies as "seen" by the nucleus of several MeV). On the galactic scale the interactions with starlight and the black body radiation (2.7 K) are not serious because the interaction lengths are very long; however in extragalactic space they do achieve importance and this aspect is considered later.

Finally, concerning electrons in the galaxy, their low mass

causes interactions by way of bremsstrahlung, Inverse Compton effect and synchrotron radiation to be important. The more important ones, away from regions of high matter density, are the latter two and it can easily be shown that the degradation of energy is severe above an energy E_c given by

$$E_c = \frac{2 \times 10^{19} mc^2}{WT},$$

where T is the lifetime in the region in question and W is the combined energy density of photons and magnetic field. Using the data of II.2 gives $W \simeq 1.2$ eV cm^{-3}. Independent measurements of T for electrons have not yet been made but there is some evidence for protons and nuclei, from the measured relative abundances referred to earlier, which suggest that the mean lifetime is 10^{14}s (this value comes from a mean path length of 5 g cm^{-2} taken together with an assumed mean interstellar density of 1 atom cm^{-3}). If electrons have the same mean life then the equation above gives $E_c \simeq 8.10^{10}$ eV.

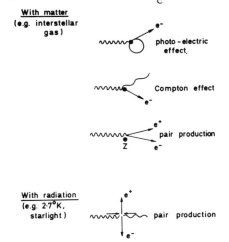

Fig. 5: Types of interaction of photons

Although this is not the place to examine this point in detail it is interesting to note that the measured electron spectrum (Figure 2) does not show any signs of the expected more rapid fall above this value of E_c; it is apparent that, unless the experimental data are inaccurate, either W or T is in error. It is probable that T is the culprit, i.e. that electrons and nuclei are not derived from the same sources and do not remain in the galaxy for the same length of time.

3. Interactions of Photons

Photons interact with matter and other photons in a variety of ways, the most important of which, are shown in Figure 5.

The photons interact with the material near their sources and the interstellar medium in general in a similar fashion to that of protons and nuclei. Confining attention to the interstellar medium the photons with energy in the region of tens of eV have comparatively short ranges whereas above about 5 keV the photons can penetrate from outside the galaxy with little attenuation.

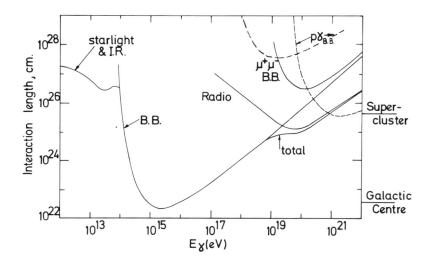

Fig. 6: Interaction length against photon energy for collisions of photons with photons of the various radiation fields (after Wdowczyk et al., 1972). The process concerned is e^+e^- production except where indicated otherwise.

The interaction length for proton - 2.7K black body photons is also shown (it is denoted $p\gamma_{BB}$).

The adopted energy densities are: starlight, 10^{-2} eV cm^{-3}; infra-red, 2.4×10^{-2} eV cm^{-3}; 2.7K, 2.4×10^{-1} eV cm^{-3} and radio, 5×10^{-8} eV cm^{-3}.

As mentioned earlier, photon-photon collisions are characterised by very long interaction lengths and are thus of much greater importance for propagation in extragalactic space. Figure 6 gives the variation of interaction length with photon energy for the photons in the various radiation fields; starlight, infra-red, 2.7K and radio. The assumed energy densities are indicated in the caption to the Figure. The attenuation lengths can be viewed against the following dimensions:

thickness of galactic disc $\simeq 400$pc $\simeq 10^{21}$cm.
distance to galactic centre $\simeq 3.10^{22}$cm.
dimension of Supercluster of galaxies $\simeq 6.10^{25}$cm.

Clearly, for energetic γ-ray propagation within the Supercluster attenuation by way of interactions with the ambient radiation (principally 2.7K radiation) must be allowed for.

At this stage, the interactions of protons with the extragalactic fields can also be mentioned. Figure 6 shows the situation for the most important interaction, that with the black body radiation, in which pions are generated. It can be seen that as with the case of energetic photons, interactions on the scale of the Supercluster and beyond are of importance.

IV. SUMMARY

The previous sections have briefly reviewed the main properties of the various primary cosmic ray components, with particular reference to their energy spectra and energy densities. Some attention has also been given to the interaction processes experienced in the Galaxy and beyond by both particles and γ-rays. Later Chapters will deal with all these matters in much greater detail, with the object of deriving as many clues as possible as to the likely sources of the components.

REFERENCES

Allen, C.W., Astrophysical Quantities (Athlone Press), 1973.

Balasubrahmanyan, V.K. and Ormes, J.F., Astrophys. J., 186, 109, 1973.

Biswas, S., Fichtel, C.E. and Guss, D.E., J. Geophys. Rev. 71, 4071, 1966.

Blandford, G.E., Proc. 12th Int. Conf. on Cosmic Rays (Hobart; Univ. of Tasmania) 1, 269, 1971.

Dodds, D., Strong, A.W., Wolfendale, A.W. and Wdowczyk, J., Nature, 1974 (in the press).

Fichtel, C.E., Roy. Soc. Meeting on Origin of Cosmic Rays, 1974 (in the press).

Fowler, P.H. et al., Proc. Roy. Soc. A318, 1, 1970.

Meyer, P., Proc. 12th Int. Conf. on Cosmic Rays (Hobart; Univ. of Tasmania) rapporteur paper, 1971.

Shapiro, M.M., Sillberberg, R. and Tsao, C.H., Proc. 12th Int. Conf. on Cosmic Rays (Hobart; University of Tasmania) 1, 221, 1971.

Strong, A.W., Wdowczyk, J. and Wolfendale, A.W., J. Phys. A., 7, 120, 1974.

Unsold, A.O.J., Science, 163, 1015, 1969.

Wdowczyk, J., Tkaczyk, W. and Wolfendale, A.W., J. Phys. A. 5, 1419, 1972.

Wolfendale, A.W., Cosmic Rays at Ground Level, Ed. A.W. Wolfendale (The Institute of Physics, London) 1, 1973.

THE GALAXY AND INTERSTELLAR MEDIUM

J.L. Osborne

Department of Physics
University of Durham, U.K.

I INTRODUCTION

The purpose of this chapter is to describe the galactic setting for the propagation and possible origin of cosmic rays. A description of the overall structure and constituents of our Galaxy is followed by an account of the physical properties and distribution of the interstellar medium. An attempt is made to solve the problem of covering such a broad subject in a brief review by concentrating on those features which are judged to have a direct bearing on the origin and propagation of cosmic rays. Even so it is not feasible to support the text fully with references. A suggested reading list is therefore given at the end.

II STRUCTURE OF THE GALAXY

1. Total Mass

The matter of the Galaxy is in the form of stars, dust and gas, distributed mainly in a flat disc. The total mass of the Galaxy is 2.10^{11} $M_\odot$. The greatest fraction of the matter is in its approximately 10^{11} stars. The total mass of gas in the form of atomic hydrogen, estimated from the 21cm wavelength emission, is about $5 \cdot 10^9$ $M_\odot$. In addition there is molecular hydrogen, helium and heavier elements, which may bring the mass of gas up to 10% of the total mass of the Galaxy. As is shown below the ratio of gaseous to stellar matter varies greatly with distance from the galactic centre. The distribution of dust, on the other hand, is similar to that of gas. The smoothed space density in dust is about 1% of that of the observed gas.

2. The Position of the Sun

The sun lies within 12pc of the galactic plane
(1pc = 3.09 10^{18}cm). The centre of the Galaxy is in the
constellation Sagittarius at 265.6° R.A., - 28.9° dec. Its
distance can be found by estimating the distance of RR Lyrae
variable stars in the central condensation. The currently
accepted value of 10 kpc has an uncertainty of about 10%. (Prior
to 1963 a value of 8.2 kpc was widely used).

A system of Galactic Coordinates (longitude l, latitude b)
centred on the sun is used to give directions in the Galaxy. The
equator of the system is the galactic plane and l = 0° is the
galactic centre. The earth's north pole points to l = 123.0°,
b = 27.4°. With respect to the average motion of nearby stars
the sun is moving at 20 km s^{-1} towards l = 60° b = 24°.

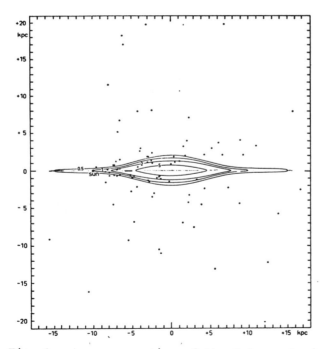

Fig. 1: A cross-section of the Galaxy showing the distribution
of matter. (After Oort (1965)).

3. Distribution of Stars

In describing the structure of the Galaxy the coordinates, R, distance from the galactic centre, and Z, perpendicular distance from the plane are used. In Figure 1 a cross section of the Galaxy is given. The contours show the distribution of mass density in stars in units of the smoothed density near the sun. The diameter of the disc is about 30kpc and the sun is thus at 2/3 of the radius, from the centre. The black spots indicate the positions of globular star clusters projected on to the plane through the sun and the galactic centre. These, the oldest stellar systems, form a roughly spherical halo of radius 15kpc. The surface density of all stars projected on to the galactic plane, is strongly peaked towards the galactic centre. Beyond R = 1kpc the surface density falls approximately as exp (-R/4kpc).

The Z-distribution of the various classes of stars and constituents of the interstellar medium differ one from another. The 'thickness' of the disc for a particular class is usually expressed as the equivalent thickness ($2Z_{eq}$) i.e. the total thickness that the disc of the objects would have if it were a slab of uniform density, or the distance between half-density points ($2Z_{\frac{1}{2}}$). The relation between these depends of course on the form of the Z-distribution. For a gaussian $Z_{\frac{1}{2}} = 0.94\ Z_{eq}$ while for an exponential $Z_{\frac{1}{2}} = 0.69 Z_{eq}$. In what follows the word 'thickness' implies $2Z_{\frac{1}{2}}$.

Class of Stars	$2Z_{\frac{1}{2}}$ (pc)
O	70
A	160
G,K,M	480
White Dwarfs	700
Pulsars	380
Supernova Remnants	125

Table I: The total thickness of the galactic disc for various classes of stars and for supernova remnants.

The thickness of the disc in the neighbourhood of the sun for some representative classes of stars are shown in Table I. The thickness increases in the sequence from the young, I stars, through the solar type (G) to white dwarfs.

To date approximately 80 pulsars are known in the Galaxy and for each the dispersion measure, $\int n_e\ dl\ cm^{-3}$ pc, has been found. The integral is along the line of sight to the pulsar and n_e is the density of thermal electrons in the interstellar medium along that line. If the average value of n_e is known the spatial distribution of pulsars follows. The thickness of the pulsar disc in units of dispersion measure is approximately $9.6 cm^{-3}$ pc. The thickness quoted in Table I corresponds to $n_e = 0.025$ cm^{-3} (see below). The Z-distribution of pulsars need not correspond to that of their parent stars because they may have been formed with high velocities. With the above value of n_e the observed pulsars lie within about 5kpc of the sun. The total number in the Galaxy is estimated to be between 10^5 and 10^6.

The thickness of the disc of supernova remnants is also given. About 100 remnants have been identified by their radio emission (17 have optical counterparts). Using those from which independent estimates of the distance are available, an empirical relation between surface brightness and diameter is obtained. This is applied to all the remnants to obtain the spatial distribution (Milne 1970, Ilovaisky and Lequeux 1972). The radial surface density distribution is flat out to R = 8kpc, falls between 8 and 12kpc and is very low beyond 12kpc. For R < 8kpc the disc of supernova remnants is half as thick as at the sun.

4. The Spiral Structure

Most external disc-shaped galaxies have a structure of a spherical or barred nucleus surrounded by spiral arms in varying degrees of development. The spiral tracers in these galaxies, at visual wavelengths, are O and B type stars, regions of ionised hydrogen gas and dust lanes. For our own Galaxy, because of absorption at visual wavelengths by dust (about 2.2 magnitudes per kpc in the plane) visual observations of the structure are limited to a few kpc. The large scale spiral structure is deduced from 21cm radio emission from neutral hydrogen. The ground state of the hydrogen atom is split into 2 levels with the electron and proton spins either parallel or antiparallel. The transition gives radiation of a frequency 1420.4MHz. The natural width of the line is negligible and it is possible to measure Doppler shifts down to 0.2 km s^{-1}. The intensity of radiation of a given frequency, I_ν, is usually expressed as a brightness temperature, $T_B = (c^2/2\nu^2 k) I_\nu$. If, over the 21cm line profile, the optical thickness is much less than unity the brightness temperature is proportional to the column density of hydrogen atoms. If the motion of the gas about the galactic centre is purely rotational and the rotation curve of the Galaxy is known the distance of a portion of the gas can be deduced from its radial velocity. The rotation curve of the Galaxy, for $R<R_\odot$, the solar radius, can be measured (in a given direction the maximum radial velocity is that of gas at $R = R_\odot \sin l$). Outside $R_\odot$ it is

deduced from a model of the Galaxy. Figure 2 shows such a curve. The rotational velocity in the neighbourhood of the sun is (250 ± 38) km s^{-1} in the direction $l = 90°$. The period of rotation at the sun is thus $2.5\ 10^8$ yr.

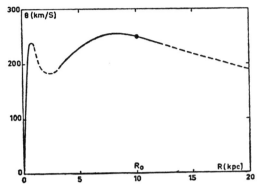

Fig. 2: Rotation curve of the Galaxy, uncertainties are typically $\pm$ 10% (After Lequeux 1969).

From 21cm surveys, maps of the spiral structure can be built up. Figure 3 is a composite map showing the position of bright ridges of emission. The next outer arm to the sun is the Perseus arm at the next inner is the Sagittarius arm. The average pitch angle of the arms is $12.5°$. The sun lies within the so-called Orion arm. Stellar and radio observations show this to have a pitch angle of $20°$. It is apparently not a major arm but an offshoot of the Sagittarius arm. The form of the arms indicates that the Galaxy is of a type intermediate between Sb and Sc.

The central region of the Galaxy has a complex structure. An interpretation of the 21cm data is shown in Figure 4. The inner, partial or complete, rings of neutral hydrogen are rotating and simultaneously expanding with the velocities shown. The present kinetic energy of expansion of the 3kpc arm is 10^{53} erg.

To produce this by a single explosion would require the ejection of $10^8 M_\odot$ 10^7 yrs ago with a total energy of $3\ 10^{58}$ erg. The nuclear disc of neutral hydrogen which has radius 500 pc is also expanding. It has within it a partial ring of molecular hydrogen. At the galactic centre is the radio source. Sag, A, a 6 pc diameter source of synchrotron emission. Infra-red radiation at 2.2 µm indicates that the density of normal stars rises to 10^6 pc^{-3} within 1 pc of the galactic centre. It has been suggested that there is a massive black hole at the centre of the Galaxy. The observed motion of the gas sets an upper limit to its mass of $2\ 10^8\ M_\odot$.

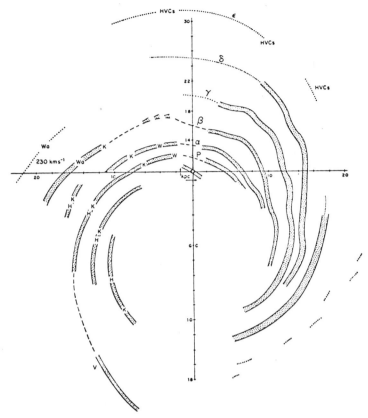

Fig. 3: A composite map of the outer spiral structure of the Galaxy from 21cm data by Verschuur (1973). Inner arms are not shown because of the large uncertainty in their distance.
Key: K = Kerr (1970), H = Henderson (1967), Wa = Wannier et al (1972), W = Weaver (1970), V = Verschuur.

Earlier radio surveys at high galactic latitudes were interpreted as evidence for a halo of dimensions comparable to the volume occupied by the globular clusters containing magnetic field and relativistic electrons. Further surveys with improved angular resolution showed that a large proportion of the emission came from features in the galactic disc. There is still some evidence for a 'radio disc' of approximate radius 10kpc and thickness up to 5 kpc.

The persistence of spiral structure in galaxies in spite of the differential rotation can be accounted for by the density wave

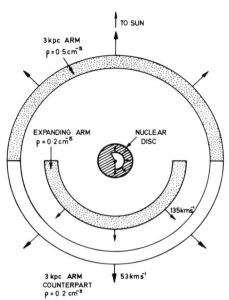

Fig. 4: Schematic diagram by Sanders and Wrixon (1973) of the central region of the Galaxy.

theory. The spiral tracers are produced due to the compression of the gas as it traverses the density wave. In the Galaxy the rotational velocity of the wave pattern, independent of R, would be 13.5 km s^{-1} kpc^{-1} (Lin 1970). At the solar position the wave then moves with about half the speed of the gas and stars. This accounts for the appearance of the dust lanes on the inner side of the bright spiral arm. It would take (3 to 10) 10^7 yr for the sun to traverse a density wave. The model requires an amplitude of variation of total mass surface density of 10% and an arm to interarm gas density ratio of 5.

III PROPERTIES OF THE INTERSTELLAR MEDIUM

1. Constitution of the Interstellar Medium

The interstellar medium consists of gas, dust and cosmic rays, which can be considered as far as the dynamics are concerned to be a hot tenuous gas. The interstellar flux of non-relativistic cosmic rays that can be important in heating and ionizing the gas cannot be measured directly because of the shielding effect of the solar wind. The chemical composition of the gas is, hydrogen 70%, helium 28%, heavier elements 2% by weight.

The commonly quoted figure for the gas density in the plane is 1 atom cm^{-3}. The gas is either neutral-atomic, ionised or molecular; these states and the variations in density are discussed below. The 'neutral' gas has a low level of ionisation (10^{-3}) due to cosmic rays and photo-ionisation of Ca and Na atoms. This is sufficient to 'freeze' the magnetic field to the gas. The dust grains will also be charged and tied to the field. The galactic magnetic field is described in the article by Thielheim. It appears to have a regular component of strength 3µG lying along the local spiral arm, or spur, and an irregular component of the same magnitude. The dust grains which have radii of the order of 0.1µm probably have no direct effect on cosmic rays. The suggestion that the soft X-ray background in the Galaxy is due to transition radiation from the grains could only be true if there were a very large energy density (10eV cm^{-3}) in low energy cosmic rays.

2. The Neutral Gas

The neutral gas, predominantly hydrogen, can be termed H_I. Because of the way that the rate of heating of the gas depends upon its density there are two stable phases of the gas: hot low density ($n_H < 10^{-1} cm^{-3}$, T > 6000K) and cool high density ($n_H > 1 cm^{-3}$, T < 300K). Observations are consistent with there being cool clouds in a hot intercloud medium. The following properties of 'standard clouds' are quoted: diameter 10pc, number along line of sight 5 kpc^{-1}, density $8cm^{-3}$. It must be stressed however that the interpretation of the 21cm data in terms of such clouds is by no means unique. Ideally 6 space and velocity coordinates of the gas are needed to define a cloud while only the directions (l and b) and radial velocity (V_r) in fact are known. For a given direction the line $T_B(V_r)$ is analysed into gaussian components and searches are made for corresponding components in a line in an adjacent direction. Typical internal velocities in the clouds are ~1 km s^{-1} while the 3 dimensional r.m.s. velocity of the clouds is ~11 km s^{-1}. The shape of the clouds, spheres or shells or sheets, is not known. Heiles (1967) has studied the cloud size spectrum. As well as a few 'standard' clouds he found many 'cloudlets' with radii 1 to 4 pc and densities ~2 cm^{-3}. From emission and absorption profiles the temperatures of the clouds are generally measured to be in the range 60K to 80K. Measurements of the intercloud medium give a few 10^3K. Falgarone and Lequeux (1973) give the following figures for the solar neighbourhood. Density in the plane: (a) for the intercloud medium 0.16 cm^{-3} (b) for a smoothed average of cloud material 0.29 cm^{-3}. Thickness of disc: (a) cloud and intercloud 360 pc (b) cloud only 310 pc (c) intercloud only 550 pc. The value of 0.45 cm^{-3} for the total density of neutral hydrogen is supported by satellite observations of the Lyman α profiles of stars. From measurements of absorption in the arm and interarm regions along the line of sight to strong radio sources the arm-interarm density ratio is estimated to be ~ 8.

A map showing the thickness of the disc of H_I gas (cloud and intercloud) by Jackson and Kellman (1973) is presented in figure 5. For 4.5 < R < 10 kpc the thickness is practically constant at ~250 pc reducing to 100 pc closer to the centre. As can be seen the thickness increases rapidly beyond 10 kpc except for an anomalous region near $l = 140°$. At high latitudes high velocity clouds are observed, the majority approaching the plane. These have been interpreted as material that is either condensing from the intergalactic medium or that is falling back after having been ejected. The radial surface density distribution of H_I

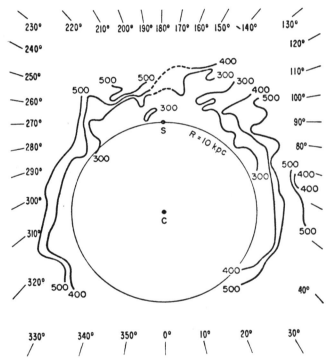

Fig. 5: A contour map of the thickness ($2Z_{\frac{1}{2}}$) of the H_I gas beyond R = 10 kpc (After Jackson & Kellman 1973).

is shown in figure 6. This peaks at 13 kpc in strong contrast to the stellar distribution. The radial distribution of the volume density of H_I at Z = 0 peaks at about R = 8 kpc but the decrease beyond 8 kpc is compensated by the increasing thickness of the H_I disc.

3. Ionised Gas

Spheres of ionised hydrogen, (H_{II} regions), surround young, hot O and B type stars. The gas is ionised completely out to a sharply defined boundary. The radius of the sphere varies as $n_H^{-2/3}$: for an O5 star it is 100 pc when n = 1cm^{-3}. The H_{II} regions can be seen optically or as thermal radio sources; the temperature of the gas is 10^4K. It is possible that they act as sources of most of the kinetic energy of the neutral gas. 'Giant' H_{II} regions (defined as those with 4 times the luminosity of the Orion Nebula) indicate sites of star formation and are useful as spiral tracers. The radial distribution of their density in the Galaxy is shown in figure 6.

The mean density of thermal electrons within a few kpc of the sun can be obtained from dispersion measures of pulsars whose distances are known independently, after discounting the effect

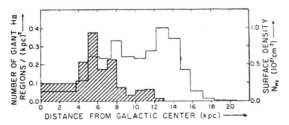

Fig. 6: Radial distributions of number density of giant H_{II} regions (Hatched) and surface density of H_I. (After Mezger 1970).

of any H_{II} region along the line of sight. Estimates in the range $n_e \stackrel{=}{} 0.025$ to 0.05 cm^{-3} have been obtained. The recombination rate of electrons and ions is roughly proportional to the square of the density. Thus at equilibrium $n_e^2 \propto n_{H_I}$. The thickness of the electron disc should then be greater than that of the neutral gas by a factor of 2 or $\sqrt{2}$ respectively for an exponential or gaussian Z-distribution. This is consistent with estimated values of 800 to 1000 pc, derived from rotation measures of extragalactic radio sources (Falgarone and Lequeux (1973)).

4. Molecular Hydrogen

Molecular hydrogen has no radio emission. It is possible that this 'invisible' hydrogen is an important constituent of the interstellar medium. Although H_2 cannot be directly photo-dissociated by radiation of longer wavelength than the Lyman limit

THE GALAXY AND INTERSTELLAR MEDIUM

there are indirect processes that limit n_{H_2}/n_H in a 'standard' cloud to 10^{-5}. In dense dust clouds, however the interior is shielded and the hydrogen should be mainly molecular. Recent observations of emission from CO molecules imply a peak in the distribution of H_2 at R = 5 kpc. This is very significant for the interpretation of γ-ray data. Details are given in the article by Stecker.

SUGGESTED READING

"The Interstellar Medium" Ed. by K. Pinkau, NATO ASI Series C6 (D. Reidel) 1974).

"Diffuse Matter in Space" by L. Spitzer (Interscience) 1968.

"The Interstellar Medium" by S.A. Kaplan and S.B. Pikelner (Harvard Univ. Press) 1970.

"The Spiral Structure of Our Galaxy" I.A.U. Symp. 38 Ed. by W. Becker and G. Contopoulos (D. Reidel) 1970.

"Interstellar Gas Dynamics" I.A.U. Symp. 39 Ed. by H.J. Habing (D. Reidel) 1970.

REFERENCES

Falgarone, E. and Lequeux, J., Astron. and Astrophys. 25, 253, 1973.

Heiles, C., Astrophys. J. Suppl. Ser. 15, 97, 1967.

Henderson, A.P., Unpublished Ph.D. dissertation, University of Maryland, 1967.

Ilovaisky, S.A. and Lequeux, J., Astron. and Astrophys. 18, 169, 1972.

Jackson, P.D. and Kellman, S.A., Astrophys. J., 190, 53, 1974.

Kerr, F.J., 'Spiral Structure of our Galaxy'(see above) p 95, 1970.

Lequeux, J., 'Structure and Evolution of Galaxies' (Gordon and Breach), 1969.

Lin, C.C., 'Spiral Structure of our Galaxy'(see above) p 377, 1970.

Mezger, P.G., 'Spiral Structure of our Galaxy' (see above) p 107, 1970.

Milne, D.K., Aust. J. Phys. 23, 425, 1970.

Oort, J.H., 'Galactic Structure' Ed. by A. Blaauw and M. Schmidt (University of Chicago) p 455, 1965.

Sanders, R.H. and Wrixon, G.T., Astron. and Astrophys. 26, 365, 1973.

Wannier, P., Wrixon, G.T., Wilson, R.W., Astron. and Astrophys. 18, 224, 1972.

Weaver, H.F., 'Spiral Structure of our Galaxy' (see above) p 126, 1970.

Verschuur, G.L., Astron. and Astrophys., 27, 73, 1973.

EXTRAGALACTIC COSMIC RAYS

G.R. Burbidge

Department of Physics
University of California, San Diego

I   INTRODUCTION

In these lectures I shall largely be concerned with the view that the bulk of the primary cosmic rays is of extragalactic origin. In order to set the stage for these ideas, a survey of the components of the extragalactic universe will be given. Then much of the discussion will centre about the various types of extragalactic non-thermal sources.

Throughout the discussion we assume that the Hubble constant $H_o = 50$ kms^{-1} Mpc^{-1}.

II   THE COMPONENTS OF THE UNIVERSE

1. Condensed Objects

(a) Galaxies. The most important component which we are certain about is simply the galaxies. They are discrete objects and appear to comprise the major mass-energy contribution. In general, they have very strong clustering tendencies. The mass-energy density present in the universe, due to galaxies, using the above value of H_o, is $\sim 10^{-31}$ g cm^{-3}.

The critical mass density required for a closed universe in which $q_o = 1/2$ is $\rho_{crit} = 5 \times 10^{-30}$ g cm^{-3} (this is also the value required in the steady state cosmology). Thus, the closure density is about 50 times the "observed" value for galaxies, and if a closed universe is accepted, the question of the missing mass is raised. None of the other components that will be mentioned and that are known to exist, give mass densities approaching that for galaxies. Some workers are therefore now suggesting, quite reasonably, that we may live in an open hyperbolic universe.

(b) <u>Quasars (= Quasi-stellar objects [QSOs])</u>. Quasars are also discrete objects which do exist and can be directly observed. It is not known what contribution they make to the mass density of the universe. There is an argument about their distances, but even if we could settle this, there is no direct way of measuring their masses. Consequently we do not know what contribution they could make, except to say that it is almost impossible that they make a very large contribution.

Large numbers have been detected through the radio and optical surveys. The total number down to about $19\overset{m}{.}5$ to 20^m probably amounts to no more than about 10^6. If they lie at cosmological distances, their space density is very low compared to that of galaxies. If they are comparatively local, their space density is very much higher, but there is every reason to believe that their masses are then quite small.

(c) <u>Highly Evolved or Collapsed Configurations</u>. A third contribution, for which we have no direct evidence, is that from various evolved or collapsed configurations. It is reasonable to suppose that some evolved galaxies and massive black holes are present. If we take the view that clusters of galaxies are bound relaxed systems, and this is likely for clusters like the Coma cluster, there must be a large amount of dark matter, because the kinetic energy of the visible galaxies is much greater than their potential energy. In one or two cases such as the Coma cluster it is difficult to argue that it is in any form other than collapsed or highly evolved matter.

(d) <u>Faint Low Mass Stars</u>. There may also be a very large population of very faint low mass star systems, such as globular clusters. Some at distances of a few hundred kiloparsecs have been detected by accident. It is not obvious that they are satellites of our Galaxy or of the Andromeda nebula. Such intergalactic globular clusters could be distributed extensively throughout the universe. However, it is hard to believe that they could make a very large contribution to the total mass of the universe.

2. Diffuse Matter

We pass on to the diffuse gas present in the universe. We are only able to set certain limits on the mass of gas that may be present in intergalactic space. It was originally thought that, if the gas is present, it is likely to be in the form of cold atomic hydrogen, a reasonable notion if we accept the idea that the universe may have expanded from an initially hot dense state.

The 21-cm line observations have been used to set a limit of $< 10^{-30}$ g cm^{-3} on the density of intergalactic matter. Attempts to detect the Lyman α absorption from quasars, assuming they are at cosmological distances, have set a lower limit of $< 10^{-35}$ g cm^{-3}. It can then be argued that, if one feels that there ought to be gas present, it must be hot gas, in which case one expects to see background bremsstrahlung X-rays. Some attribute part of the background X-ray flux to this, and it has been argued that one can reach the closure density in this way. It is also possible that the background flux is the integration of discrete sources. These sources may be clusters of galaxies, Seyfert galaxies or a new class of X-ray galaxies. If clusters dominate, the X-rays may either be due to Compton scattering or from hot gas in the clusters. But in any case the discrete source explanation for the background if accepted necessarily means that no evidence is available on intracluster gas.

We have no idea how to estimate the contribution of intergalactic dust to the mass-energy of the universe, except to say that dust cannot form unless comparatively heavy elements are available (at least carbon) and this would seem to imply that there cannot be an appreciable contribution.

3. The Electromagnetic Fields

The energy density of starlight is estimated to be $\sim 10^{-35}$ g cm^{-3}. (All energy densities of radiation will be quoted in mass units [1 eV cm^{-3} = 1.8x10^{-33} g cm^{-3}].) The contribution from the ultraviolet region is certainly not bigger than this by more than a factor of ~ 10. The ultraviolet part of the electromagnetic spectrum is one which has not been surveyed for faint discrete sources and it is likely that when this is done new classes of discrete ultraviolet sources will be discovered.

Radio waves contribute $\sim 10^{-36}$ g cm^{-3} due to the integration of the radio emission of all the discrete radio sources throughout the universe. The X-ray and γ-ray backgrounds are expected to contribute 10^{-37} g cm^{-3}. The microwave background is the largest component $\sim 10^{-33}$ g cm^{-3}. This assumes we are looking at a black-body spectrum of 2.7 K. The microwave measurements follow the black-body curve fairly well, but we may still be in for surprises as we come down to the shorter wavelength side of the curve. A discrete source component may be present, making a comparatively minor contribution. It is important to realise, however, that such a component would almost certainly arise from discrete non-thermal sources. Such sources would therefore generate cosmic electrons and protons.

4. Neutrino Flux

The limits that can be set on the low-energy neutrino flux are exceedingly poor. The possibility that the bulk of the mass-energy in the universe is due to neutrinos cannot be eliminated. The limits that can be set by looking at the end of the β-decay spectrum are about a factor of 10^3 to 10^4 higher than the value corresponding to the closure density.

5. Gravitational Waves

If it is assumed that the gravitational wave phenomenon that Weber is discussing is real and that the gravitational waves come from our Galaxy, then if all galaxies are radiating in a similar fashion, it is clear that a very large amount of mass-energy is present in the universe in the form of gravitational waves.

Almost certainly some gravitational waves are present, and they do make a contribution. However, without definite results from the experimental standpoint, one cannot make definite arguments.

6. Cosmic Rays

The final component to be mentioned is cosmic rays. Here we can go from the extreme universal hypothesis which says that the energy density is 10^{-33} g cm^{-3} to the extreme non-universal hypothesis which says only the highest energy cosmic rays are extragalactic giving an energy density value as low as 10^{-37} g cm^{-3}.

It is the generation of the cosmic rays in the galaxies and their interaction with some of the other components listed above that we shall be concerned with here.

III RADIO SOURCES

1. Normal Galaxies

It is generally assumed that the detection of non-thermal radio sources tells us that relativistic electrons are present and this is therefore a pointer for cosmic ray physics. Where an electron component is present, it is reasonable to assume that there is also a proton component.

In normal spiral galaxies, like our own, we see radio emission from the disk, and possibly from more extended regions. Recently a number of spirals have been studied with the high resolution Westerbork array (e.g. Van der Kruit, 1973) showing very interesting structures.

There are a number of elliptical galaxies which show compact, weak, radio sources in their nuclei. This phenomenon appears different from that found in the spirals, in that in the ellipticals which show such continuum emission, it is in all cases confined to their nuclei, and is associated with high excitation gas which is also confined to the nucleus. In general, not much gas is found in ellipticals (no 21-cm emission has ever been seen) which sets quite severe limits on the amount of neutral atomic hydrogen ($< 10^{-4}$ M_{galaxy}) in some of the nearest ellipticals. However, in some cases compact sources can be seen and this appears to be confined to the situation where there are comparatively small amounts of ionized gas. The electron temperature is in the range 3×10^4 K to 4×10^4 K and optical lines of the kind one would expect (O III, O II, N II, S II, Na V, etc.) are observed. The amount of gas that is excited need only be $\sim 10^2$ $M_\odot$, this reflects on the nuclear activity involved on this comparatively small scale.

2. Powerful Extended Sources

(a) Radio galaxies with compact components. The powerful radio sources tend to be double with the optical object (elliptical galaxy, N-system or QSO) lying between the two radio components. The calculations of the minimum total energy in the sources in the form of relativistic particles and magnetic flux gives values lying in the range $\sim 10^{58}$-10^{60} ergs. These are conservative estimates and require that there is rough equipartition between energy in particles and magnetic field, i.e. $E_p = (4/3)E_m$. For a long time I have argued that equipartition is unlikely and that probably $E_p \gg E_m$. Some of the strongest sources have been studied in detail, and at high resolution. I shall very briefly mention some of the individual cases which have been studied in detail. One of the newer results is that it is found that in quite a large fraction of the double sources, a weak compact component is associated with the central optical object. This shows that the violent activity is going on continuously.

NGC 1275 - Perseus A[‡]. The galaxy NGC 1275 is a strong radio source in the Perseus cluster. The radio emission from the cluster is extended and several galaxies appear to be involved. The cluster

---

[‡]In the lectures, slides of this and other objects were shown.

also contains an extended X-ray source with possibly a compact component on NGC 1275. The extended X-ray source may be connected with the radio phenomenon, if indeed the X-rays are generated by the Compton effect and are not bremsstrahlung of hot gas. NGC 1275 has a Seyfert nucleus. This nucleus also contains a very compact, variable high-frequency radio source. The galaxy also is ejecting a large mass of gas at velocities $\geq$ 3000 km s^{-1}. The total kinetic energy in this gas alone is of the order of 10^{59} ergs.

NGC 5128 - Centaurus A. This is the nearest powerful extragalactic radio source to us, and it lies at a distance of about 4 Mpc. It is not impossible that some of the cosmic rays we observe at the earth are coming from this object. Photographs show a giant elliptical galaxy with a very broad dust band across it. The extended radio source is double, with a dimension $\sim$ 800 kpc. There is an inner pair of small sources close to the optical object. Most recent observations show there is an extremely small source in the nucleus of the galaxy. Thus a range of time scales of activity is apparent. There is also an X-ray source in the nucleus. The X-rays may be generated by the Compton collisions of radio electrons with the microwave radiation.

M87 = NGC 4486 - Virgo A. This source has an extended halo of radio emission. There is also an extended X-ray source. This galaxy contains the famous optical synchrotron jet and the weaker counterjet. In the nucleus a very tiny (high frequency) radio source with a dimension of only a few light months has been found. The existence of the jet indicates that fluxes of particles with energies in the 10^4 GeV range are being generated or reaccelerated continuously, while the compact radio source indicates that activity must be going on on time scales of months or years.

Cygnus A. This is the most powerful of the sources with small redshifts. It is a classical double, and the early calculations showed that it posed a severe energetic problem. Recently, Hargrave and Ryle (1974) have mapped it with high resolution at 5 GHz using the Cambridge 5-km telescope. There are a pair of extended radio sources reaching out to approximately 90 kpc from the central optical object. Both contain a bright compact source at the furthest point from the central object. Hargrave and Ryle calculate the minimum total energies, assuming equipartition between relativistic electrons and the magnetic field, to be 3×10^{57} erg and 4×10^{57} erg for the two compact components. The equipartition magnetic field is 3×10^4 gauss for both. The lifetimes of the particles giving rise to the radio emission are 5×10^4 years and 4×10^4 years, respectively. These lifetimes are very short indeed, much less than the light travel time from the central object. The authors concluded that there must be a continuous injection or re-acceleration of electrons with a power of 10^{45} erg s^{-1}. These equipartition magnetic fields are very large indeed, and it appears likely that the true values of the fields are much less. If this is the case, then the total energies go up accordingly. If the fields are $\sim 10^{-6}$ gauss then the total energy in relativistic electrons will be about 5×10^{61} ergs. The particle lifetime will then be increased to $\sim 10^8$ years. However, the existence of the highly compact components suggests that they contain particle generators which are continuously active. Unless it is proposed that the mechanism generates electrons only, radio sources such as these will be very powerful sources of primary cosmic rays.

(b) <u>Large Radio Sources</u>. Our understanding of extragalactic radio sources is not very good and the new observations suggest two things. One is that very compact sources exist at very considerable distances from the places where they were generated. This raises many questions concerning the propagation of the particles. Secondly, from the observations of very large radio sources at Westerbork, we can now see that extragalactic cosmic rays, at least as far as the electron component is concerned, <u>do</u> exist and fill very large volumes of space, larger even than the usual volumes of clusters of galaxies. These last remarks are made following the discovery by Willis et al. (1974) of the faint and very extensive radio components of 3C 236 and DA 240. The dimensions of these sources are 5.7 Mpc and 2.0 Mpc, respectively. The equipartition magnetic field strengths for both approach 10^{-6} gauss (assuming the energy of cosmic-ray protons to be negligible). The extended source 3C 236 contains a minimum energy in electrons of $\sim 10^{60}$ ergs in a volume $\sim 10^{74}$ cm^3. The electron energy density of $\sim 10^{-14}$ erg cm^{-3} is therefore similar to that in the disk of our Galaxy. The smaller source DA 240 has a slightly higher energy density of electrons. Despite their energy losses (for the parameters used here, Compton scattering is more important than synchrotron radiation) the electrons are able to propagate over very large distances. We are looking then at systems with electrons which are truly extragalactic, in the sense that the containment values are considerably larger than the volumes of clusters of galaxies.

3. Highly Compact Variable Sources

A third category of objects which are probably important particle generators are the extremely compact radio sources, which are often found to be variable. Included here are the nuclei of galaxies, the BL Lacertae objects, QSOs and N-systems. Their very small angular sizes (10^{-4} and 10^{-3} arc seconds) imply linear dimensions of a few parsecs or less. Also if one argues that the light travel time across the object is less than the observed period of fluctuations, sizes of the order of light years are deduced. These scales do seem highly relevant to the whole phenomenon of explosive events in galaxies, whether they involve multiple supernovae or the collapse of massive objects or something more exotic.

BL Lacertae itself was originally classified as a variable star. It is not a galaxy but may be a violent event superimposed on a galaxy*. It may be closer to the quasi-stellar phenomenon than any event we know. The number of objects of this type being discovered is increasing rapidly.

The simplest model that can be considered for the very compact radio sources is a model in which one supposes there is a cloud of relativistic electrons and a magnetic field giving rise to an incoherent synchrotron source. Some part of the variability may be due to expansion of this cloud as was first proposed by Shklovsky for galactic supernova remnants and by van der Laan for extragalactic sources. In some cases there may be several such components. In the simplest case the radio spectrum of a non-expanding source follows a power law, with a turn-over which is almost certainly due to synchrotron self-absorption. If we make this assumption, the frequency of the turnover and the angular size of the source gives directly a measure of the magnetic field. From this one can look at the models and decide what are the energetics of a cloud of this type. It is found that in nearly all cases the energy in the particles, E_p, is very much greater than the energy in the magnetic field, E_m. (In general, as I described earlier, without knowing anything about the magnetic field, one assumes $E_p = E_m$.) But in these compact sources a conservative estimate of the ratio $E_p/E_m \simeq 10^4$ to 10^6 (cf. Burbidge, Jones and O'Dell 1974). It turns out then, that in sources of this type one expects, in general, 10^{52} to 10^{56} ergs per outburst in relativistic electrons. It appears that nature has a way of making particles in a situation that is very far from equipartition. Thus, outbursts occur on time scales of years or less, so that additively these types of sources are prolific generators of relativistic particles.

IV REQUIREMENTS FOR AN EXTRAGALACTIC THEORY OF COSMIC RAYS

I have concentrated so far on describing the properties of extragalactic non-thermal sources and in so doing I have shown that there is evidence that powerful generators of extragalactic cosmic rays exist. I now turn to a summary of the requirements for an extragalactic theory of cosmic-ray origin.

---

*Recent observations by Baldwin, Burbidge, Robinson and Wampler (1975), however, do not confirm the claim made recently by Oke and Gunn (1974) that a galaxy underlies the central object.

EXTRAGALACTIC COSMIC RAYS

By extragalactic origin, I mean the idea that the bulk of the energetic particles, i.e. the nucleons in the cosmic rays, come from outside the galaxy. These ideas have been developed in detail in the paper by Brecher and Burbidge (1972).

The requirements are summarized under several headings.

1. Cosmic Rays Outside the Galaxy

Evidence is required that cosmic rays exist outside the Galaxy with an energy density equal to that inside. We do not have this evidence. However, Ginzburg (1972) has proposed the test of looking for γ-rays from nearby galaxies (from the Magellanic Clouds for example) which arise from the decay of neutral pions produced in pp collisions. He points out that if one does not find γ-rays, it is evidence against extragalactic cosmic rays. If we do find γ-rays, this does not necessarily mean it is evidence in favour, since the external galaxy may be a source of cosmic rays itself.

2. Evidence of the Existence of Powerful Extragalactic Cosmic Ray Sources

Evidence is required for the existence of powerful extragalactic sources of cosmic rays. If one accepts that the mechanism that gives rise to the radio emission is an incoherent synchrotron process, then such evidence does exist. I have summarized it in the previous lectures. Colgate has suggested an alternative mechanism in which the radio emission is produced by an oscillating plasma process. However, most of us believe at present that the synchrotron mechanism is responsible for most of the non-thermal radiation generated in the universe.

3. Evidence for the Escape of Cosmic Rays from Sources

As I have described earlier, evidence does exist that the electron component does escape from such sources, as electrons are found at very great distances from the places in which they must have originated. If protons are also generated, they too will escape.

4. Energy Density Requirement

The energy density required for a completely universal theory is 10^{-12} erg cm^{-3}. The space density of galaxies is about one per 10^{75} cm^3, so each galaxy must on average give rise to 10^{63} ergs in cosmic rays over a time scale of H_o^{-1} ($\sim 10^{10}$ yrs). If each galaxy has a mass of 10^{11} $M_\odot$, the rest mass energy is 2×10^{65} erg and about 0.5% of the rest mass energy of the galaxy must be converted by some mechanism into relativistic particles to give an energy density of 10^{-12} erg cm^{-3}. In view of the fantastic energetics that are observed, and the rate at which power is poured out in non-thermal processes in the universe, this number does not seem particularly unreasonable.

In the modified extragalactic theory (Brecher and Burbidge 1972), it was suggested that the cosmic rays cannot easily propagate over very large distances to fill the whole universe. Galaxies are largely confined to clusters, and if the view is taken that the propagation characteristics of extragalactic cosmic rays are such that they cannot travel distances greater than the dimensions of clusters in 10^9 to 10^{10} yrs, the overall energy requirements are reduced in the following sense: the clusters occupy only about 1% of the volume of space and our energy requirement is that one fills up only this volume. The requirement then becomes $\sim 10^{61}$ ergs per galaxy. The clusters will slowly fill up and the particles will spill out. Recent radio observations by Willis et al. (1974) are, however, against this notion. Cosmic electrons are present in volumes large compared with those occupied by clusters of galaxies. This may force one to return to the universal hypothesis. In the case of our own Galaxy, we are situated about 10 Mpc from the centre of the Virgo cluster. We may lie on the edge of the supercluster, but in any case, if galactic cosmic rays are extragalactic in origin and arise in systems like M87, they must have travelled 10 Mpc or more to reach us. As was mentioned earlier, closer objects like Centaurus A may contribute.

For the 3C radio sources the equipartition magnetic fields are 10^{-5} to 10^{-4} gauss and minimum energies are 10^{58}-10^{60} ergs. However, if the situation is far from equipartition with weaker magnetic fields, the energy in particles will be much greater. For example, if the assumed magnetic field is reduced by a factor of 10, the particle energy density is increased 30 times. To obtain a large extragalactic flux of particles, $10^7 M_\odot$ to $10^9 M_\odot$ per galaxy are required to be converted to relativistic particles. Whether these numbers are reasonable depends on our future understanding of galactic explosions. If electrons and protons are generated in each outburst in a fixed ratio such that $E_{protons}/E_{electrons} \geq 1$, it is clear that while the electrons are rapidly reduced in energy by synchrotron losses and Compton scattering, the proton flux will steadily build up.

5. Cosmic-Ray Propagation

Evidence is required that cosmic rays can propagate to us without being destroyed or giving rise to components which are not observed. For most energies of interest, electrons cannot propagate to us, since interaction with the universal radiation field will cut them off. Thus, we know that the electrons that we do see are largely Galactic in origin, and the extragalactic theory is for a nuclear component only.

An intergalactic proton flux will give rise to γ-rays due to collisions with the intergalactic gas. If the universe were closed, by the intergalactic gas, the predicted γ-ray flux would be greater than that observed. However, as was described earlier, there is little evidence that there is an appreciable density of truly intergalactic gas, and there is no contradiction with observation at present.

6. Cosmic Rays Entering the Galaxy.

For an extragalactic theory we would like to have evidence that cosmic rays can enter the Galaxy. Parker (1973) has suggested that if the Galactic magnetic field configuration is closed, it could be difficult for cosmic rays to enter. He has also suggested that only sufficiently energetic cosmic rays ($\gtrsim 10^{17}$ eV) would be expected to enter. The field structure is not known, however, and there are explosive events which could blow holes in the magnetic field allowing particles to flow out. Certainly particles could then flow in.

I suspect that this problem, if it exists, is partly man-made. So far, nearly all of the theoretical investigations have been made on the assumption that cosmic rays are Galactic and are trying to get out, not in. If the Galactic magnetic field structure is open, as it may be, particles will get in and flow through. Moreover, if we can show that the Galaxy is bathed in a sea of externally generated particles, I would have no doubt that they would eventually get in.

V. CONCLUSION

I have tried to make a reasonable case for the extragalactic cosmic-ray hypothesis, and to show you why someone like myself has been interested in it for a long time. It started with the discovery of powerful extragalactic objects which, over the past 15 to 20 years, have been shown to be very important indeed. This theory has a similar genesis to the modern development of supernova origin theory which really started when Ginzburg, Shklovsky and Pikelner investigated the Crab Nebula and put forward the argument that here was direct evidence for the generation of galactic cosmic rays. In my view, a similar case can be made for the extragalactic theory. Let us hope that the observations will enable us eventually to decide whether violent events inside our Galaxy, or outside are more important.

Extragalactic research at UCSD is supported in part by the

National Science Foundation and by NASA.

REFERENCES

Baldwin, J., Burbidge, E.M., Robinson, L. and Wampler, E.J., 1975, Ap. J. (Letters). To be published January, 1975.

Brecher, K. and Burbidge, G.R., 1972, Ap. J., 174, 253.

Burbidge, G.R., Jones, T.W. and O'Dell, S, 1974, Ap. J., 193, 43.

Felten, J.E. and Morrison, P., 1966, Ap. J., 146, 686.

Ginzburg, V.L., 1972, Nature Physical Sciences, 239, 8.

Hargrave, P.J. and Ryle, M., 1974, M.N.R.A.S., 166, 305.

van der Kruit, P.C., Astron. and Astrophys. 1973, 29, 249 and 263.

Oke, J.B. and Gunn, J.E., Ap. J. (Letters), 189, L5, 1974.

Parker, E.M., 1973, Ast. and Space Sci., 24, 279.

Willis, A.G., Strom, R.G. and Wilson, A.S., 1974, Nature, 250, 625.

SIDEREAL DAILY VARIATIONS IN COSMIC RAY INTENSITY AND THEIR
RELATIONSHIPS TO SOLAR MODULATION AND GALACTIC ANISOTROPIES

T. Thambyahpillai

Physics Department, Imperial College of Science and
Technology, London SW7.

I. INTRODUCTION

From the earliest days of cosmic ray intensity recording it has been recognised that with a fixed instrument one could use the spin of the earth to scan different parts of the sky and to measure anisotropies by the diurnal periodic variations that should be induced in the data. However, significant daily variations occur in both solar and sidereal times and because of the extreme closeness of the two periodicities (365 solar days and 366 sidereal days in one earth year) one requires unbroken data spanning one complete year or preferably a series of complete years in order to separate the two periodic variations. Even then a direct setting of the data in solar or sidereal time will not result in the separation of the solar or sidereal diurnal variations unless each type of periodic variation retained a constant amplitude and phase throughout the year. This can easily be seen by examining the situation on a harmonic dial. A harmonic dial is a method of representing a sinusoidal variation by means of a vector and is a diagram in the form of a clock-dial which is graduated in 24 hours for displaying diurnal variations. A vector radiating from the centre of the dial has a length equal to the amplitude of the sine wave and points to the time of day at which maximum intensity occurs.

Let us consider the behaviour of a solar diurnal variation plotted on a sidereal time dial. Let the data be sub-divided into 12 monthly groups and let the solar diurnal vector for each group be plotted on the dial. Because the solar clock loses about 4 minutes per day or about 2 hours per month relative to the sidereal clock the solar vectors for successive months will rotate clockwise, the angular separation between successive vectors being $30°$. Let us consider the special case in which the solar vector retains a constant amplitude and phase throughout the year. Here we can select pairs of months which are six months apart such as January and July, February and August and so on and find that these pairs have equal amplitude and are diametrically opposed to each other. Hence if we find the average variation in sidereal time over the year the six pairs will average to zero. If, however, the amplitude and/or the phase did not remain constant throughout the year, the average of the 12 monthly vectors will not reduce to zero. Thus, it appears that if there is an annual variation in amplitude and/or phase of the solar diurnal variation, this will result in the generation of a spurious diurnal variation in sidereal time and vice versa. Therefore, one has to isolate and eliminate spurious sidereal variations of solar origin before one can relate the remaining sidereal variation (if any) to an anisotropy of galactic cosmic rays. The theoretical methods necessary for tackling this problem are already available in radio communications theory and were first adapted for the cosmic ray field by Farley and Storey (1954). It is well known that a radio frequency carrier which is amplitude modulated by an audio-frequency can give rise to signals at the carrier frequency as well as at two side-bands which correspond to the sum and difference of the radio and audio frequencies. In the cosmic ray case, the solar diurnal variation corresponds to the carrier and when amplitude modulated at the rate of one cycle per year gives rise to variations in the sidereal (366 cycles per year) and anti-sidereal (364 cycles per year) side-bands. In a similar way one can show that phase modulation of the solar diurnal variation would generate appreciable variations in a series of side-bands falling on either side of the carrier and differing from the carrier by an integral multiple of the one cycle per year frequency. The important difference between amplitude and phase modulations appears to be that whereas only two side-bands are excited by amplitude modulation, a series of side-bands should be excited by phase modulation. Our analysis of data from equipment located at sea-level as well as at moderate depths underground shows that no significant variations are observable outside the anti-sidereal and sidereal side-bands suggesting that phase modulation is not an important phenomenon in cosmic ray diurnal variations and hence we shall not consider phase modulation any further. Although Farley and Storey considered one component of the solar diurnal variation which undergoes an annual amplitude modulation, their method can very easily be adapted to cover the more general case of two components, one of which is due to atmospheric temperature changes and the other caused by primary anisotropy. We shall now consider this adapted version of the Farley and Storey analysis.

II. THE MODIFIED FARLEY AND STOREY ANALYSIS

We shall use the subscripts atm and ani to denote the atmospheric and anisotropic components respectively of the solar diurnal variation. Let $\bar{A}$ represent the annual mean amplitude and δ the depth of modulation. It will be assumed that both solar components undergo amplitude modulation. It is convenient to measure time t in units of 1 year and let N denote the number of solar days per year. The percentage changes of cosmic ray intensity I can then be represented by the equation:

$$100 \frac{\delta I}{I} = \bar{A}_{atm} \left[1 + \delta_{atm} \cos(2\pi t - \Phi_{atmy})\right] \cos(2\pi N t - \Phi_{atm})$$

$$+ \bar{A}_{ani} \left[1 + \delta_{ani} \cos(2\pi t - \Phi_{aniy})\right] \cos(2\pi N t - \Phi_{ani})$$

$$+ A_{sid} \cos\left[2\pi(N+1)t - \Phi_{sid}\right]$$

This can then be resolved into components in solar, sidereal and anti-sidereal times as follows:

Solar component = $\bar{A}_{atm} \cos(2\pi N t - \Phi_{atm}) + \bar{A}_{ani} \cos(2\pi N t - \Phi_{ani})$

Sidereal component = $A_{sid} \cos\left[2\pi(N+1)t - \Phi_{sid}\right]$

$$+ \tfrac{1}{2}\bar{A}_{atm} \delta_{atm} \cos\left[2\pi(N+1)t - (\Phi_{atm} + \Phi_{atmy})\right] +$$

$$+ \tfrac{1}{2}\bar{A}_{ani} \delta_{ani} \cos\left[2\pi(N+1)t - (\Phi_{ani} + \Phi_{aniy})\right]$$

Anti-sidereal component

$$= \tfrac{1}{2}\bar{A}_{atm} \delta_{atm} \cos\left[2\pi(N-1)t - (\Phi_{atm} - \Phi_{atmy})\right]$$

$$+ \tfrac{1}{2}\bar{A}_{ani} \delta_{ani} \cos\left[2\pi(N-1)t - (\Phi_{ani} - \Phi_{aniy})\right]$$

The important relationships which Farley and Storey pointed out can be illustrated by considering one of the components, say the atmospheric, subject to amplitude modulation. Firstly, the amplitudes of the spurious sidereal and anti-sidereal variations are equal. Secondly, the phase constants ($\Phi_{atm} + \Phi_{atmy}$) for the sidereal and ($\Phi_{atm} - \Phi_{atmy}$) for the anti-sidereal variations show that the two vectors are symmetrical about the generating solar vector with a phase constant Φ_{atm}. Thus the corresponding spurious sidereal vector can be determined by 'reflecting' the anti-sidereal vector in the solar vector. A similar symmetrical relationship obtains with regard to the anisotropic component of the solar vector. Unfortunately, when we combine the two solar components and the two anti-sidereal components, this symmetrical relationship no longer exists as far as the resultants are concerned unless very special circumstances prevail. It is therefore regrettable that some workers continue to apply the Farley and Storey correction to the observed (resultant) vectors without studying the suitability of the method in any particular case and there is some danger that the results so obtained may not be quite correct. It may therefore be worthwhile to examine briefly the possible ways in which the various components of the solar diurnal variation arise.

III. THE ORIGIN OF THE SOLAR DIURNAL VARIATION

As seen earlier, a solar diurnal variation can originate partly from atmospheric causes and partly from an anisotropy existing outside the earth's magnetosphere. The effects which fall within the first category are the barometer effect, the muon decay or negative temperature effect and the positive temperature effect, the last arising from a competition between capture and decay processes of the pion. Two types of anisotropy following solar time are known, viz. the co-rotation anisotropy and the earth's orbital motion anisotropy. These effects will be examined in greater detail in the following, mainly with a view to familiarising the reader with the magnitudes of the variations involved.

1. The Atmospheric Effects

(a) <u>The Barometer Effect.</u> This describes the changes of cosmic ray intensity associated with changes in atmsopheric pressure. This effect arises partly as a mass absorption effect and partly from the finite life-time of the muon. The barometer coefficient can be calculated if the muon momentum spectrum and the rate of energy loss in the air are known. (Dutt and Thambyahpillai, 1965). Using the readings from a barograph, the effects of pressure changes can be easily eliminated from the data.

(b) <u>The Muon Decay Effect.</u> Muons are believed to be produced mainly near the 100 mb pressure level and have to travel from the point of production to the recorder at sea-level or underground without decaying. If the atmospheric temperature increases, the muons have to travel a longer distance in order to reach the recording instrument and thus a lower muon intensity is recorded because of a larger number of muons decaying in flight. The decay coefficient is about 3.5% per Km. for a sea-level recorder and has decreased to 0.7% per Km. for a recorder at a depth of 60 metres water equivalent (m.w.e. underground). It diminishes still further as the depth of observation increases. In order to calculate the diurnal variation caused by the decay effect, it is necessary to know the diurnal variation in atmospheric temperature throughout the atmosphere. It is this information that is usually either not available or, if available, is open to serious doubts about its reliability. These doubts arise because in atmospheric soundings made during daylight hours there is uncertainty as to whether the readings given by the temperature elements reflect the true atmospheric temperature or some higher temperature produced by the absorption of solar radiation. However, indirect methods have been devised to estimate the annual mean diurnal temperature effect in sea-level muon data and these suggest that the temperature vector has an amplitude of about 0.1% and a time of maximum around 06 hr. Some estimates made on the basis of radio-sonde data give a vector which has a nearly double amplitude and a phase different from the previously given value by 4 hours. By using the indirect estimate, one can deduce that the temperature vector for a recorder at 60 m.w.e. depth yielded by the muon decay effect should have an amplitude of 0.02%. The vector would be still smaller at larger depths underground.

(c) <u>The Positive Temperature Effect.</u> This effect occurs only in the upper atmosphere well above the 200 mb. pressure level and as mentioned earlier arises because of the competition between decay and nuclear capture of the pions. At any pressure level, an increase in atmospheric temperature reduces the density and thus reduces the probability of nuclear capture. The decay probability is increased resulting in more muons reaching the detector. Thus a positive correlation between recorded muon intensity and stratospheric temperature results and this gives the effect the name of positive temperature effect. This effect cannot be calculated accurately because the absorption length of the pion may not be known precisely. Making simplifying assumptions, Trefall (1955) obtains the expression for the positive coefficient applicable to muons of energy E as

$$\frac{E}{E + E_0} \cdot \frac{1}{T}$$

where E_0 is about 80 GeV and T represents stratospheric temperature. For observations at any particular depth, this expression has to be weighted using the muon energy spectrum but it is clear that the positive coefficient increases with increasing muon energy until is asymptotically approaches the value of 0.46% per °C. It is worthy of note that this effect, unlike the two effects described earlier, increases with increasing depth. Because the present trend is to make observations at greater depths underground than hitherto, it should be stated that this atmospheric effect can contribute to the observed solar diurnal variation and possibly add a spurious component in the sidereal side-band at these greater depths. Once again, the amplitude of the diurnal variation generated by the positive effect cannot be calculated directly because of the lack of reliable information about atmospheric temperature changes.

2. Solar Diurnal Anisotropies

There are two known effects which give rise to anisotropic distriubtions which follow solar time. These can be called the co-rotation anisotropy and the oribital motion effect. These will now be considered in brief in the following:

(a) <u>The Co-rotation Anisotropy.</u> In order to maintain a fixed number of cosmic ray particles of a rigidity P in the inner solar system, the number of cosmic rays convected out radially by the solar wind should be replaced by the number of particles undergoing anisotropic diffusion inwards. The application of this condition leads to the result that the cosmic ray gas is subject to an azimuthal streaming with the same velocity as it would have if the gas rotated as if it was rigidly attached to the rotating sun and interplanetary magnetic field. This co-rotation velocity which is close to 400 km/sec at the orbit of the earth when substituted in the Compton-Getting expression $(2 + \gamma)v/c$ where γ is the exponent of the differential rigidity spectrum, yields an amplitude of 0.6% outside the magnetosphere. The time of maximum outside the magnetosphere is 18 hr. The amplitude has been found experimentally to be rigidity-independent up to an upper limit of about 100 GV beyond which it falls to zero. A sea-level muon detector measures this corotation anisotropy as a diurnal variation of amplitude about 0.2% and the amplitude diminishes to 0.02% in a detector at a depth of 60 m.w.e. underground. This corotation anisotropy should not be observable at much greater depths exceeding, say, 200 m.w.e.

(b) <u>The Earth's Orbital Motion Effect.</u> The relative motion between an observer and an isotropic flux of cosmic radiation should result in an apparent anisotropy to the observer with an amplitude $(2 + \gamma)v/c$ where v is the velocity. The maximum flux would appear to arrive from the direction in which the observer is moving and the minimum from a diametrically opposite direction. Because the earth has a velocity of about 30 km/sec in its orbital motion around the sun, there should be an apparent anisotropy to an observer on the earth with an amplitude of 0.046% and maximum intensity in the 06 hr. direction. Although this orbital motion effect is too small to be perceptible at sea-level its effects are quite evident at a depth of

60 m.w.e. underground and because there is no upper limit beyond which this anisotropy vanishes, it should be observable even at much greater depths than 60 m.w.e.

IV. IS THERE AN ANNUAL AMPLITUDE MODULATION?

The question as to whether the solar diurnal variation generated in the various ways listed above is subject to amplitude modulation with a period of one year will be considered in two separate groups corresponding to atmospheric and anisotropic origins of the diurnal variation.

1. Diurnal Variation of Atmospheric Origin.

The barometer effect need not be included here because this effect can be eliminated satisfactorily from the data. It has already been stated that even the average diurnal variation caused by the muon decay and positive temperature effects cannot be calculated directly because of the difficulties associated with the measurement of diurnal temperature variations in the atmosphere. Similar difficulties are encountered when it is desirable to know whether these atmospheric diurnal variations are subject to an annual amplitude modulation. In fact, it may be possible at the present time to reverse the problem and to use the anti-sidereal variations in cosmic ray intensity to make deductions regarding the amplitude modulation in atmospheric diurnal variations. However, it is known that in middle latitudes, the amplitude of the temperature variation in the lower atmosphere (where insolation errors of the temperature elements are minimal) tends to be largest during local summer and smallest during local winter. It would be very interesting to know whether a similar trend exists at higher altitudes as well.

2. Solar Diurnal Anisotropies.

The reasoning to be given in this section will be illustrated by applying it to the case of the co-rotation anisotropy but it is equally valid for the orbital motion effect, the only differences being in the time of maximum of the anisotropy and in the time of year at which the geometrical configuration occurs.

The only serious suggestion for providing an amplitude modulation of the co-rotation anisotropy that has been advanced hitherto is that the tilt of the earth's axis by 23° from the normal to the ecliptic is responsible. The argument advanced can best be understood by reference to Figure 1. The co-rotation anisotropy is assumed to be caused by a flux of additional particles flowing from left to right in the ecliptic with maximum intensity from the 18 hr. direction. During the (northern) vernal equinox, the axis of the telescope of a mid-latitude station in the northern hemisphere makes the smallest angle with the direction of the flux while the axis of a southern telescope makes the largest angle. Six months later, at the (northern) autumnal equinox, the axis of the northern telescope makes the largest angle with the flux, this angle being 46° more than in the previous case. On the other hand, the southern telescope now makes the smallest angle. Hence one expects an amplitude modulation of the co-rotation anisotropy which is six months out of phase between the northern and southern hemispheres.

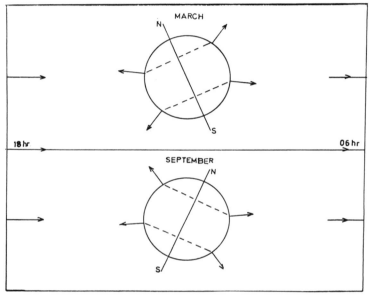

Fig. 1: A section through the centre of the earth normal to the earth-sun line. The 18 hr. direction is on the left and the 06 hr. direction on the right. The positions of the axes of telescopes in the Northern and Southern hemispheres at 06 and 18 hr. are shown for the months March and September.

Unfortunately, there is a fundamental objection to the above reasoning which is that the co-rotation anisotropy is observed experimentally to be <u>not</u> a peak in the 18 hr. direction but a sine wave with a <u>maximum</u> in the 18 hr. direction and a minimum in the 06 hr. direction. The use of the Compton Getting expression in the theoretical derivation of the co-rotation amplitude also supports this view. Therefore we have to examine the geometry associated not only with the peak but also the trough.

Reference to Figure 1 will show that at the northern vernal

equinox when the axis of a northern telescope makes the smallest angle with the direction of maximum intensity, the axis makes the largest possible angle with the direction of minimum intensity. Six months later when the axis of the northern telescope makes the largest angle with the direction of maximum intensity the same axis makes the smallest angle with the direction of minimum intensity. From considerations of symmetry we would expect the first harmonics measured in both situations to be the same, the only difference being in the higher harmonics generated because the telescope measures a distorted sine wave. It appears, therefore, that a careful examination of the geometry does not lead us to expect an annual amplitude modulation of either the co-rotation or orbital motion anisotropies. There may be a modulation with a period of six months but this does not generate signals in the sidereal and anti-sidereal side-bands and will therefore be ignored.

There is considerable experimental evidence which contradicts the suggestion that an annual amplitude modulation of the co-rotation anisotropy generates appreciable anti-sidereal variations. Firstly, the theory predicts maximum amplitudes in March in the northern hemisphere and during September in the southern hemisphere. The observed phases of the anti-sidereal vectors at Budapest (40 m.w.e. depth, N. hem.) and Hobart (30 m.w.e. depth, S) are such that maximum amplitudes occur in June in the northern hemisphere and in January in the southern hemisphere. Secondly, (because the depth of modulation is determined only by geometrical factors) the explanation predicts a strong positive correlation between the amplitudes of the solar and anti-sidereal vectors which is clearly absent at least in the Hobart data. It is concluded, therefore, that there is no theoretical or experimental evidence to support the suggestion that an annual amplitude modulation occurs in the co-rotation or orbital motion anisotropies.

3. Additional Method of Generating an Anti-sidereal Variation.

Because one relies heavily on the anti-sidereal variation in the data in order to infer the spurious sidereal variations generated by changes in the solar diurnal variations it is worth considering whether an anti-sidereal variation can be produced in any other way. As pointed out by Elliot et al. (1970) the configuration of the interplanetary magnetic field is such that a genuine galactic anisotropy can be amplitude modulated with a periodicity of six months. This type of amplitude modulation would result in an anti-sidereal variation but this would be accompanied by an equal vector in the USV2 (ultra-sidereal variation of order 2 with a frequency of 368 cycles per year) side-band. Also the anti-sidereal and the USV2 vectors will be symmetrical about the sidereal vector undergoing amplitude modulation. We can therefore determine the USV2 vector in order to obtain an estimate of the magnitude and phase of the anti-sidereal vector generated by the changes of the sidereal vector. At the surface and depths up to 50 m.w.e. the USV2 vector appears to have a negligibly small amplitude (less than 0.004%) and in the London data at 60 m.w.e. depth it has a barely significant amplitude of 0.008 $\pm$ 0.004%. Thus, with the exception of the London data, it does not seem to be necessary to consider this method of anti-sidereal vector generation as important.

V. EXPERIMENTAL RESULTS AND THEIR INTERPRETATION

1. Experimental Results

It emerges from the foregoing discussion that the observed anti-sidereal vector is in all likelihood generated by atmospheric temperature effects and that this vector should be reflected in the atmospheric component (and not the observed resultant) of the solar diurnal variation in order to determine the spurious sidereal vector. However, the phase of this atmospheric vector is not known except at sea-level where it is believed to have a maximum between 05 and 06 hr. Very little is known about the temperature vector (annual mean) for underground recorders.

In order to obtain some information about the temperature effects in the London underground station, directional measurements using inclined telescopes pointing in the north and east directions were carried out between 1965 and 1969. The axis of the north telescope was made nearly parallel to the earth's rotation axis so that this telescope cannot register any anisotropies in primary particles of sufficiently high rigidity (above about 100 GV) so as not to suffer appreciable deflection in the geomagnetic field. Those primaries which enter the north telescope and have asymptotic latitudes substantially smaller than 90° will cause a diurnal variation in the north telescope due to primary anisotropy. Therefore the north telescope will register only a fraction of the diurnal variation caused by anisotropy and which will be registered by the east or vertical telescopes. Temperature effects, however, should be the same in all telescopes. Reasoning of this kind has led Thambyahpillai and Speller (1974) to conclude that the annual mean temperature vector at 60 m.w.e. depth is not significantly different from zero. The anti-sidereal and (apparent) sidereal vectors measured by the north telescope, however, were quite large and would mostly have originated from temperature effects. This is a bizarre situation in which there is no signal at the carrier frequency but substantial signals are present in the two sidebands. Moreover, the sidereal and anti-sidereal vectors of the

(a) Northern Hemisphere Stations

Station	Depth	Sidereal Amplitude %	Time of Max. Hrs.
Cheltenham	Sea-level	0.024 ± .006	22.0
Budapest	40 m.w.e.	0.023 ± .007	24.0
London	60 m.w.e.	0.016 ± .006	22.5

(b) Southern Hemisphere Stations

Station	Depth	Sidereal Amplitude %	Time of Max. Hrs.
Christchurch	Sea-level	0.029 ± .006	08.2
Hobart	30 m.w.e. (nominal)	0.028 ± .006	09.0

Table I: Corrected Sidereal Variations at Various Stations.

north telescope have nearly the same amplitude and are symmetrical about the 18 hr. (or 06 hr.) direction. This leads to the result that the spurious sidereal vector can be obtained by reflecting the anti-sidereal vector in a vector with a time of maximum of 18 hr. Thus the direction of the base vector in which reflection is to take place has not changed from sea-level to a depth of 60 m.w.e. underground and for lack of more definite information the phase of the base solar vector will be assumed to be 06 hr. (or 18 hr.) for all depths between 0 and 100 m.w.e.

Corrected sidereal diurnal variations measured at sea-level and various depths underground are presented in Table I.

It will be seen from Table I that the time maximum of the corrected sidereal variation is around 23 hr. in the northern hemisphere and around 09 hr. in the southern hemisphere. The explanations offered for the remarkable 10 hr. difference in phase between the two hemispheres will be considered now.

2. North-South Streaming and the Swinson Model.

As will appear later, trajectory calculations of cosmic rays moving in the interplanetary magnetic field show that the magnetic field is completely opaque to particles with rigidities less than roughly 100 GV so that any galactic anisotropies existing in these low rigidity particles cannot be observed from the earth. Yet small but significant sidereal variations are registered by recorders at sea-level and shallow depths underground. Because the major part of the counting rate of these recorders is caused by these low rigidity primaries it appears necessary to invoke a spurious mechanism of solar origin to explain these sidereal variations. However, these sidereal variations could not have arisen from amplitude modulation of the solar vector because corrections based on the observed anti-sidereal vectors have been applied. This property has been independently verified by using north and south-pointing telescopes at near-equatorial stations where a 12 hr. phase difference between north and south directions was observed even though amplitude modulation effects should be identically the same in both directions. Such a mechanism for a sidereal variation of solar origin was first suggested by Swinson (1969). This mechanism relies on the streaming of cosmic rays in north-south direction, this streaming originating from the existence of a radial density gradient in the cosmic rays at the earth's orbit coupled with the tangential component of the inter-planetary magnetic field. It can be shown that when the inter-planetary magnetic field is directed inwards (towards the sun) there will be a larger intensity flowing from the northerly direction and when the IPMF is directed away from the sun, maximum intensity will arrive from the southerly direction. Swinson pointed out that this north-south streaming will generate diurnal variations in sidereal time because of the $23°$ tilt of the earth's rotation axis from the normal to the ecliptic. Because of this tilt, the axis of a middle latitude telescope in the northern hemisphere will make the smallest angle with the north-south direction at 18 hr. sidereal time right through the year. Similarly the axis makes the largest angle with the flow direction at 06 hr. sidereal time. Consequently an excess of cosmic rays from the north would cause a sidereal diurnal variation with a maximum in the 18 hr. direction. This happens in telescopes in both the northern and southern hemispheres but because the IPMF is stronger inside the orbit of the earth than it is outside, the northern amplitude will be larger than the southern amplitude. Similarly when the IPMF reverses sign, there is an excess of particles from the south and this results in sidereal time variations with a maximum in the 06 hr. direction. However, the larger amplitude will now be registered in the southern hemisphere. When the data are averaged over a year when the earth had spent a roughly equal number of days in fields of either sign, the data in the northern hemisphere will show a residual sidereal variation with maximum at 18 hr. and a southern telescope one with a maximum at 06 hr.

Although the physical principles underlying the Swinson model are very well understood, no serious attempt has, as yet, been made to calculate the magnitude of the effect theoretically. The reasons for this state of affairs are two-fold: firstly, the quantities which enter the calculation and are necessary to be known have not been determined accurately and secondly an easier alternative method involving only available experimental data can be employed for verifying the Swinson model. Swinson himself, as well as several other researchers, have pursued the second alternative and the results of these efforts will be detailed after considering a simple relationship which is obtained at very low rigidities.

Provided the radius of gyration of the primaries is very much smaller than 1 A.U., the motion of the particles can be treated as though the IPMF was uniform at the earth's orbit and inclined at $45°$ to the helio-centric radius vector. We assume that there is a radial density gradient of the cosmic radiation of magnitude $40/p$ % per A.U. where p is the rigidity of the particle. The amplitude of the north-south anisotropy is then given by the product of the radius of gyration and the component of the density gradient in a direction perpendicular to the magnetic field. Because the radius of gyration is directly proportional to p, the amplitude becomes about 0.3% and is independent of the rigidity of the cosmic ray particles. Since the tilt of the earth's rotation axis is only $23°$, a recorder at the earth will observe only about a third of the full amplitude, i.e. 0.1%. In order to make order of magnitude calculations, it is convenient to extrapolate this rigidity independent amplitude to higher rigidities up to some upper limit beyond which the amplitude falls to zero. By comparison with the co-rotation anisotropy which has a similar rigidity dependence, but a free-space amplitude of 0.6%, one would expect a sea-level recorder to measure a Swinson amplitude of 0.03% and an underground recorder at 60 m.w.e. to measure less than 0.01%. It is worth noting that these values which apply for a particular direction of the IPMF are about equal to the observed average sidereal amplitudes (for both directions of the IPMF).

Now we note that according to the Swinson model when the IPMF is directed inwards (towards the sun) the observed sidereal variation in both hemispheres should have a maximum at 18 hr. and when the IPMF reverses sign the time of maximum should change to 06 hr. Because of the sector structure characteristic of the IPMF whereby alternate sectors have magnetic fields of opposite sign, the earth is immersed alternately in opposite magnetic fields as the sectors, sharing the rotation of the sun, rotate past the earth. Thus, if the cosmic ray data are grouped according to the sign of the IPMF at the earth, the 18 hr. maximum should occur when the field is inward and the 06 hr. maximum when the field is directed outwards. The necessary IPMF information has been obtained from artificial satellites from

1964 onwards and, for the earlier periods, has been inferred from polar magnetometer records by Svalgaard (1972).

Swinson (1973) has grouped data from Chacaltaya (20 m.w.e.), Embudo (40 m.w.e.) and Socorro (80 m.w.e.) according to the sign of the IPMF and finds the times of maxima strongly dependent on the sign of the magnetic field. He concludes that his data can be fully explained as spurious sidereal variations generated by north-south streaming.

Data from Hobart (30 m.w.e; southern hemisphere) have been analysed in the same way by Humble et al. (1973). These data, corrected for the anti-sidereal variation as outlined in section II are presented in Figure 2. The tips of the vectors with standard deviation circles are shown labelled with the corresponding direction of the IPMF. It can be seen from the diagram that a line joining the two end-points does not pass through the origin as is to be expected if the spurious variation caused by north-south streaming were the only sidereal variations being observed. In order to separate out the residual sidereal variation, it is necessary to know the ratio of the amplitudes of the IN and OUT vectors but if this ratio lay between 0.5 and 1.0, the residual vector would have an amplitude of about 0.03%. Humble et al.

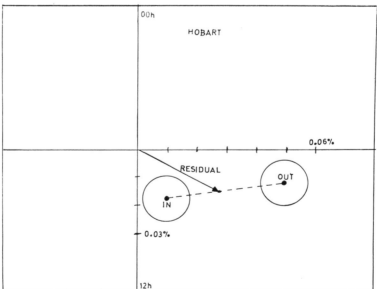

Fig. 2: Harmonic dial showing the end-points of the corrected sidereal vectors obtained at Hobart when the data are grouped according to whether the IPMF is inwards or outwards. The residual vector remaining after allowing for the Swinson effect is also shown.

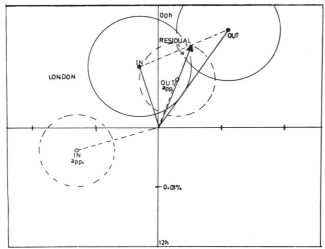

Fig. 3: Harmonic dial showing the end-points of the apparent (app.) and corrected sidereal vectors obtained at London when data are grouped according to the direction of the IPMF. The residual vector remaining after allowing for the Swinson effect is also shown.

---

have expressed their belief that this residual vector represents a genuine galactic anisotropy. Since Swinson's data apply mainly to the Northern hemisphere, presumably this genuine sidereal variation, if Humble et al. are correct, is observable only in the southern hemisphere.

Data from the London (60 m.w.e.) station which extend over a period of 10 years have been analysed according to the sign of the IPMF and the results are plotted on a harmonic dial in Figure 3. Only the end-points of the vectors together with standard deviation circles are plotted. It is evident from an examination of the diagram that although the uncorrected apparent sidereal vectors separate out in the manner found by Swinson, the correction for the anti-sidereal vectors in the manner described earlier alters the state of affairs radically. The corrected vectors show a barely significant separation which can be explained by the Swinson model and even if this separation is treated as real, there appears to be a residual vector exceeding 0.01% in amplitude which can be interpreted as a genuine galactic anisotropy.

Summarizing, it appears that the Swinson model is successful in explaining the sidereal variations observed at the surface and shallow depths underground, but it does not exclude the existence of variations which require other explanations, including that of a possible galactic anisotropy, particularly in the Hobart and London data.

3. Current Trends in the Observation of Sidereal Variations

Many workers have come to accept the Swinson thesis that almost the entirety of the sidereal diurnal variations observed below a depth of 100 m.w.e. is caused by solar modulation. One self-evident solution is to make measurements at much higher rigidities where one can be confident that solar modulation is no longer operative. Groom at Utah, Bercovitch in Ottawa and Fenton at Hobart are making measurements at depths of over 200 m.w.e. A somewhat different approach has been adopted by the London group because the London data can be interpreted as showing a possible galactic anisotropy in addition to any effects of north-south streaming. This anisotropy has been attributed by Elliot et al. (1970) to the motion of the sun relative to the Galactic rotation frame. It turns out that if this interpretation is correct, one would expect an equatorial telescope to observe over twice the sidereal amplitude of a vertical telescope operating in London. On the other hand the Swinson mechanism and the earlier two-way sidereal anisotropy advocated by Jacklyn predict zero sidereal amplitudes for an equatorial telescope. Therefore, highly inclined telescopes with the axes of their viewing cones pointing into the earth's equatorial plane are being operated at the London underground station. The experiment has been running for about two years and the results are preliminary but the results seem to indicate the presence of a finite sidereal variation in these telescopes.

VI DERIVATION OF SIDEREAL VARIATION AT THE EARTH DUE TO A GALACTIC ANISOTROPY

In order to calculate the sidereal diurnal variation measured by an underground recorder at the earth in the presence of a Galactic anisotropy which exists outside the solar cavity, it is necessary to know the response function (also called coupling constant) of the recorder, the smoothing effect of the interplanetary magnetic field (also any systematic deflection) and the rigidity dependence of the amplitude of the anisotropy. A brief review of the current state of affairs regarding our knowledge of these various factors will be given in the following.

1. Response Functions

Let a number dN of the recorder count rate be produced by primary particles with energy lying between E and $E + dE$ striking the top of the earth's atmosphere. Then, if $D(E)$ is the differential energy spectrum and $m(E)$ is the effective multiplicity of muons produced by primaries of energy E, the following relationship holds:

SIDEREAL DAILY VARIATIONS IN COSMIC RAY INTENSITY

$$dN = D(E) \cdot m(E) \, dE$$

The total count rate is given by

$$N = \int_{E_{min}}^{\infty} D(E) \cdot m(E) \, dE$$

$$= \int_{E_{min}}^{\infty} W(E) \, dE$$

where E_{min} is the minimum energy required by a primary to generate a muon that can penetrate the absorber to reach the recorder and $W(E)$ is the differential response function. Usually $W(E)$ is normalised so that $N = 100$. The usefulness of $W(E)$ is illustrated by the following example.

Let the differential spectrum suffer a change $\delta D(E)$ which results in a change δN in the counting rate. Then

$$\delta N = \int_{E_{min}}^{\infty} \delta D(E) \cdot m(E) \, dE$$

$$= \int_{E_{min}}^{\infty} \frac{\delta D(E)}{D(E)} \cdot D(E) \, m(E) \, dE$$

$$= \int_{E_{min}}^{\infty} \frac{\delta D(E)}{D(E)} \cdot W(E) \, dE$$

The fractional (or percentage) change is given by:

$$\frac{\delta N}{N} = \int_{E_{min}}^{\infty} \frac{\delta D(E)}{D(E)} \cdot W(E) \, dE \Big/ \int_{E_{min}}^{\infty} W(E) \, dE$$

Thus if a theory predicts a particular functional form for $\delta D(E)/D(E)$ it is possible to predict the percentage change in the count rate of the recorder if the differential response function is known. Although the above example refers to scalar changes, the response function can also be used to evaluate the sinusoidal variations in which the phase may be rigidity dependent because of magnetic deflection. In such a case, the integration is replaced by vector summation to evaluate the resultant sine wave.

At times, instead of W(E), an integral response function may be evaluated. This function is merely the integral of W(E) dE between E_{min} and some higher energy E, shown as a function of E.

In order to evaluate W(E), indirect we well as direct methods have been employed. The indirect method was first devised by Dorman (1957) who derived the response function for an underground recorder from that of a sea-level muon detector. Dorman suggested that the portion of a W(E) vs E plot which lay above a threshold energy E_T was applicable to an underground recorder provided the curve was re-normalised to have an area under the curve equal to 100, E_T was the product of an effective multiplicity of muons (about 5) and the minimum energy required by a muon to traverse the atmosphere and the rock or other material above the recorder. Dorman has noted two criticisms that could be made about his method of derivation. Firstly, the sharp rise of W(E) to a high value at the threshold E_T was not realistic and the rise can be expected to be more gradual. Secondly, W(E) for a sea-level muon detector was not known accurately because only 8% of the counts are produced by latitude-sensitive primaries. That portion of the W(E) curve above the equatorial geomagnetic threshold which accounts for 92% of the count rate was assumed to have a functional form which may not be accurate. Recently, Ahluwalia and Ericksen (1971) have revised the Dorman calculations reducing E_T by reducing the muon multiplicity to about 2. As these authors have not attempted to make revised calculations in those areas where Dorman has indicated weaknesses, their derivation is open to the same criticisms as were noted above.

Direct calculations which involve a knowledge of the differential spectrum D(E) and the multiplicity function m(E) have recently been made by Groom (1973), Gaisser (1974) and Erlykin et al. (1974). The calculation of m(E) can be made much more confidently through the availability of ISR data from the European Organization for Nuclear Research and the use of Feynman scaling. Essentially, a proton entering the atmosphere undergoes a series of interactions with air nuclei generating pions at each interaction. These pions can interact further with air nuclei producing further pions. Many of the pions decay into muons and these muons have to survive decay and energy loss in the atmosphere and in the solid earth in order to reach the recorder. All these processes enter into the calculation of m(E). Erlykin et al. have given a comparison of the response function calculated by themselves with one calculated by Gaisser and the one given by Ahluwalia and Ericksen. The differences between the first two response curves are small but these two differ by a larger amount from the third. However, all authors agree that when transforming from energy to rigidity, the energy scale has to be increased by about 25% because of the presence of alphas and heavier nuclei in the primary radiation. Erlykin et al. have given the median primary energy for a recorder at 60 m.w.e. as 230 ± 30 GeV as compared to 240 GeV obtained by Gaisser. These values are substantially higher than the value 130 to 140 GeV found by Ahluwalia and Ericksen.

2. Effects of the Interplanetary Magnetic Field

The average properties of the IPMF have been found by space probes travelling between the orbits of Jupiter and Mercury and close to the ecliptic plane to be consistent with the spiral configuration predicted by the Parker solar wind model. Although large deviations of direction and field strength are known to occur on a daily basis, it is a smooth field which is constant in time and independent of heliographic longitude that has been adopted for trajectory calculation purposes. The field components used in the calculation were

$$B_r = 3.5 \times 10^{-5}/r^2$$

$$B_\theta = 0$$

$$B_\phi = 3.5 \times 10^{-5} \cos\theta/r$$

where r is the distance from the sun measured in AU and θ and ϕ are heliographic latitude and longitude respectively. It was further assumed that the IPMF possessed a sector structure with alternate positive and negative sectors having field directed respectively away from and towards the sun. Four equal sectors, each extending over 90° of heliographic longitude were assumed for purposes of the calculation. Using this field model the equations of motion for negatively charged particles emitted from the earth in various directions have been integrated and the asymptotic directions of the trajectories determined at r = 50 AU.

The starting conditions were varied for direction, position of the earth relative to the sector structure and particle energy so as to produce a representative selection covering the motion of the earth during the year and the relevant part of the primary spectrum. Asymptotic directions were determined at two hourly intervals throughout the sidereal day for primaries with rigidity 100, 150, 200, 300, 500 and 9000 GV. In each case the computations were carried out for the solstitial and equinoctical positions of the earth assuming first that it lay at the mid-point of a negative sector and then at the mid-point of a positive sector. In all eight asymptotic directions were determined for each even hour of the sidereal day.

The reader is referred to Figure 4 in the paper by Speller et al. (1972) for a presentation of the results of the calculation. The results are shown separately for each value of the rigidity specified earlier. In each case, cos α is plotted as a function of sidereal time where α is the angle between the asymptotic direction and the direction of maximum intensity assumed to be that of the solar apex at RA = 18 hr. and δ = 34°N. Eight separate points corresponding to the eight starting conditions as well as curves fitted through the means of the eight values are given in the diagram. It can be seen from the figure that the scatter is so large at a rigidity of 100 GV that the mean value of cos α is not appreciably different from zero at any time of day. This implies that the smearing out of any galactic anisotropy by the deflecting action of the IPMF is fairly complete at this low rigidity. The situation improves rapidly as the rigidity increases until at a rigidity of 300 GV the amplitude of the mean curve has risen to be over 80% of the amplitude observed at a rigidity of 9000 GV. Thus the reduction of amplitude by the action of the IPMF is less than 20% above about 300 GV. It is thus possible to construct a graph of the ratio of the amplitude at earth to that outside the solar cavity as a function of primary rigidity. This curve rises rapidly from nearly zero at 100 GV to reach 80% at 300 GV and flattens out so that the rise is very gradual at higher rigidities. It is this curve that is very useful in deriving the sidereal amplitude at the earth caused by a galactic anisotropy. The results obtained by Speller et al. are essentially similar to those obtained by Barnden (1973) who has considered both 4-sector and 2-sector structures for the IPMF.

3. Application to the Case of Solar Motion

Even if the cosmic radiation were completely isotropic in the Galactic rotation frame, a small anisotropy is expected at the earth due to motion of the sun relative to this frame with a speed of about 18km/sec. in the direction of the constellation Hercules. The Compton-Getting expression then shows that the anisotropy outside the heliosphere will have an amplitude and phase which are independent of primary rigidity. The amplitude is about 0.028% and the time of maximum near 18hr. Because of the tilt of the earth's axis with respect to the Galactic equatorial plane, this full amplitude cannot be observed and it is the variation of cos α (where α is the angle between the telescope axis and the direction of solar motion) that determines the amplitude of the observed anisotropy.

Because of the symmetrical sector structure whereby the earth spends roughly equal times in the inward and outward spiral fields, there will be no systematic phase difference between the variation at the earth and that outside the solar cavity. Unlike the case of geomagnetic deflection, it is not necessary to carry out a vector summation in order to determine the anisotropy at the earth. Thus, if A is the amplitude of the anisotropy for the 9000 GV particles, the observed amplitude at the earth is given by

$$a = A \int_{E_{min}}^{\infty} r(E) \cdot W(E) \, dE$$

where r(E) is the ratio of amplitude at the earth to that outside the heliosphere. In practice, the integration is replaced by summation over about ten suitable rigidity intervals. In the case of the detector at 60 m.w.e. depth in London, the observed amplitude is about 60% of that which would be observed in the absence of the IPMF.

V.II DISCUSSION AND AUTHOR'S COMMENT

It was noted by Meyer that the value 40/p used for the radial density gradient in the calculation of the Swinson effect was unacceptably high in the light of recent spacecraft measurements. This fact merely makes the difficulties encountered by the Swinson model almost insuperable.

In reply to a question as to why the calculations of the sidereal anisotropy at the earth ignored the effects of the geomagnetic field, it was stated that the geomagnetic field had practically no effect on those primaries which are capable of traversing the IPMF without suffering large deflections.

Meyer stated that the IPMF at the earth's orbit undergoes large fluctuations and inquired what effect these would have on trajectory calculations which assumed a smooth and constant value for the IPMF. It was stated in reply that a smooth field was more efficient in deflecting cosmic rays than a turbulent one. At Denver, Elliot pointed out that since the deflection increases as square root of B, the errors of the assumption of a constant field are not serious. This assertion is supported by the results of Barnden who used two different values for the IPMF which differed by 50% and could not find any appreciable differences in the attenuation factor for the amplitude of a galactic anisotropy at the earth.

It was pointed out by Wolfendale that if the attenuation factor introduced by the IPMF was close to 2 at 60 m.w.e depth, it is likely to be considerably higher at 30 m.w.e. which is the depth of the Hobart station, perhaps as much as 3. This coupled with the large residual vector of 0.03% at Hobart would imply that the amplitude of the anisotropy observable in the Southern Hemisphere could approach 0.1% outside the heliosphere. In reply it was stated that this was certainly true if the residual vector represented a Galactic anisotropy but since this separation depends crucially on the validity of the Swinson model (and doubts had been cast on the validity of the Swinson model) it is not possible to be certain about the large amplitude of the anisotropy. However, the equatorial scan telescopes operating in London show the wrong phase and therefore fail to confirm the existence of a large anisotropy in the Southern Hemisphere.

ACKNOWLEDGEMENTS

The author is indebted to his colleagues at Imperial College, particularly Professor H. Elliot, F.R.S., for many stimulating discussions. Special thanks are due to Dr. R.D. Speller and Dr. J.C. Dutt who rendered valuable assistance in the computation and numerical reduction of the data. The trajectory integrations in the interplanetary magnetic field were carried out by Dr. R.D. Speller. We wish to acknowledge the financial assistance given by the Science Research Council towards the acquisition of the recording equipment used in the London underground experiment.

REFERENCES

Ahluwalia, H.S. and Ericksen, J.H. (1971), J. Geophys. Res. 76 6613.

Barnden, L.R. (1973), Proc. Int. Conf. on Cosmic Rays, Denver, 2, 963.

Dorman, L.I., (1957), Cosmic Ray Variations (Moscow: State Publishing House for Tech. and Theor. Lit.)

Dutt, J.C. and Thambyahpillai, T., (1965), J. Atmos. Terrest. Phys., 27, 349.

Elliot, H., Thambyahpillai, T. and Peacock, D.S. (1970), Acta Phys. Hung. 29 Suppl.1, 491.

Erlykin, A.D., Ng. L.K. and Wolfendale, A.W. (1974) J. Phys. A. (in the press).

Farley, F.J.M. and Storey, J.R. (1954), Proc. Phys. Soc. A67, 996.

Gaisser, T.K. (1974), J. Geophys. Res. 79, 2281.

Groom, D.E. (1973), Proc. 13th Int. Conf. on Cosmic Rays, Denver, 2, 851.

Humble, J.E., Fenton, A.G. et al. (1973), Proc. 13th Int. Conf. on Cosmic Rays, Denver, 2, 976.

Speller, R.D., Thambyahpillai, T and Elliot, H., (1972), Nature, Lond. 235, 25.

Svalgaard, L. (1972), Danish Meteorolog. Inst. Geophys. Papers, paper R-29.

Swinson, D.B., (1969), J. Geophys. Res. 74, 5591.

Swinson, D.B., (1973), Proc. 13th Int. Conf. on Cosmic Rays, Denver, 2, 970.

Thambyahpillai, T. and Speller, R.D., (1974), Planet. Sp.Scien. (in the press)

Trefall, H., (1955), Proc. Phys. Soc. A68, 625.

ENERGY SPECTRUM AND MASS COMPOSITION OF COSMIC RAY NUCLEI FROM 10^{12} TO 10^{20} eV

A.A. Watson

Department of Physics, University of Leeds, Leeds 2.

I. INTRODUCTION

A view of the cosmic ray energy spectrum above 10^{12} eV which has been popular during the last ten years is one in which the spectral index above 10^{12} eV changes fairly sharply every 3 decades in energy. Between 10^{12} and 10^{15} eV the primary intensity was believed to fall with an integral exponent of -1.6; from 10^{15} to 10^{18} eV the fall became significantly steeper (exponent = -2.2) but above 10^{18} eV the, earlier, gentler, slope was resumed. It was often held that the steepening above 10^{15} eV was associated with an enrichment of heavy nuclei in the cosmic ray beam consequent upon the onset of leakage of protons from the galaxy, while the feature at 10^{18} eV has been associated with the appearance of extragalactic particles, probably protons, as the dominant component in the primary flux.

Evidence is accumulating which suggests the need for a revision of this picture. At the lower energies (10^{12} to 10^{15} eV) the prospects of identifying specific astronomical objects, for example pulsars, as the sources of cosmic rays have been heightened by claims that there are differences between the spectral indices of different nuclei at energies of about 10^{10} eV/nucleon, and near 10^{15} eV the details of the spectral break have begun to appear rather complex. At greater energies a major question is whether or not the cosmic rays are extragalactic; the flattening of the spectrum above 10^{18} eV has been seriously questioned and at the highest energies (>5.10^{19} eV) attention is focussed on the problem of seeking spectral features which may have cosmological interest.

J. L. Osborne and A. W. Wolfendale (eds.), Origin of Cosmic Rays, 61-95. All Rights Reserved.
Copyright © 1975 by D. Reidel Publishing Company, Dordrecht-Holland.

In these lectures evidence relevant to the primary energy spectrum and mass composition is discussed. The two problems are most intimately linked in the energy range considered here as nearly all experiments have to be made deep in the atmosphere on secondary phenomena so that interpretation is by way of iterative procedures in which changes in the energy, the primary mass and the nuclear physics are successively made in attempts to reach a consistent viewpoint. Although the extrapolation of the nuclear physics from machine to cosmic ray energies is an important aspect of the interpretation of cosmic ray data it will not be discussed specifically; a useful review of this topic has been made by Wdowczyk and Wolfendale (1973).

II COMPOSITION AND SPECTRUM MEASUREMENTS IN THE RANGE $10^{12}-10^{14}$ eV

1. Direct Measurements

Amongst recent results relevant to the question of the origin of cosmic rays are those of the group from the Goddard Space Flight Centre (Ramaty et al. 1973) which suggest that the spectra of Fe nuclei, in the range 3 to 50 GeV/nucleon is significantly flatter than that of carbon, nitrogen and oxygen. For Fe the differential spectral index, γ, is -2.12 ± 0.13 while for C, N and O it is -2.64 ± 0.04. The greater slope is also a good fit to the proton and α-particle spectra in the same energy range, although there is evidence that both of these spectra become slightly steeper, $\gamma = -2.7$, at higher energies for which there are no comparable data for the heavier nuclei. Ramaty et al. propose that these data imply that iron nuclei could have a different origin from the rest of the primary cosmic rays. While there is not complete agreement about the accuracy of the GSFC results (Golden et al. 1974, Webber 1973) it is important to examine evidence on the spectral slope and composition above 10^{12} eV to see if any similar trends are apparent at higher energies.

The only direct method of measurement available for studying nuclei above 10^{12} eV is the ionization calorimeter technique pioneered by Grigorov et al. (1965) in which the particle of interest interacts in an absorber interspersed with ionization detectors. With a sufficiently large device most of the ionization produced by the primary particle and its secondaries is absorbed enabling the energy of the incoming particle to be estimated. The charge of the primary can be found using a system of proportional counters and Cerenkov light detectors. The ionization calorimeter of the GSFC group (Ryan et al. 1972, Balasubramanyan and Ormes 1973) is a more sophisticated instrument than that of Grigorov et al. (1971) but unlike the latter it has not yet been flown in a satellite. Results from the two instruments can be compared for protons and α-particles and the agreement as to slope and intensity is good below 10^{12}eV.

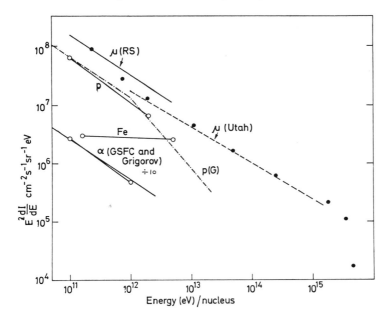

Fig. 1: Energy spectrum measurements, from direct and indirect methods $10^{11} \div 10^{15}$eV.
μ(RS) - Ramana Murthy and Subramanian (1972); μ(Utah) - Elbert et al. (1973); o———o p, α and Fe data from Goddard Flight Center (GSFC) - Ryan et al. (1972), Ormes and Balasubramanyan (1973); p(G) and ● Grigorov et al. (1971).

Above 10^{12}eV/nucleus the only direct measurements are those of Grigorov et al. (1971) and these data exhibit features which, if substantiated in an independent experiment, would be of extraordinary importance. It was found, using apparatus carried by PROTON 1, 2 and 3 satellites, that the proton spectrum steepened from an integral spectrum index, γ + 1, -1.7 to about -2.5 at 10^{12}eV. The "all particle" spectrum measured in these experiments showed no such steepening and data from an instrument on board the PROTON 4 satellite have been used to make a confirmatory measurement. Grigorov et al. claim that their results indicate the rapid disappearance of ~35% of the primary cosmic rays. the PROTON 4 results also show a 'cut-off' in the all-particle spectrum just above 10^{15}eV but this might arise because of the limited volume of their calorimeter.

2. Indirect Measurements

(a) <u>Conclusions based on muon data.</u> Amongst the best measured quantities in cosmic ray physics are the muon intensity and the μ^+/μ^- charge ratio observed at sea level. A number of attempts have been made to use these data to investigate the behaviour of nuclear processes above machine energies and to deduce details about the slope and mass composition of the primary cosmic ray spectrum.

Ramany Murthy and Subramanian (1972a) have used information on the sea-level muon momentum spectrum and the slope of the primary nucleon spectrum to test the validity of Feynman scaling. The muon spectrum extends to 4000 GeV/c so that the upper limit of the range of energies tested corresponds to about 2.10^{13} eV. They conclude that scaling is indeed valid to this energy, irrespective of the <u>magnitude</u> of the fluxes of the primary nucleons provided the <u>differential</u> exponent of the nucleon spectrum is approximately -2.67, which is close to the value derived by Grigorov et al. for the 'all particle spectrum' and to that of Ryan et al. for protons and α-particles below 10^{12} eV.

It is not feasible to determine the scaling parameters at these energies by comparing the measured nucleon flux with sea-level muon measurements since the primary nucleon flux is insufficiently well known, but Ramana Murthy and Subramanian (1972b) have attempted to derive a precise nucleon flux <u>assuming the spectral index to be -2.67,</u> by adopting the scaling parameters determined at machine energies. Their flux estimates, multiplied by 1.55 to convert to energy/nucleus, are plotted in figure 1 together with the 'all-particle' data of the Grigorov group and the approximate agreement of these independent results lends some support to the Grigorov conclusions. The indirect measurements of the nucleon flux would change by $\pm$ 8% for a change in γ of $\pm$ 0.05 and is claimed to be uncertain by an additional $\pm$ 18% because of uncertainties in the scaling parameters. The approach of these authors assumes that the scaling parameters are not significantly changed due to the difference between proton-proton and proton-light nucleus collisions and they claim that this uncertainty plus their adoption of the superposition model to describe heavy nucleus - air nucleus collisions may lead to a systematic over-estimate of the nucleon flux of about 15%. Work by the Utah group (Elbert et al. 1973) along similar lines suggests that this systematic error is considerably underestimated. Their nucleon flux estimates are about 1.8 times lower than those of Ramana Murthy and Subramanian (Figure 1) and the discrepancy appears to lie partly in a different choice of scaling parameters and partly in different mean free paths for proton and pion interactions.

The analyses just described do not give any information on the mass composition but measurements on the ratio of positive to negative muons can be used as this quantity is sensitive to the ratio of neutrons to protons in the primary nuclei (Pal and Peters 1964). At muon energies of about 10^{10}eV the muon charge ratio, μ^+/μ^-, is about 1.28 and remains more or less constant with energy up to a few TeV where measurements are due to the Utah group (Ashley et al. 1973). An increase in the percentage of the Fe nuclei present at a few TeV/nucleon would cause a decrease in the μ^+/μ^- ratio at high energies because of the increased number of neutrons which would be present in the incoming cosmic beam. Calculations along these lines have been carried out by the Durham (Daniel et al. 1974) and Utah (Elbert et al. 1973) groups; the Utah calculations seem to be the most detailed. Elbert et al. find that if the results of Ramaty et al. (1973) are extrapolated to the range 0.4 to 20 TeV/nucleon then the charge ratio would decrease by about 0.13 which is excluded by the experimental results, and so their work argues against the extrapolation of the new, lower energy, composition data. The Utah charge ratio results are also inconsistent with the increase in the proton spectrum exponent observed in the PROTON 1, 2 and 3 satellite experiments.

(b) <u>Conclusions based on EAS data.</u> All of the spectral and composition information at energies above about 10^{14}eV/nucleus has to be based on the rather indirect technique of studying the properties of the extensive air showers (EAS) created by the successive interactions of the progeny of the primaries. At energies below 10^{14}eV the information available from EAS is considerably less direct than that from muons but some interesting preliminary results have been obtained.

First we consider results from techniques using Cerenkov light produced by EAS. Many of the particles in the EAS produced by primary cosmic rays are sufficiently energetic to produce Cerenkov light in the atmosphere. The first detection of this light was achieved some time ago (Galbraith and Jelley, 1953) but there has been a renaissance of interest in the field. At the energies of concern here the light signal is rather weak and search light mirrors and good 'seeing' at high altitudes are required to obtain useful results.

A particularly exciting development is the method of Grindlay (1971) and Grindlay and Helmken (1973) used at Mt. Hopkins (2260m) which is claimed to be capable of measuring the muon/electron ratio in 10^{12}eV EAS. Studies of the angular distribution of the detectable atmospheric Cerenkov radiation have revealed a distinct component which is identifiable with locally penetrating muons (4.5 GeV) close to the EAS axis. This portion of the distribution is rich in the ultra violet by comparison with light from the electrons present at shower maximum (about 330 g cm^{-2} at 10^{12}eV) because of atmospheric absorption. Two search light mirrors, spaced at ∼100m and pointing at the correct angle are used to detect the electron maximum, while a third reflector at one end of the baseline, is aligned to detect the same EAS at its muon 'core'. The ratio of the two pulse heights is proportional to the muon/electron ratio and thus the measurements may reflect the cosmic ray mass distribution. To extract detailed mass information requires careful comparison of the results with Monte Carlo calculations and Grindlay and Helmken (1973) claim that they find evidence of an enhanced iron abundance at about 2.10^{12}eV/nucleus. Their comparison is not entirely convincing but support for the principle of the method comes from showers with effectively zero μ/e ratio (γ-ray primaries?) detected in the direction of the Crab Nebula pulsar, synchronous with the periodic (33ms) light flash (Grindlay et al. 1973). The method is potentially very powerful as it combines a large area detector (10^4 m^2) of high angular resolution (10^{-4} sr) with the possibility of identifying major mass groupings.

A novel approach to the problem of determining the energy spectrum above 10^{12}eV has been begun by Gerdes, Fan and Weekes (1973). They are using search light mirrors of 1.5 m and 10 m diameter at Mt. Hopkins to measure the photon density spectrum from which the primary energy spectrum can be deduced. A very preliminary result indicates that the differential energy spectrum slope is about -2.4.

We now turn to results from conventional EAS detection methods. The particles of EAS produced by primaries of energy 10^{12}-10^{14}eV can only be detected directly by operating an array of particle detectors at high altitudes and at Tien Shan (3333m) the number spectrum of EAS has been measured in association with very high energy (> 0.3 TeV) muons detected with an underground calorimeter (Erlykin et al. 1973). The ratio of the intensity of showers greater than a certain size, and containing a muon above a given energy, to the intensity of muons above that energy is dependent upon the composition of the primary beam and on the nuclear physics and fragmentation features of the collisions of primary nuclei with air nuclei. For a model in which an incoming nucleus is completely fragmented in the first nucleus - air nucleus collision the muon energy is proportional to E_p/A while the shower size N, is proportional to $A^{(1-\alpha)}E_p$, where α is greater than 1. Hence if the composition is enhanced with heavy nuclei the ratio $I(N > 3.10^4)/I(N > 0)$ will be increased. The data shown in Figure 2 apparently exclude an enhanced heavy composition but application of the scaling model and of a more appropriate fragmentation picture for heavy nuclei (e.g. Waddington and

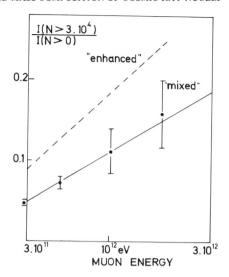

Fig. 2: From Erlykin et al. 1973. Ratio of intensity of showers with $N > 3.10^4$ particles to intensity of 'muon-only' events as a function of energy of detected muon. "Mixed" implies a composition as commonly given at lower energies and "enhanced" a composition deficient in protons as given by Grigorov et al. (1971).

Frier 1973) would tend to reduce the predicted difference. Further calculations are needed before drawing definite conclusions from these interesting data which refer to primaries in the range 10 - 100 TeV.

III SPECTRUM AND MASS MEASUREMENTS IN THE RANGE 10^{14} to 10^{17} eV

All information on the energy spectrum and mass composition of cosmic rays above 10^{14} eV has to be deduced from the EAS created by these particles in the atmosphere. Study of the range 10^{14} to 10^{17} eV has been particularly intense since early indications that a sudden steepening or 'knee', in the energy spectrum occurs at about 10^{15} eV were interpreted (Peters 1960) as implying evidence for leakage of cosmic rays from the galaxy. This hypothesis leads naturally to the idea that the mass composition might change fairly rapidly and become significantly richer in heavy nuclei at energies greater than 10^{15} eV.

1. Estimations of the Primary Energy Spectrum.

Of the two distinct methods used to derive the primary energy

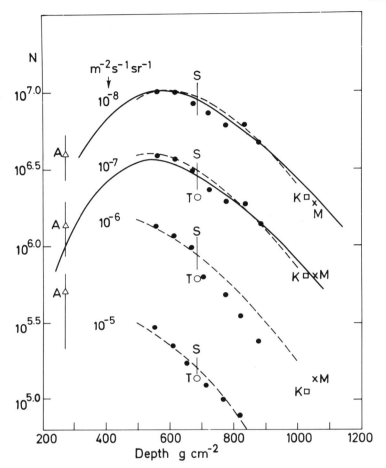

Fig. 3: Comparison of theoretical longitudinal development curves with experiment. ⌀ Antonov et al. 1971 (at intensities 10^{-5}, 10^{-6} and 10^{-7} m^{-2} s^{-1} sr^{-1}). ● Chacaltaya data (Bradt et al. 1965, La Pointe et al. 1968).
M × Moscow data (Khristiansen et al. 1973).
K □ Kiel data (Hillas 1974).
T ○ Tien Shan data (Aseiken et al. 1971).
full line: Hillas (1974) model E and primaries of mass 10.
dash line: Gaisser (1974) scaling model and Fe primaries.
'S' marks a possible irregularity in the Chacaltaya data.

spectrum from EAS data, the most direct, in that it is less dependent on the uncertainties of models of shower development, is to measure the total energy deposited by the shower particles in the atmosphere, adding to this estimate the small fraction of energy remaining unabsorbed at the ground; over 80% of the energy of the primary should be dissipated in ionization of the atmosphere. The other less direct method is based on a comparison of the observed electron and/or muon number spectra with that expected under certain assumptions about the shape of the primary spectrum and the mass composition. In principle this approach is similar to that based on the single muon spectrum at lower energies (section II.2(a)) but it is more susceptible to error than that work because of the greater experimental uncertainties and because of the extrapolation of the nuclear physics which is involved. We will consider results from both of these methods.

(a) <u>Energy Deposition in the Atmosphere.</u> The growth and decay of the number of charged particles, N, in a shower as a function of atmospheric depth, x, is used to find the total energy deposited in ionization from the area $\int N(x)dx$ under such a development curve, multiplied by the rate of energy loss per particle per g cm^{-2}. This method is analogous to the use of an ionization calorimeter at lower energies. A development curve is established by measuring the size of showers at a given integral rate per unit solid angle as a function of atmospheric depth, the depth being varied either by using an aeroplane (Antonov et al. 1971) or by studying EAS at various inclinations at a mountain site (Bradt et al. 1965, La Pointe et al. 1968). Data from these experiments are shown in figure 3. Those for intensities of 10^{-7} and 10^{-8} m^{-2} s^{-1} sr^{-1} are well fitted by a model calculation in which the mass of the primaries is assumed to be constant and equal to 10 (Hillas 1974), except for the aeroplane results which

Intensity, I_s m^{-2} s^{-1} sr^{-1}	Energy (eV) (Hillas)	$\gamma+1$	Energy (eV) (Gaisser)	$\gamma+1$
10^{-5}	-	-	5.9×10^{14}	1.64
10^{-6}	-	-	2.4×10^{15}	2.39
10^{-7}	5.9×10^{15}	2.31	6.3×10^{15}	2.47
10^{-8}	1.6×10^{16}	1.86	1.6×10^{16}	2.10
10^{-9}	5.5×10^{16}	2.04	4.8×10^{16}	2.15
10^{-10}	1.7×10^{17}	1.96	1.4×10^{17}	1.97
10^{-11}	5.5×10^{17}		4.5×10^{17}	

$\gamma + 1$ = integral slope between I_j and I_{j+1} where $I = kE^{-(\gamma+1)}$

Table I

are of rather low statistical weight and zenith angle accuracy and may additionally be influenced by geomagnetic effects. Hillas finds that 'any model so far tested which fits the points beyond the maximum of the shower gives virtually the same area before maximum' and estimates an energy of 5.9×10^{15}eV to correspond to 10^{-7} m^{-2} s^{-1} sr^{-1}. Other energy assignments are listed in Table I.

Data from the unique high altitude laboratory at Chacaltaya (550 g cm^{-2}) have deservedly received particular attention in many discussions and it has been noted for some time that the development curves for the higher intensities (10^{-5} and 10^{-6} m^{-2} s^{-1} sr^{-1}) show a more rapid absorption of the showers with depth than expected. A common interpretation of this observation (e.g. Kalmykov et al. 1973) is that there must be a marked increase (perhaps $n_s \propto E^{\frac{1}{2}}$) in the dependence of the multiplicity of secondary particles upon the energy released in nuclear interactions at about 10^{13} - 10^{14}eV, the aeroplane data of Antonov et al. (1971) being cited to support the very high shower maximum which would be expected under such circumstances. Hillas (1974), however, has argued against such a change as he finds agreement with a wide variety of EAS data at energies above 10^{16}eV using a more traditional model with a multiplicity variation more like $E^{\frac{1}{4}}$; he has suggested that many of the showers near 10^{15}eV are produced by 'very heavy nuclei', which are not dominant above 10^{16}eV but which may be linked to the high flux of Fe nuclei seen at 10^{12}eV/nucleon. Gaisser (1974) has tried to fit the development curves of the Chacaltaya experiment using Fe primaries and Feynman scaling and his results are also shown in Fig. 3 and Table I. While his calculations agree reasonably with the data for 10^{-7} and 10^{-8} m^{-2} s^{-1} sr^{-1} and the energy assignments are also close to those of Hillas, the fit with the curves is exceedingly poor at 10^{-6} m^{-2} s^{-1} sr^{-1} though slightly better at 10^{-5} m^{-2} s^{-1} sr^{-1}.

There have been few critical discussions of the Chacaltaya longitudinal development curves, perhaps because of the paucity of information which has been published about their methods of derivation, but two points may be worth noting. Firstly, data from Tien Shan, Kiel and Moscow (Fig. 3) do not, taken on their own, provide compelling evidence for very rapid absorption of EAS. Secondly, the Chacaltaya curves show an irregularity in the region marked S in figure 3, between depths of 665 and 705 g cm^{-2} which may be indicative of a zenith angle dependent systematic error in the size spectra from which the development curves are derived.*

---

* Earlier data from Chacaltaya (Clark et al. 1963) show stronger evidence of this irregularity and zenith angle measurements may have been less accurate during this epoch (Barker et al. 1965).

(Continued on next page ....)

A reduction of the sizes or intensities measured at depths less than 705 g cm^{-2} or an increase of the sizes or intensities at greater depths would smooth the development curves and there are a number of factors which might be worth checking in these respects.

It is well known (Murzin 1965, Edge et al. 1973) that random errors in size measurement can lead to an overestimate of the intensity at a given size. Far from the triggering threshold the intensity is enhanced by a factor

$$F = \exp(\tfrac{1}{2} \sigma^2 (\gamma-1)^2)$$

where γ is logarithmic standard deviation and γ is the modulus of the differential spectral index. For a differential slope of about -2.7 the size estimates would need to be uncertain by a factor of more than 3 for the near vertical showers, while retaining an accuracy of better than 60% in very inclined events, to explain the irregularity. Such a large random error seems unlikely.

A systematic error may arise in near vertical showers due to the influence of the geomagnetic field. Chacaltaya is close to the geomagnetic equator and showers in near vertical directions will be distorted from circular symmetry.

The effect of errors in the measurement of the zenith angles of showers must be considered. Barker et al. (1965) have reported that angles determined from the timing array are <u>systematically</u> smaller (by about 2.5°) than those measured in the cloud chamber. If the cloud chamber measurements are correct the systematic error is in the right sense and of such a magnitude that the measured longitudinal development curves will be significantly too steep. To evaluate this effect accurately** requires a detailed knowledge of the source of the systematic error and its variation as a function of θ and shower size: complete information is not available and the work of Barker et al. was restricted to zenith

---

* (Continued from previous page)
It may or may not be significant that data at 10^{-5} m^{-2} s^{-1} sr^{-1} are not included in the final publication from the Chacaltaya group (La Pointe et al. 1968).

** An approximate calculation assuming $\sigma_\theta = 2.5°$ and a systematic error of 2.5° in the sense observed shows, for an intensity distribution of showers varying as $\cos^5\theta$, that $I_{obs}(\theta < 20°)/I_{exp}(\theta < 20°) = 1.2$ while $I_{obs}(50°-60°)/I_{exp}(50°-60°) = 0.79$. I_{obs} and I_{exp} are the observed and expected intensities respectively.

angles less than $30°$ and shower sizes in the range $10^5 < N < 10^7$.

The above factors are likely to be more important for small showers ($N < 10^6$) but quantitative estimates of their magnitude require detailed information about the Chacaltaya data.

(b) <u>Indirect Estimates of the Energy Spectrum.</u> Many groups have measured the number spectrum of showers produced by primary cosmic rays of energies 10^{14} to 10^{16}eV and in general the observation first made by Kulikov and Khristiansen (1958) that the spectrum steepens around an electron sea-level size of about 4.10^5 particles has been strongly confirmed. A similar steepening in the number spectrum of muons has been found both at sea-level (Vernov et al. 1968) and at mountain altitudes (Aseikin et al. 1971). That the break in the electron number spectrum does not arise due to systematic effects has been argued cogently by Vernov and Khristiansen (1968), and there seems little room for doubt that both the electron and muon "knees" are real features. Estimates of the energy of the primaries which produce showers near these spectral breaks are mostly close to 4.10^{15}eV, and, at least among air shower workers, it is commonly believed that the primary energy spectrum undergoes a rather marked change of slope at this energy.

One of the more detailed attempts to fit the electron data using an assumed primary spectrum and a specific shower development model has been that of Hillas (1974), shown, in part, in figure 4. In this calculation it is assumed that the primaries have a mass of 10 and that the primary spectrum slope changes from -2.5 to -3.3 at 4.10^{15}eV. The fits to Chacaltaya results and sea-level data (from Moscow and Kiel) are quite impressive. The intermediate altitude measurements (Miyake et al. 1971) and Aseiken et al. (1971) are not so well explained but the internal consistency of these two experiments is rather poor. Hillas points out that if the primary spectral index was constant above 4.10^{15}eV then the sea-level number spectra would tend to be convex while it is observed to be concave (figure 4 and Khristiansen et al. 1973). This implies a flattening of the energy spectrum and Hillas has likened the spectral shape to that first reported by Linsley in 1963, except that the levelling off of the spectrum is at an energy more than an order of magnitude lower than was indicated then.

A further feature of this analysis, and that of other workers (Vernov and Khristiansen 1968, Aseiken et al. 1971, Kempa et al. 1974), is that the primary spectrum slope before the 'knee' at 4.10^{15} appears to be less than the proton and alpha particle slope measured by Ryan et al. (1972). Nevertheless it must be stressed that although the increase in slope above 4.10^{15}eV seems well established the value before the

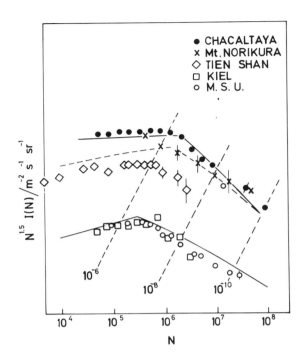

Fig. 4: Adapted from Hillas (1974). Integral size spectrum of showers at various altitudes compare with predictions based on a primary spectrum in which the exponent changes from −2.5 to −3.3 at 4.10^{15}eV. The primaries were assumed to have mass 10.

---

knee is only poorly known as the summary of intensity measurements in Table II, and the uncertainties about the Chacaltaya data, indicate. What is clear is that at 5.9×10^{14}eV the intensities deduced from EAS experiments are significantly higher than those derived by extrapolation from lower energies.

(c) <u>Summary of Spectral Information, 10^{14} to 10^{17}eV.</u> There is considerable evidence for complex structure in the energy spectrum from 10^{14} to 10^{17}eV. The independent analyses of Hillas (1974) and Gaisser (1974) are summarized in Table I and there is agreement that the integral spectral index from $2.4\ 10^{15}$eV to $1.6\ 10^{16}$eV is -2.39 ± 0.08, which is substantially greater than the index for energies less than 10^{14}eV. For at least one decade above $1.6\ 10^{16}$eV the spectral index falls to about −2.0 and there is evidence (discussed later) to suggest that this value is sustained up to near 10^{20}eV. There are indications that the spectral index from about 10^{14} to 10^{15}eV may be smaller than

Authors	5.9×10^{14} eV	5.9×10^{15} eV	Comments
Grigorov et al. (1971)	4.10^{-6}	4.10^{-8}	Satellite experiment
Hillas (1974)	–	10^{-7}	Derived from fits to Chacaltaya development curves
Gaisser (1974)	10^{-5}	$1.2 \, 10^{-7}$	
Vernov and Khristiansen (1968)	4.10^{-6}	$7.5 \, 10^{-8}$	Reviews of data of EAS
Nikolsky (1970)	4.10^{-6}	$4.5 \, 10^{-8}$	
Kempa et al. (1973)	$1.7 \, 10^{-5}$	$2.5 \, 10^{-7}$	
Ramana Murthy et al. (1972)	$2.7 \, 10^{-6}$	$6.3 \, 10^{-8}$	Extrapolations from μ-analysis (section 2)
Elbert et al. (1973)	$1.6 \, 10^{-6}$	3.10^{-8}	
Ryan et al. (1972)	6.10^{-7}	1.10^{-8}	Extrapolation of proton spectrum.

Integral intensity estimates ($m^{-2} \, s^{-1} \, sr^{-1}$)

Table II: Primary energy spectrum intensity estimates at two energies, above and below the 'knee' at 4.10^{15} eV.

---

the value of -1.7 found at lower energies. These results are probably the most accurate available as they are obtained by the calorimetric method.

An interesting proposal which explains these features in an approximate manner has been made by Karakula et al. (1974). They suggest that the flattening in the spectrum at energies less than 10^{15} eV occurs because of the presence of cosmic rays which have been accelerated in pulsars and they have made a rather successful attempt to calculate the spectrum to be expected on the basis of one specific model of a pulsar cosmic ray acceleration process (Gunn and Ostriker 1969). The spectral shape outlined in (a) and (b) is also explained and it is predicted that above 10^{16} eV a large fraction of heavy nuclei might be present over a narrow energy range which just precedes the region in which non-pulsar accelerated cosmic rays again becomes dominant. Gold (1974) has concluded independently that pulsars might be important sources of galactic cosmic rays of energy 10^{14} to 10^{16}eV.

2. The Mass Composition from 10^{14} to 10^{17} eV

It is worth recalling that definite charge identifications have been made of most nuclei between protons and iron between 10^{14} and 10^{15}eV using nuclear emulsions (Teucher et al. 1959) but otherwise the mass resolution of all experiments is so poor that the only questions to which answers can be sought relate to changes relative to the composition at low energies (Ginzburg and Syrovatsky 1963).

(a) <u>Studies Close to the Cores of EAS.</u> Because of the relatively high intensity of primaries of about 10^{15}eV it has been possible to attempt to deduce their nature by studying the central regions of EAS, and for a shower of a given electron size it is expected that the density of electrons near the core will be higher for showers initiated by protons than for those initiated by heavier nuclei. This approach was first suggested by the Sydney group (Bray et al. 1964); distributions of the central density, Δ, have been measured as a function of shower size, N, by them and by workers at Kiel (Samorski et al. 1971). While the Sydney workers (McCusker et al. 1969) find a decrease in $(\Delta/N)_{av}$ and $\sigma(\Delta/N)$ with shower size which they interpret as indicating that the mean mass becomes progressively greater above 3.10^{15}eV than at lower energies, Samorski et al. claim that (Δ/N) and $\sigma(\Delta/N)$ remain substantially constant and are consistent with a proton or 'mixed' composition up to nearly 10^{16}eV.

A mixed composition, $(\bar{A} = 10)$, as normally understood contains at a given energy per nucleus 37% protons and 27% alpha particles, but in studies <u>at a fixed shower size</u> about 75% of showers detected at sea level will have been produced by protons, from a range of energies, and only about 15% by alpha particles (Samorski et al. 1971). This 'over-dominance' of protons, due to upward fluctuations in the shower sizes produced at sea-level by protons of lower energies, makes it hard to detect small changes in the mean mass composition as a function of energy. The Kiel group claim that the observed fluctuations are such that an enhancement of heavy primaries with $A \gtrsim 10$ can be excluded in direct contradiction to the Sydney interpretation. It is not clear where the discrepancy lies.

Above 10^{16}eV only the Sydney experiment (McCusker et al. 1969) has significant data and it is claimed that most of the observed events are created by heavy nuclei (A > 16) but a dramatic change in the transverse momentum of the very high energy particles in the cores of showers must occur at about the same energy (Bakich et al. 1971). It could be that the Sydney results can be interpreted as arising from proton primaries together with a change in features of nuclear interactions detectable only in the very central regions of EAS: this possibility has not been thoroughly explored.

(b) <u>Studies on Nuclear Active Particles</u>. Theoretical studies of nuclear active particles (nap) produced in showers by primaries of 10^{14} to 10^{16} eV have indicated that it would be very difficult to gain information on the mass problem from the spatial distribution of these particles close to the core (Bradt and Rappaport 1968, Thielheim and Biersdorf 1969) but it is claimed that fluctuations of the energy flow of nap might yield mass information. Accordingly Rappaport and Bradt (1969) and Trumper et al. (1970) have measured the fluctuations in the nuclear active particle component of EAS at 550 g cm^{-2} and at sea-level respectively. Rappaport and Bradt used the 60m^2 mosaic of 15 x 4m^2 scintillators, shielded by 320 g cm^{-2} of absorber, at the centre of the Chacaltaya array while Trumper et al. employed at 14m^2 neon hodoscope shielded by 800 g cm^{-2} concrete with the Kiel EAS array. At 10^{15}eV the Chacaltaya group find fluctuations which are consistent with either a mixed composition or a relatively pure beam with an average mass of about 12. The Kiel group find, at 4.10^{15}eV, that their data are consistent with a pure proton beam or a mixed composition, while at 10^{16}eV a mixed composition seems likely though a pure iron beam cannot be excluded. In a similar type of experiment Fomin and Khristiansen (1970) have measured the variation of the energy flux of nap > 10 GeV as a function of distance from the shower axis. By comparing their results with model calculations they conclude that the average atomic mass of the primaries at 10^{15}eV is "not large".

(c) <u>Cerenkov Light Studies of EAS</u>. Theoretical studies of the total flux of Cerenkov light associated with an extensive air shower suggest that it is nearly proportional to the track length integral of electrons above the observation plane, which should depend in part upon the nature of the primary particle for a shower of constant size. Consequently if observations are carried out close to shower maximum more Cerenkov light per electron should be observed for EAS initiated by heavy nuclei than for EAS produced by protons. Hence the width of the distribution of the Cerenkov light signal at a constant shower size will depend upon the mass distribution in the primary beam.

Cerenkov light experiments made in the Pamir mountains (3860m) by Nesterova (1968) and at Mt. Chacaltaya (5000m) by Kreiger and Bradt (1969) have yielded measured distributions of the Cerenkov light per electron which are more consistent with a

mixed composition than either a pure proton or pure heavy beam.
It is to be hoped that further measurements of the Cerenkov light
component of EAS will be made in this energy range and in particular that attempts be made to measure the distribution of muon
number in showers of fixed Cerenkov light signal as this is
expected to yield accurate information on the mass distribution
(Vernov and Khristiansen 1968).

(d) <u>Lateral Distribution Function of EAS</u>. Qualitatively,
from predictions of the differences likely to be observed between
proton and heavy nucleus initiated showers, one expects that the
lateral distribution of electrons will be flatter in the latter
than in the former. In early EAS experiments insufficient detectors were available to detect such differences as might or might
not have existed between individual showers but accurate results
have now become available from the complex installation operated
by the Moscow State University Group (Khristiansen et al 1971 and
earlier work). These workers find that there are considerable
fluctuations in the lateral distribution of EAS, which are much
larger than would be expected due to instrumental effects and
which they believe indicate a dominance of protons up to 10^{17}eV.

A similar experimental result has been obtained by Catz et al.
(1973) from studies of showers produced by primaries of about
10^{15}eV. These authors have compared the fluctuations observed
with what would be expected on an isobar model for primaries of
mass 1, 4 and 10 and conclude that the fluctuations are consistent
with a dominance of protons in the primary beam.

(e) <u>Study of Muons in Inclined EAS</u>. The Durham group
(Rogers et al 1969) have studied the frequency of multiple muons
as a function of zenith angle and have compared their results
with a model in which a mean transverse momentum of ~ 0.4 GeV/c
and mean multiplicity varying as $E^{\frac{1}{4}}$ are assumed for the pions
produced in nucleon-nucleus collision. They find that the
observed frequencies are in much better agreement with a pure
proton or mixed composition than with a composition in which
heavy nuclei become dominant above 10^{15}eV.

Although there appears to be a rather strong consensus that
the mass composition does not undergo any very rapid change in
the region of the 'knee' and much of the data are consistent with
a mass distribution similar to that at 10^{10}eV it must be noted
that neither the effect of the scaling hypothesis nor the consequences of a possible increase of the proton-proton crosssection have been fully explored in relation to the very extensive
data on the composition problem. Interpretation of these data
requires constant revision until a better knowledge of the
appropriate nuclear physics is available.

IV. SPECTRUM AND MASS MEASUREMENTS ABOVE 10^{17}eV

1. The Primary Energy Spectrum

The very low flux of particles with energies above 10^{17}eV hampers their study: at 10^{17}eV the rate is about 10^{-2} m^{-2} y^{-1} sr^{-1} falling to 0.5 km^{-2} y^{-1} sr^{-1} at 10^{19}eV and perhaps to 100 times lower at 10^{20}eV where intensity and energy assignments are very uncertain. To compensate for these rates, arrays of detectors, spread out over areas of many square kilometres, have been operated by a number of groups (Table III).

Array	Altitude	Type of detector	shower size parameter	Enclosed area km^2	Exposure (above 10^{19}eV used in published data) km^2 y	Reference
Volcano Ranch	2km	8 gcm^{-2} plastic scintillators	No. of electrons, N_e	8	30	Linsley (1963)
Haverah Park	sea level	120 cm deep-water Cerenkov detectors	Density at 600m, $\rho(600)$	12	30	Edge et al (1973)
Sydney	sea level	15 gcm^{-2} liquid scintillator shielded by 375 gcm^{-2} earth	No. of muons, N_μ	45	175	Bell et al (1974)
Yakutsk	sea level	8 gcm^{-2} plastic scintillator	No. of electrons N_e	3.3	2	Kerschenholz et al (1973)

Table III

ENERGY SPECTRUM AND MASS COMPOSITION OF COSMIC RAY NUCLEI 79

Although these experiments have been carried out with different types of detector (and at altitudes other than sea level) there is sufficient information available to enable internal consistency checks to be made so that a reliable estimate of the primary spectrum can be reached <u>within the limits of the nuclear physics used to derive it.</u>

The Yakutsk and Volcano Ranch groups measure as their size parameter the number of charged particles in the shower at ground level. This is not a straight forward procedure since approximately half of the particles lie inside 50m where measurements are rarely possible. In attempts to surmount this problem Linsley (1973c) has contracted the area covered by his detectors and increased their number thus improving the probability of recording signals close to the shower axis, while the Yakutsk group (Dyakanov et al. 1973) have supplemented estimates of the shower size with measurements of the atmospheric Cerenkov light signal which, between 300 and 500 m of the shower axis, is

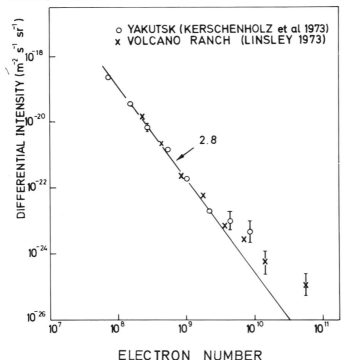

Fig. 5: Comparison of differential number spectrum as measured at Yakutsk (sea level) and Volcano Ranch (834 g cm^{-2}).
X   Volcano Ranch (Linsley 1973)
O   Yakutsk (Kerchenholz et al. 1973)

believed to be independent of shower development. The number spectra from these experiments are compared in figure 5 and the agreement is striking although at the higher differential intensities (10^{-21} m^{-2} s^{-1} sr^{-1}) the Volcano Ranch sizes would have been expected to be about 2.3 times larger than those seen at Yakutsk. The lateral distribution function used to find the shower sizes is nearly identical in these experiments and is not well controlled by data close to the axis but guidance towards an understanding of the anomalous agreement should come from the further measurements being made by Linsley (1973c) and from the accumulation of additional data at Yakutsk. For both experiments a single spectral index ($\sim$ -2.8) defines the data although there are indications of a reduction of this index above 3.10^9 particles.

A useful check on the consistency of the Volcano Ranch, Haverah Park and Sydney experiments can be made by comparing the muon signals measured there. Linsley (1973a) has given details of the muon lateral distribution function for showers recorded at an intensity of 10^{-11} m^{-2} s^{-1} sr^{-1}. Such showers contain $5.7^{+1.1}_{-0.7} \times 10^6$ muons above an energy of about 220 MeV (10 cm lead shielding) and might be expected to contain about 5.3×10^6 muons at sea level. The Sydney muon size at the same intensity is 4.2×10^6 muons above a threshold of about 750 MeV. The sizes are in the ratio 1.26, while for a wide range of models Hillas et al. (1971b) give $N_\mu(> 320 \text{ MeV})/N_\mu(> 770 \text{ MeV})$ in the range 1.14 to 1.19. Sydney and Volcano Ranch results are thus very consistent at this intensity. At Haverah Park the total number of muons is not measured but the muon density at 300m above two energy threshold levels (Armitage et al. 1973, Dixon et al. 1974) is compared in figure 6 with data from other experiments. Haverah Park results agree well with alternative muon energy spectra fitted to the Volcano Ranch and Sydney data; measurements of Armitage et al. are perhaps a little low. At an intensity of 10^{-11} m^{-2} s^{-1} sr^{-1}, therefore, one may conclude that each experiment is recording showers of the same energy to within about 30%; the energy assignments from the Chacaltaya work are in the range 4.5 to 5.5×10^{17}eV (Table I). Using Cerenkov light measurements the Yakutsk group ascribe an energy of 5.5×10^{17}eV to showers of this intensity, which at their array have an electron size of 7.9×10^7.

Such a detailed check on consistency between the basic data of the various experiments cannot yet be made at substantially greater energies but comparison of the size spectrum of muons measured at Sydney (Bell et al. 1974) with the spectrum of densities recorded at 600m from the shower axis in the Haverah Park experiment (Edge et al. 1973) is useful. The Leeds group at Haverah Park have adopted $\rho(600)$, the deep (120 cm) water-

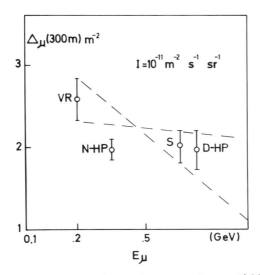

Fig. 6: Comparison of density of muons above different threshold energies, E_μ, measured at 300m in vertical showers recorded at an intensity of $10^{-11}m^{-2}$ s^{-1} sr^{-1}.
V R Linsley (1973a) corrected to sea level. ($E_\mu \sim 220$ MeV)
N-HP Armitage et al. (1973) ($E_\mu \sim 300$ MeV)
S Bell et al. (1974) ($E_\mu \sim 750$ MeV)
D-HP Dixon et al. (1974) ($E_\mu \sim 1$ GeV)
The dashed lines represent muon spectra fitted to the Volcano Ranch and Sydney data which are consistent on an independent test.

Cerenkov response at 600m from the shower axis, as their size parameter because extensive model calculations (Hillas et al. 1971a) suggest this parameter to be relatively independent of shower development features. These calculations also show that the slopes of the muon number spectrum and the $\rho(600)$ density spectrum should be closely similar (within 5%) for a wide range of models and this is in fact found (see Edge et al. 1973 for a detailed discussion). The slopes of the size parameter spectra may therefore be deemed to be consistent between these two experiments over the energy range 3.10^{17} to 10^{19}eV, with an average differential index of -3.22 ± 0.02.

Hillas (1974) has discussed how the shape of the energy spectrum may be derived from the various measured size spectra assuming that the energy E is related to the shower size parameter, S as $E \propto S^\alpha$. Then, if the differential spectrum is proportional to $E^{-3.0}$, a plot of j $S^{2\alpha+1}$ against S would be a horizontal line, and deviations in the energy spectrum from a slope of -3.0 will be readily detectable. (j is the differential flux with respect to the variable S). For the experiments under discussion

Haverah Park	$S = \rho(600)$	and $\alpha = 1.04$	(Edge et al. 1973)
Sydney	$S = N_\mu$	and $\alpha = 1.07$	(Bell et al. 1974)
Yakutsk	$S = N_e$	and $\alpha = 0.90$	(Kerschenholz et al. 1973)
Volcano Ranch	$S = N_e$	and $\alpha = 0.96$	(Linsley 1973b)

and there is little variation between models. Figure 7 reproduces the result of Hillas (1974) and it is seen that a slope of -3.0 is consistent with all experiments from 5.10^{17} to 10^{19}eV although a straight line is not followed exactly. The presence of a decrease in the spectral index near 10^{18}eV is excluded.

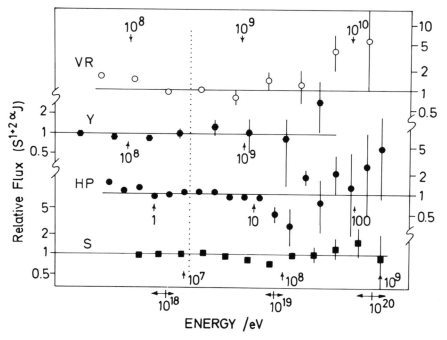

Fig. 7: From Hillas (1974). Differential shower size spectra at energies above 3.10^{17}eV for Volcano Ranch, Yakutsk, Haverah Park and Sydney. Energies estimated on the horizontal axis are for guidance only.

Using the model E of Hillas et al. (1971b) the data of the Leeds group (Edge et al. 1973) and of Sydney (Bell et al. 1974) lead to the following representations of the energy spectrum.

Leeds (Haverah Park):

$$J(E)\,dE = 6.17 \times 10^{-30} \left(\frac{E}{10^{18}\,\text{eV}}\right)^{-3.18 \pm 0.02} \text{m}^{-2}\,\text{s}^{-1}\,\text{sr}^{-1}$$

Sydney:

$$J(E)\,dE = 7.94 \times 10^{-30} \left(\frac{E}{10^{18}\,\text{eV}}\right)^{-3.07 \pm 0.02} \text{m}^{-2}\,\text{s}^{-1}\,\text{sr}^{-1}$$

Because of the difficulties already discussed it is not at this time feasible to deduce reliable estimates of the energy spectrum from charged particle size experiments but we note (a) that there is good agreement in the energy estimates at the intensity $10^{-11}\,\text{m}^{-2}\,\text{s}^{-1}\,\text{sr}^{-1}$ and (b) the derived differential energy spectrum slope may be close to -3.0.

2. The Cosmological Significance of the Energy Spectrum.

If the highest energy cosmic rays are protons from extragalactic sources - and the isotropy of the particles above $1.5.10^{19}$ eV supports this view (Linsley and Watson 1974) - interactions with the 2.7K black body radiation are expected to modify the source spectrum significantly above 5.10^{19} eV because of the incidence of photopion production at these energies (Greisen 1966, Zatsepin and Kuzmin 1966), while at lower energies ($\sim 5.10^{18}$ eV) losses due to electron pair production will change the source spectrum slightly. Even if the particles are heavy nuclei a steepening in the spectrum is expected in the region between 10^{19} and 10^{20} eV because of the effect of photodisintegration. Consequent upon these expectations, comparisons have been made by several authors of the data from the Sydney and Haverah Park experiments with the spectra predicted for various models of cosmic ray production in the sources. A typical comparison (partly reproduced in figure 8) is that of Hillas (1974) in which the protons are assumed to have been produced with a spectrum $E^{-2.75}\,dE$ and to have experienced losses due to photopion production, red-shifts and pair production (Hillas 1968). It is apparent that the experimental results can only be reconciled with a universal flux of protons if current energy estimates of the biggest showers are reduced. The experimental flux cannot be reproduced even if the cosmic ray sources were stronger in the past epochs. A similar conclusion is reached by

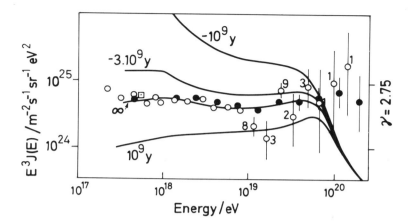

Fig. 8: From Hillas (1974). The energy spectrum derived from observations of the Leeds group at Haverah Park (open circles) and Sydney (black circles) using the Leeds shower model. The square indicates a calorimetric estimate (see text). The curves are the predicted spectral shapes for protons produced with spectral exponent -2.75 and subject to interaction with the 2.7K radiation. The times attached to the curves are leakage and absorption lifetimes.

Strong et al. (1974) who point out that random errors in the energy estimates of about 80% are required to bring the experimental data and the predictions of an $E^{-2.75}dE$ source spectrum into agreement. Wolfendale (1974) has also discussed how an enhanced intensity from nearby extragalactic sources (for example the Virgo cluster) may be present and would improve the agreement.

Both experimental groups have made careful assessments of the errors involved in their measurements of the size parameters of the largest showers upon which the estimates at the highest energies are based. Bell et al. (1974) find for all showers that the errors and uncertainties of measurement compound so that the size of showers and, hence the primary energy, is underestimated by about 45%. However they also find that the logarithmic standard deviation falls from 0.40 to 0.23 over their energy range, and a detailed test shows these two effects approximately cancel leaving the input spectrum shape scarcely distorted. For the Leeds experiment (Edge et al. 1973) the estimates of the size parameter uncertainties are rather small (30% at about 10^{19}eV) mainly because determination of the deep water-Cerenkov density at 600m, $\rho(600)$, requires little extrapolation to radial distances in which there are no measurements. Tests have been made to see if the measurement errors would mask a 'cut-off' if it existed and it has been concluded that they would not.

Although the checks made by both groups using simulated EAS data have been carefully done, neither have considered uncertainties which could arise if the primary particles are protons and fluctuations in the lateral distribution function of muons arise at large distances as predicted by Dedenko and Dimova (1973). Additionally the Sydney simulations have not yet included the effect of the diffuse nature of the shower front on the triggering probabilities of their detectors. Both these factors will be most significant at the highest energies.

Particular examples of large showers seen at Volcano Ranch, Haverah Park and Sydney have been discussed in detail (Linsley 1962, Andrews et al. 1968, Bell et al. 1974) and it seems likely that more than 10 events have been recorded with energies greater than the 'cut-off' predicted by Hillas, although this last statement presupposes that the shower models presently favoured are accurate representations of the nuclear physics at the highest energies. Detailed comparisons of the model predictions with experiments have been possible up to about 10^{18}eV and the models of Hillas appear to provide adequate descriptions of shower data at least at large radial distances. However the marked differences between the "Andromeda" and "Centauro" events observed by the Japan-Brasil collaboration (1971, 1973) near 10^{16}eV, caution against too definite an assignment of energies at 10^{20}eV, and at this time an honest view must be that no definite discrepancy between experiment and cosmological predictions has been established.

3. Composition Above 10^{17}eV

A number of the difficulties of interpretation of the data above 10^{17}eV would be more tractable if reliable information existed on the mass composition. Additionally the implications of the isotropy of particle arrival directions for the origin of high energy cosmic rays would be more clear, but unfortunately the problem of determining even the grossest features of the mass distribution has proved very difficult.

The work of Linsley and Scarsi (1962), originally presented as evidence for a high proportion of protons in the flux above 10^{17}eV, has recently been re-evaluated (Linsley 1973a) and held, but not strongly, to be consistent with a heavy composition. At similar energies Orford and Turver (1968) have reported measurements on high energy muons at large axial distances as suggesting a primary flux of average mass greater than 10: from more refined model calculations by the same group, it now appears that muon momentum spectrum measurements are not capable of distinguishing between masses as disparate as 1 and 56! Kawaguchi et al. (1971) have proposed that the absence of fluctuations in muons observed at Chacaltaya combined with evidence for muon fluctuations in showers recorded at Tokyo from primary particles considered to be of the same energy, indicate a high proportion of protons at about 10^{18}eV. However, details of this work currently available are insufficient for critical evaluation. Earlier indications that radio emission from EAS would provide information on the primary mass (Allan et al. 1971, 1973) appear to have been over-optimistic and Allan and Jones (1974, private communication) now regard the technique as being too insensitive to give much information except at about 10^{18}eV where the rate of suitable events is prohibitively low.

A number of lines of investigation are in progress which look promising. Following a suggestion by Fomin and Khristiansen (1971), the groups at Yakutsk (Efimov et al. 1973) and at Haverah Park (Orford and Turver 1974 private communication) are measuring the width of the air-Cerenkov light signal at several hundred metres from the shower axis; the width is believed to reflect the longitudinal development of EAS. At the Haverah Park array progress is well advanced on an experiment to look for fluctuations in the lateral distribution of EAS at distances from the axis less than 100m. Such fluctuations are believed to be measurably large in proton initiated EAS and indeed may already have been observed in a preliminary experiment by Edge (1974).

Very significant fluctuation effects have been found by Watson and Wilson (1974) in the spread of arrival time of particles at large distances from the shower axis. The principle of this experiment is illustrated schematically in figure 9; at a given distance from the shower axis the rise time of the signal due to a late developing shower is expected to be slower than for an early developing shower. Although detailed model calculations are not yet available to interpret the experimental results, the authors are of the opinion that some of the observed fluctuations at about 10^{18}eV are so large as to require proton primaries for their explanation. The aim of this experiment, and of those mentioned in the previous paragraph, is the development of methods to enable the percentage of protons in the range 10^{17}eV to 5.10^{18}eV to be estimated. Hopefully by the time the highest energy end of the spectrum has greater statistical significance - and this will require extended running by the largest arrays at Yakutsk (Diminstein et al. 1973) and Sydney - more will be known about the primary mass composition above 10^{17}eV.

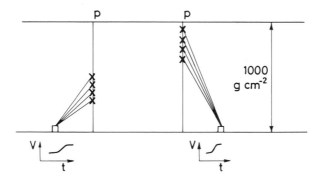

Fig. 9: Schematic illustration of the principle of the experiment of Watson and Wilson (1974). The "shower" at the left has developed "early" and the "shower" at the right has developed "late". Hence at similar axial distances the rise time of the detector signal will be faster in the former shower than in the later.

V. SUMMARY

At all energies the spectrum slope is very much better known than is the intensity and the spectral indices quoted seem unlikely to change drastically even if fairly dramatic changes take place in the nuclear physics at energies not yet accessible to machines. At the lowest energies, $\sim 10^{12} \div 10^{13}$ eV, there is disagreement between different estimates of the primary nucleon flux derived from muon data which are as large as a factor of 2, but it appears that the differential spectral slope is close to -2.65 ± 0.05, the error indicating the range of uncertainty as between experiments. At around 10^{16} eV there is agreement between workers using different models and different assumptions about the primary mass as to the energy to be ascribed to showers of a given intensity, but at energies near 10^{15} eV the situation is less clear. Although it must be considered well established that just above 10^{15} eV the slope is significantly steeper than at 10^{12} eV (-3.39 ± 0.08 compared with the lower value just quoted) there are doubts about the true value of the intensity at energies less than 10^{15} eV because of uncertainty about the interpretation of some of the longitudinal development curves seen at Chacaltaya. This is particularly unfortunate as the extrapolations from the low energy data hint at a flattening of the spectral slope somewhere before 10^{15} eV which is what might be expected were pulsars to provide the bulk of the cosmic rays in this region.

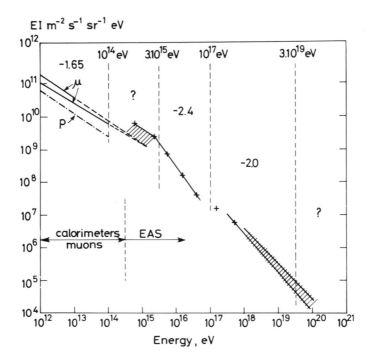

Fig. 10: Composite integral energy spectrum from 10^{12}eV to 10^{20}eV. Integral slopes are indicated which seem well established in certain energy ranges. Shading indicates where greatest uncertainty exists as regards the slope and intensity. A discussion of the details of the figure has been given in the text.

The steep slope above 10^{15}eV appears to give way to a slightly flatter one at energies near 10^{17}eV and there is encouraging evidence of consistency in the results of rather difficult measurements around 5.10^{17}eV. It seems probable that all of the EAS data can be reconciled with a slope of -3.0 from this energy to 10^{20}eV. Near 10^{20}eV there is still room for doubt as to whether or not any particles have been seen beyond the cut-off which is expected to be imposed on the spectrum if the particles are protons of extragalactic origin. The ground-level measurements on these particles are probably correct to 40% but no very realistic estimate of the accuracy of the model calculations used to describe the EAS development and hence to deduce the energy can be made. An attempt to summarise some of the data on the energy spectrum which has been discussed in detail above is made in figure 10.

The spectral intensity would be considerably more certain if better information was available on the mass composition. There is no conclusive evidence for any departure from the canonical mixed composition at least up to 10^{15}eV, although there are slight indications that in the following energy decade the mean mass may become rather higher. There is little evidence to support the hypothesis that the trend towards an iron spectrum which is flatter than the proton spectrum observed at low energies is sustained in the $10^{12} \div 10^{15}$ range. At intermediate energies, 10^{15}-10^{17}eV, there are large quantities of data available and it may be that efforts to interpret them using the most modern ideas on nuclear physics and carefully matching the results of calculations to the experiments will be rewarded.

Above 10^{17}eV the situation is not at all clear and although model calculations used to deduce the primary spectrum in this region often suppose the primaries to be protons this should not be taken as evidence that there are hard data to support the assumption. At the present time there is promise of progress on the problem from studies of fluctuation in EAS which are under way at Yakutsk (U.S.S.R.) and Haverah Park (U.K.) but because of a lack of detailed understanding of many factors concerning shower development interpretation of these data will require extraordinary care.

ACKNOWLEDGEMENTS

It is a pleasure to thank Dr. A.M. Hillas for a number of clarifying discussions and Dr. R.J.O. Reid for a critical reading of part of the manuscript.

REFERENCES

Allan, H.R., Jones, J.K., Mandolesi, N., Prah, J.H. and Shutie, P., 1971, Proc. 12th Int. Conf. on Cosmic Rays, (Hobart), 3, 1102-1107.

Allan, H.R., Shutie, P.F., Sun, M.P. and Jones, J.K., 1973, Proc. 13th Int. Conf. on Cosmic Rays, (Denver), 4, 2407-2413.

Andrews, D., Evans, A.C., Reid, R.J., Tennant, R.M., Watson, A.A. and Wilson, J.G., 1968, Nature, 219, 343-346.

Antonov, R.A., Ivanenko, I.P., Samosudov, B.E. and Tulinova, Z.I., 1971, Proc. 12th Int. Conf. on Cosmic Rays, (Hobart), 6, 2194-2200.

Aseiken, V.S., Benko, D., Kabanova, N.V., Nesterova, N.M., Nikoskaja, N.M., Romakhin, V.A., Stamenov, I.N., Stanev, T.S. and Janminchev, V.D., 1971, Proc. 12th Int. Conf. on Cosmic Rays, (Hobart), 6, 2152-2164.

Armitage, M.L., Blake, P.R. and Nash, W.F., 1973, Proc. 13th Int. Conf. on Cosmic Rays, (Denver), 4, 2545-50.

Ashley, G.K., Elbert, J.W., Keuffel, J.W., Larson, M.D. and Morrison, J.L., 1973, Phys. Rev. Letters, 31, 1091-1094.

Balasubramanyan, V.K. and Ormes, J.F., 1973, Astrophysical Journal, 186, 109-122.

Bakich, A., McCusker, C.B.A. and Winn, M.M., 1970, J. Phys. A., 3, 662-688.

Barker, P., Hazen, W. and Hendel, A., Proc. 9th Int. Conf. on Cosmic Rays, (London), 2, 718-719.

Bell, C.J., Bray, A.D., Denehy, B.V., Goorevich, L., Horton, L., Loy, J.G., McCusker, C.B.A., Nielsen, P., Outhred, A.K., Peak, L.S., Ulrichs, J., Wilson, L.S. and Winn, M.M., 1974, J. Phys. A., 7, 990-1009.

Bradt, H.V. and Rappaport, S.A., 1967, Phys. Rev., 164, 1567-1583.

Bradt, H.V., Clark, G., La Pointe, M., Domingo, V., Escobar, I., Kamata, K., Murakami, K., Suga, K. and Toyoda, Y., 1965, Proc. 9th Int. Conf. on Cosmic Rays, (London), 2, 715-7.

Bray, A.D., Crawford, D.F., Jauncey, D.L., McCusker, C.B.A., Poole, P.C., Rathgeber, M.H., Ulrichs, J., Wand, R.H. and Winn, M.M., 1964, Nuovo Cimento, 32, 827-845.

Catz, P.H., Hochart, J.P., Maze, R., Zawadzki, A., Gawin, J. and Wdowczyk, J., 1973, Proc. 13th Int. Conf. on Cosmic Rays, (Denver), 4, 2494-2499.

Clark, G., Bradt, H., La Pointe, M., Domingo, V., Escober, I., Murakami, K., Suga, K., Toyoda, Y. and Hersil, J., 1963, Proc. 8th Int. Conf. on Cosmic Rays, (Jaipur), 4, 65-76.

Daniel, B.J., Hume, C.J., Ng, L.K., Thompson, M.G., Whalley, M.R., Wdowczyk, J. and Wolfendale, A.W., 1974, J. Phys. A., 7, L20-24.

Dedenko, L.G. and Dimova, I.A., 1973, Proc. 13th Int. Conf. on Cosmic Rays, (Denver), 4, 2444-2448.

Diminstein, A.S., Glushkov, A.V., Kaganov, L.I., Maximov, S.V., Mikhailov, A.A., Pravdin, M.I., Sokurov, V.F. and Yefimov, N.N., 1973, Proc. 13th Int. Conf. on Cosmic Rays, (Denver), 5, 3232-3236.

Dixon, H.E., Earnshaw, J.C., Hook, J.R., Smith, G.J. and Turver, K.E., 1973, Proc. 13th Int. Conf. on Cosmic Rays, (Denver), 4, 2473-2488.

Dixon, H.E., Machin, A.C., Pickersgill, D.R., Smith, G.J. and Turver, K.E., 1974, J. Phys. A., 7, 1010-1016.

Dyakonov, M.N., Kolosov, V.A., Krasilnikov, D.D., Kulakovskaya, V.P., Lischenjok, F., Orlov, V.A., Sleptsov, I. Ye and Nikolsky, S.I., 1973, Proc. 13th Int. Conf. on Cosmic Rays, (Denver), 4, 2384-2388.

Edge, D.M., 1974, Ph.D. Thesis, University of Leeds.

Edge, D.M., Evans, A.C., Garmston, H.J., Reid, R.J.O., Watson, A.A., Wilson, J.G. and Wray, A.M., 1973, J. Phys. A., 6, 1612-1634.

Efimov, N.N., Krasilnikov, D.D., Khristiansen, G.B., Shikalov, F.V. and Kuzmin, A.I., 1973, Proc. 13th Int. Conf. on Cosmic Rays, (Denver), 4, 2378-2383.

Elbert, J.W., Keuffel, J.W., Lowe, G.H., Morrison, J.L. and Mason, G.W., 1973, Proc. 13th Int. Conf. on Cosmic Rays, (Denver), 1, 213-219.

Erlykin, A.D., Kulichenko, A.K. and Macharariani, 1973, Proc. 13th Int. Conf. on Cosmic Rays, (Denver), 4, 2500-2505.

Firkowski, R., Grochalska, B., Olejniczak, W. and Wdowczyk, J., 1973, Proc. 13th Int. Conf. on Cosmic Rays, (Denver), 4, 2605-2609.

Fomin, Yu A. and Khristiansen, G.B., 1970, Acta Phys. Acad. Sci. Hung., 29, (Supp), 3, 435-438.

Fomin, Yu A. and Khristiansen, G.B., 1971, Soviet J. Nuclear Phys., 14. 360-362.

Gaisser, T.K., 1974, Nature 248, 122-124.

Galbraith, W. and Jelley, J.V., 1953, Nature, 171, 349-350.

Gerdes, C., Fan, C.Y. and Weekes, T.C., 1973, Proc. 13th Int. Conf. on Cosmic Rays, (Denver), 1, 219-224.

Ginzburg, V.L. and Syrovatskii, S.I., 1963, The Origin of Cosmic Rays, (Pergamon Press), p45.

Gold, T., 1974, Royal Society Discussion Meeting, February 1974, Proc. Roy. Soc. A. (to be published).

Golden, R.L., Adams, J.H., Badhwar, G.D., Deney, C.L., Lindstrom, P.J. and Heckman, H.H., 1974, Nature, 249, 814-815.

Greisen, K., 1966, Phys. Rev. Letters, 16, 748-50.

Grigorov, N.L., Nesterov, V.G., Papaport, I.D., Savenko, I.A. and Skuridin, G.A., 1965, Proc. 9th Int. Conf. on Cosmic Rays, (London), 1, 50-52.

Grigorov, N.L., Gubin, Yu V., Rapaport, I.D., Savenko, I.A., Akimov, V.V., Nesterov, V.E. and Yakovlev, B.M., 1971, Proc. 12th Int. Conf. on Cosmic Rays, (Hobart), 5, 1746-1751.

Grindlay, J.E., 1971, Nuovo Cimento, 2B, 119-138.

Grindlay, J.E. and Helmken, H.F., 1973, Proc. 13th Int. Conf. on Cosmic Rays, (Denver), 1, 202-207.

Grindlay, J.E., Helmken, H.F., Weekes, T.C., Fazio, G.G. and Boley, F., 1973, Proc. 13th Int. Conf. on Cosmic Rays, (Denver), 1, 36-40.

Gunn, J.E. and Ostriker, J.P., 1969, Phys. Rev. Letters, 22, 728.

Hillas, A.M., 1968, Can. J. Phys., 46, S623-6.

Hillas, A.M., 1974, Royal Society Discussion Meeting, February 1974, Proc. Roy. Soc. A. (to be published).

Hillas, A.M., Marsden, D.J., Hollows, J.D. and Hunter, H.W., 1971a, Proc. 12th Int. Conf. on Cosmic Rays, (Hobart), 3, 1001-6.

Hillas, A.M., Hollows, J.D., Hunter, H.W. and Marsden, D.J., 1971b, Proc. 12th Int. Conf. on Cosmic Rays, (Hobart), 3. 1007-1012.

Kalmykov, N.N., Fomin Yu A. and Khristiansen, G.B., Proc. 13th Int. Conf. on Cosmic Rays, (Denver), 4, 2633-2638.

Karakula, S., Osborne, J.L. and Wdowczyk, J., 1973, Proc. 13th Int. Conf. on Cosmic Rays, (Denver), 5, 3092-3097.

Kempa, J., Wdowczyk, J. and Wolfendale, A.W., 1974, J. Phys. A., 7, 1213-1221.

Kreiger, A.S. and Bradt, H.V., 1969, Phys. Rev., 185, 1629-35.

Kulikov, G.V. and Khristiansen, G.B., 1958, Zh Eksper Teor Fiz, 35, 635-640.

Japan - Brasil Emulsion Chamber Collaboration, 1971, Proc. 12th Int. Conf. on Cosmic Rays, (Hobart), 7. 2775-2780.

Japan - Brasil Emulsion Chamber Collaboration, 1973, Proc. 13th Int. Conf. on Cosmic Rays, (Denver), 4, 2671-2675.

Kauagughi, S., Sakuyama, H., Suga, K., Takano, M., Uchino, K., Hara, T., Ishikawa, F., Nagano, M. and Tanahashi, G., 1971, Proc. 12th Int. Conf. on Cosmic Rays, (Hobart), 7, 2736-41.

Kerschenholz, I.M., Krasilnikov, D.D., Kuzmin, A.I., Orlov, V.A., Sleptsov, I. Ye, Yegorov, T.A., Khristiansen, G.B., Vernov, S.N. and Nikolsky, S.I., 1973, Proc. 13th Int. Conf. on Cosmic Rays, (Denver), 4, 2507-2512.

Khristiansen, G.B., Vedeneev, O.V., Kulikov, G.V., Nazarov, V.I. and Solovjeva, V.I., 1971, Proc. 12th Int. Conf. on Cosmic Rays, (Hobart), 6, 2097-2108.

Khristiansen, G.B., Kulikov, G.V. and Solov'eva, V.I., 1973, JETP Letters, 18, 207-209.

Linsley, J., 1963, Proc. 8th Int. Conf. on Cosmic Rays, (Jaipur), 4, 77-99.

Linsley, J., 1973a, Proc. 13th Int. Conf. on Cosmic Rays, (Denver), 5, 3202-3206.

Linsley, J., 1973b, Proc. 13th Int. Conf. on Cosmic Rays, (Denver), 5, 3207-3211.

Linsley, J., 1973c, Proc. 13th Int. Conf. on Cosmic Rays, (Denver), 5, 3212-3219.

Linsley, J. and Scarsi, L., 1962, Phys. Rev. Letters, 9, 123-125.

Linsley, J. and Watson, A.A., 1974, Nature, 249, 815-817.

La Pointe, M., Kamata, K., Gaebler, J. and Escobar, I., 1968, Can. J. Phys., 46, S68-71.

McCusker, C.B.A., Peak, L.S. and Rathgeber, M.H., 1969, Phys. Rev., 177. 1902-1920.

Miyake, S., Ito, N., Kawakami, S., Hayashida, N. and Suzuki, N., 1971, Proc. 12th Int. Conf. on Cosmic Rays, (Hobart), 7, 2748-2752.

Murzin, V.S., 1965, Proc. 9th Int. Conf. on Cosmic Rays, (London), 2 (London: The Institute of Physics and The Physical Society), 872-4.

Nesterova, N.M., 1968, Can. J. Phys., 46, S92-94.

Nikolsky, S.I., 1970, P.N. Lebedev Physical Institute Preprint, No. 35.

Orford, K.J. and Turver, K.E., 1968, Nature, 219, 706-708.

Pal, Y. and Peters, B., 1964, Mat. Fys. Medd. Dan. Vid. Selsk, 33, 1-55.

Peters, B., 1960, Prox. 6th Int. Conf. on Cosmic Rays, (Moscow), 3, 157-160.

Ramaty, R., Balasubrahmanyan, V.K., Ormes, J.F., 1973, Science, 180, 731-733.

Ramana Murthy, P.V. and Subramanian, A., 1972a, Phys. Letters, 39B, 646-648.

Ramana Murthy, P.V. and Subramanian, A., 1972b, Proc. Ind. Acad. Sci., 76A, 1-11.

Rappaport, S.A. and Bradt, H.V., 1969, Phys. Rev. Letters, 22. 960-963.

Rogers, I.W., Thompson, M.G., Turner, M.J.L. and Wolfendale, A.W., 1969, J. Phys. A., 2, 365-373.

Ryan, M.J., Ormes, J.F. and Balasubrahmanyan, W.K., 1972, Phys. Rev. Letters, 28, 985-8 and 1497.

Samorski, M., Staubert, R., Trumper, J. and Bohm, E., 1971, Proc. 12th Int. Conf. on Cosmic Rays, (Hobart), 3, 959-964.

Strong, A.W., Wdowczyk, J. and Wolfendale, A.W., J. Phys. A., (to be published).

Thielheim, K.O. and Biersdorf, R., 1969, J. Phys. A 2, 341-353.

Tuecher, M.W., Hohrmann, E., Haskin, D.M. and Schein, M., 1959, Phys. Rev. Letters, 2, 313-315.

Trumper, J., Bohm, E., Fritze, R., Samorski, M. and Staubert, R., 1970, Acta. Phys. Acad. Hung., 29, Suppl. Vol. 3, 447-450.

Vernov, S.N. and Khristiansen, G.B., 1968, Proc. 10th Int. Conf. on Cosmic Rays, (Calgary), Part A. 345-396.

Vernov, S.N., Khristiansen, G.B., Abrosimov, A.T., Atrashkevitch, V.B., Beljaeva, I.F., Kulikov, G.V., Mandritskaya, K.V., Solovjeva, V.I. and Khrenov, B.A., 1969, Can. J. Phys., 46, S197-200.

Waddington, J. and Frier, P.S., 1973, Proc. 13th Int. Conf. on Cosmic Rays, (Denver), 4, 2449-2454.

Watson, A.A. and Wilson, J.G., 1974, J. Phys. A., 7, 1199-1212.

Wdowczyk, J. and Wolfendale, A.W., 1973, J. Phys. A., 6, 1594-1611.

Webber, W.R., 1973, Proc. 13th Int. Conf. on Cosmic Rays, (Denver), 5, 3568-3614.

Wolfendale, A.W., 1974, Royal Society Discussion Meeting, February 1974, Proc. Roy. Soc. A (to be published).

Zatsepin, G.T. and Kuzmin, V.A., 1966, Soc. Phys. JETP, 4, 114-7.

NUCLEAR MASS COMPOSITION AT 'LOW' ENERGIES (i.e.< 10^{12} eV/NUCLEON)

Ib Lundgaard Rasmussen

Danish Space Research Institute, Lundtoftevej 7,
DK-2800 Lyngby, Denmark.

I. INTRODUCTION

The fully ionized atomic nuclei of all the elements from hydrogen up to at least uranium, which constitute the nuclear component of the galactic cosmic radiation, are the only sample of matter which reach us from outside the solar system. We believe that these nuclei were created in the nuclear burning processes that take place inside stars during their evolution, and that the composition of the nuclear cosmic rays is one of the most important sources of information about the nucleosynthesis in the stars and the composition of matter in important regions of our galaxy.

Studies of the cosmic ray charge composition during the last 25 years have shown that between the time of creation deep inside a star and the time of détection here in the solar system, the cosmic rays have gone through many processes in widely different environments. After they are created they must be moved out of the star, accelerated to high energies and propagated through interstellar space. Each of these processes change the composition of the cosmic rays and a study of this composition will, therefore yield information on all these processes.

A knowledge of the charge composition is not sufficient to allow us to trace the cosmic rays back to the source and learn what the source composition of the cosmic rays was. In order to do this, and thereby obtain information on what and where the cosmic ray sources are, we must also know the isotopic composition of the cosmic rays.

In this paper I will give a review of present and proposed methods for the measurement of the masses of cosmic rays, and discuss some of the results already obtained. In order to provide a suitable background for a more detailed discussion of instruments in section III and results in section IV, section II will contain a brief review of the charge composition.

II COSMIC RAY CHARGE COMPOSITION

An overall picture of the composition of the cosmic radiation can be obtained from figure 1 which shows the measured relative abundance of nuclei with a mean kinetic energy of a few hundred MeV/nuc, at the top of the atmosphere compared to the solar system abundances. Although many differences can be noted, the figure shows the striking similarity between the cosmic ray and solar system values. The abundance values cover more than ten orders of magnitude and the differences are less than one order of magnitude, except in a few cases which I will explain later.

It is important at this point to emphasize the similarity of the two samples. Later, when I discuss details of the composition I will concentrate on the differences, but these differences can only be interpreted when they are taken as indicators of the different development of cosmic rays and solar system material from the interstellar gas out of which the stars have condensed.

The rapid drop in abundance beyond the iron group shows that the cosmic ray flux is almost exclusively composed of elements below charge 30. At the time when the data for this figure was compiled, a couple of years ago, a total of only ~200 nuclei of charge greater than 40 had been detected. This number of events indicates that the absolute flux of heavy cosmic ray nuclei are extremely low, and it will not be possible to measure the isotopic composition of elements heavier than the iron group for a long time in the future. I will, therefore, concentrate on elements below charge 30 in this review.

Figure 2 shows this region in more detail. The data are the cosmic ray abundances as known at the time of the Denver Conference (Aug. 1973) compared to the solar system values from Cameron's 1973 table (Cameron 1973). The mean kinetic energy of the cosmic rays is a few GeV/nucleon. The cosmic ray abundances shown here are by now well established although the uncertainties are still rather large for the less abundant, odd-Z elements.

The agreement for the elements most prominent in the nuclear burning stages in stars (H, He, C, O, Ne, Mg, Si, Fe) is fairly good and must be taken as an indication that information on the source composition is still present in the cosmic rays. The ratio

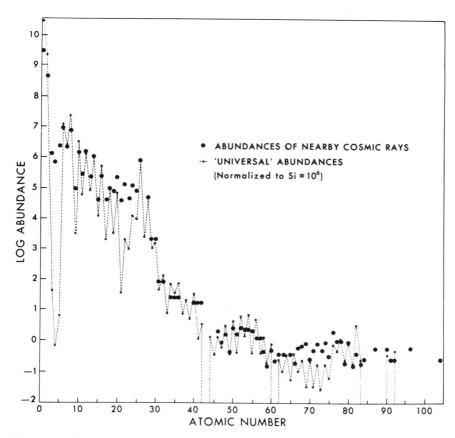

Fig. 1: The Cosmic Ray Composition measured in the vicinity of the earth. Universal abundances Cameron 1968 (Courtesy P.B. Price).

between even and odd Z elements is lower in the cosmic ray data, and Li, Be, B and the sub-iron group elements (Z = 19-25) are grossly overabundant. These large differences are caused by changes in the composition due to nuclear interactions taking place during the propagation of the cosmic rays through interstellar space, Li, Be, B being spallation products mainly from C, N, O and the sub-iron group products from iron.

If the actual mechanism causing changes is simple and well established, other aspects of the propagation are not so well known. Questions like: how long does it take the cosmic rays to reach the solar system from the sources and where is that time spent, do not have clear answers at the moment. The answers to these questions

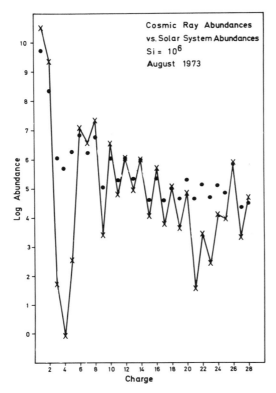

Fig. 2: The charge composition of arriving cosmic rays (Garcia-Munoz 1973) compared to solar system composition (Cameron 1973).

• cosmic ray abundances

x solar system abundances

will affect the composition we can infer by tracing the cosmic rays back to the source regions.

A detailed knowledge of the cross sections for the various nuclear interactions which can occur in interstellar space is essential when we want to determine the composition of the cosmic rays leaving the source region, from observations in the neighborhood of the sun. Although many relevant cross sections have been measured experimentally, mainly by the Orsay group, the majority must still be estimated from semiempirical formulae (Silberberg and Tsao 1973a, 1973b). Using this knowledge one can solve the diffusion equations that govern the propagation and thereby determine how much matter the cosmic rays must pass through in order to produce the observed secondary abundances. It can be shown that a wide range of pathlengths is necessary in order to obtain a good fit to the observed abundances. A recent example of such a pathlength distribution is

$$\frac{dN}{d\lambda} = \left[1 - \exp(-2.8\lambda^2)\right] \exp(-0.23\lambda)$$

where N is the fractional flux intensity of particles having a path length $\lambda(g\ cm^{-2})$ (Shapiro and Silberberg 1974)

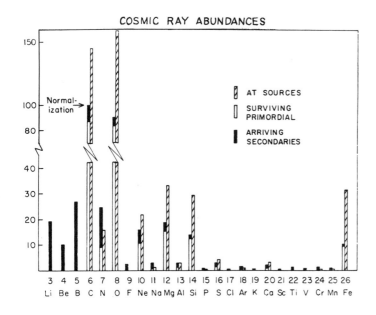

Fig. 3: Relative abundances of cosmic rays at the sources and near the earth. Arriving carbon is normalized to 100. (Shapiro and Silberberg 1974).

Figure 3 shows the result of a calculation of this type made by taking a trial source composition and propagating it forward until the composition resembles the observations. It can be seen clearly from figure 3 that the interstellar interactions cause substantial changes in the composition and that the produced secondaries constitute a sizeable fraction of even the most abundant elements. It must be noted, however, that in a calculation of this type only the source abundances of C, N, O, Ne, Mg Si, and Fe are determined with a precision of better than 20%.

It is obvious that this type of calculation does not tell us anything about where the sources are, what is their confinement region, or what is the lifetime of individual high energy particles. However, a model of confinement and propagation must predict a suitable pathlength distribution in order to be acceptable. Lately, interest has been focussed on the so called 'leaky box' model (Jokipii and Parker 1969, Ramaty and Lingenfelter 1971), that indeed does predict an exponential path length distribution. The model assumes that only one type of source is necessary to produce all the elements, that the sources are evenly distributed in the galactic disk, that the cosmic rays are confined to the disk by the galactic magnetic field, and that the particles have an energy independent probability of leaking out of the confinement region.

The model picture of the propagation as a diffusion process appears reasonable when we compare path lengths of 5 g cm^{-2} to the size of the galactic disk, 0.05 g cm^{-2} from rim to rim in the galactic plane, and 10^{-3} g cm^{-2} perpendicular to the plane. The assumption that the probability of leaking out of the confinement region is energy independent must break down for the very high energies, where the gyroradius in the galactic magnetic field ($\sim 3.10^{-6}$ gauss) becomes comparable to the dimensions of the galaxy.

To test the above assumption, one must study the composition of the arriving cosmic rays as a function of energy. Many groups are active in this field, and although there are still great uncertainties in the measurements a picture is emerging. The abundance ratios of secondary to primary elements and of C+O to Fe seems to be decreasing with increasing energy in the range 10 to 100 GeV/nuc, indicating that the probability of leakage is increasing with energy (a more detailed discussion of these results will be found in the review talks by M. Garcia-Munoz and W.R. Webber at the Denver Cosmic Ray Conference and references contained therein).

In appendix A one of the possible interpretations of the energy dependence is used to give a prediction of the charge composition at energies greater than 10^{12} eV/nucleon.

In order to obtain information about the confinement region we note that because most cosmic ray nuclei are moving with nearly the speed of light, the confinement time is inversely proportional to the average matter density in the confinement region. In case of confinement to the galactic disk a density of 1 atom cm^{-3} would result in a propagation time around 3.10^6 years. The importance of long-lived radioactive isotopes, the "cosmic ray clocks" was realized a long time ago (Hayakawa et al. 1958, Peters 1963). Attempts have been made to estimate the abundances of these isotopes from the charge composition, but it has recently been shown (Raisbeck and Yiou 1973) that due to uncertainties in the production cross sections this is still impossible, even for the most abundant of the radioactive nuclei, Be^{10}.

A measure of the relative abundance of radioactive isotopes in the arriving cosmic rays will thus tell us the "age" of the

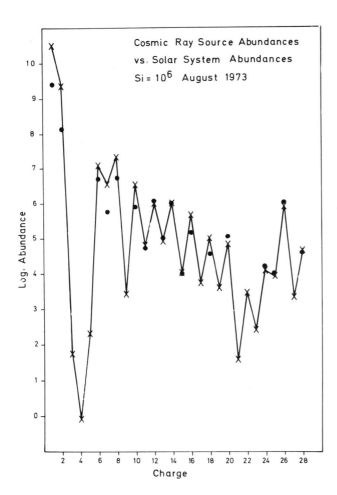

Fig. 4: Cosmic ray composition near the sources (Shapiro et al. 1973) compared to solar system composition (Cameron 1973)

- • Cosmic ray abundances
- x Solar system abundances

cosmic rays, however, the isotopic composition in general will tell us much more. The interactions taking place during the traversal of the interstellar space are interactions between isotopes, which produce new isotopes and a study of the isotopic composition will give a much more detailed picture of the propagation and provide answers to questions like: do all the elements come from the same type of source, i.e. do all elements have the same pathlength distribution. We will also be able to get a much clearer picture of the composition of the cosmic rays when they leave the source region.

At present all we know about the source composition comes from calculations like the one shown in figure 3. The result of such a calculation is compared to the solar system abundances in figure 4. The overall agreement is excellent with the underabundance of the light element cosmic rays as the only major difference.

Although the composition shown on figure 4 is often loosely referred to as the source composition, it is important to realise that it is really the composition immediately after acceleration.

It is also important to bear in mind that when we speak about cosmic ray sources, we really require two kinds of sources, one source that supplies the cosmic ray particles, and one that supplies the energy necessary to accelerate the particles to the high energies seen. These two sources could be far apart both in time and space although some possible source candidates, like supernovae, might satisfy all the requirements.

Figure 5 shows the processes that might take place in the source region. The nuclei are made inside the matter source by nucleosynthesis and then ejected into the surrounding space in a process (e.g. supernova explosion) that might change the composition from the source composition into the ejected composition. If there is a great gap between ejection and acceleration, in space or time, the ejected nuclei will be mixed with the interstellar material in the region surrounding the source, which will give a resulting mixed composition. If ejection and acceleration occur simultaneously no such mixing will take place. The nuclei are then accelerated, a process which could also introduce changes in the composition. The nuclei may then have to move through a, possibly dense, cloud of material surrounding the source region before they finally escape into the galaxy and propagate to the solar system.

With this in mind let us look at the explanations for the underabundances of the light elements. It has recently been found that the depletion of the cosmic rays shows some correlation

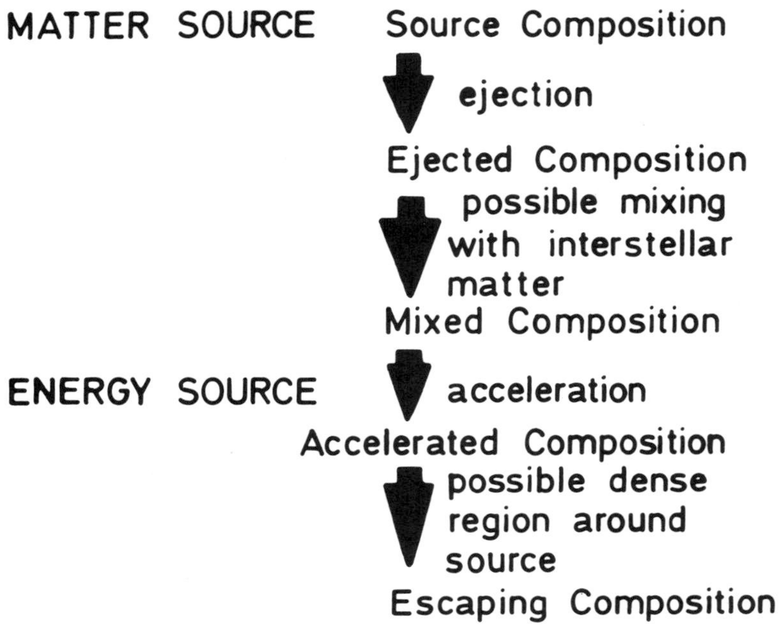

MATTER SOURCE   Source Composition
 ejection

Ejected Composition
 possible mixing
  with interstellar
  matter

Mixed Composition

ENERGY SOURCE   acceleration

Accelerated Composition
  possible dense
  region around
  source

Escaping Composition

Fig. 5: Schematic representation of important features of cosmic ray source regions. Note especially the two different kinds of sources.

---

with both the first ionization potential (Havnes 1973, Cassé and Goret 1973), and the ionization cross section (Kristiansson 1971, 1972) of the elements considered. These authors thus claim that the differences are caused by the acceleration mechanism. Another explanation recently proposed (Schramm and Arnett 1973) is that the composition of ejecta from the explosions of massive stars (with masses greater than 8 solar masses) can give a reasonable match to the cosmic ray data.

If the underabundances are caused by the acceleration process then the composition before acceleration must be very like the solar system composition, and this explanation thus implies that a considerable amount of mixing must have taken place between ejection and acceleration, because we do not at present believe that the ejected composition is similar to the solar system composition. If on the other hand the ejected composition is similar to the accelerated one, no mixing has taken place and the ejection and acceleration will most probably occur simultaneously.

That acceleration process can change the composition can be seen from measurements of energetic particles ejected from the sun

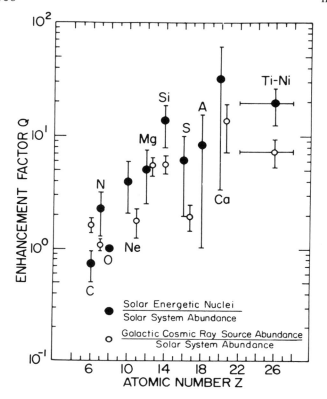

Fig. 6: Enhancement factor as a function of Z determined for solar particles averaged over seven flares (Mogro-Campero and Simpson, 1972) and for galactic cosmic ray source abundances.

All abundances are measured relative to oxygen (Figure from Price, 1973).

---

as shown in figure 6. These particles have much lower energies than the galactic cosmic rays, and we do not know whether similar acceleration mechanisms can work on the much larger scale of cosmic ray acceleration, but it is clear from these measurements that acceleration can introduce composition changes.

Therefore the experimental data obtained so far do not allow us to decide which of the proposed theories is the correct one. However, it would be possible to get a much clearer picture of the processes taking place in the source region from measurements of the isotopic composition and it is important to note that although the charge composition can be changed by the acceleration, the ratios between the isotopic abundances within each element will not be altered significantly.

It can be seen from this very brief review of the charge composition that the additional information we can obtain from measurements of the isotopic composition of the cosmic ray nuclei will be of great importance for the study of the regions of the universe where the cosmic rays are created, or through which they travel.

III BASIC TECHNIQUES

Before I begin to discuss some of the methods proposed for or used in the study of the mass composition I would like to point out some features of the cosmic ray flux that greatly influence the experimental design.

Compared to the clean and well controlled measurement conditions of accelerator experiments the conditions for cosmic ray measurements are really dirty. The particles that arrive at the instrument can have any combination of energy, charge, and mass. This means that everything we want to know about a particle must be measured by the instrument. Furthermore, the cosmic ray flux is very isotropic . If we consider energies so high that the particles are not deflected by the earth's magnetic field then the particles can reach the instrument from all directions not obscured by the earth. This means that the instrument must have a large opening angle in order to collect a sufficient number of particles because the total flux is very low.

In fact one of the major problems in abundance studies is how to collect enough particles. In figure 7 I have shown the time it takes to collect 1000 particles of a given element when particle kinetic energy accepted is above 500 MeV/nuc, 2 GeV/nuc, and 20 GeV/nuc. The 1000 particles will give 10% uncertainty on the abundance of an isotope if the ratio between the elemental abundance and the isotopic abundance is 10. The left hand scale shows the exposure time needed for a collecting area of 1 cm^2 ster typical of many early and present satellite experiments, the right hand scale corresponds to 1 m^2 ster an area characteristic for many balloon experiments and future satellite experiments.

The figure clearly shows that the heavier charges ($Z \gtrsim 40$) are out of reach for a long time to come as they require 10^5 times more exposure time in order to collect the necessary number of particles. The figure shows also the effect of the energy dependence of the spectrum. Note, however, that the times presented are based on the integral fluxes, if an instrument only detects particles in the range 500 MeV/nuc to 1 GeV/nuc the times must

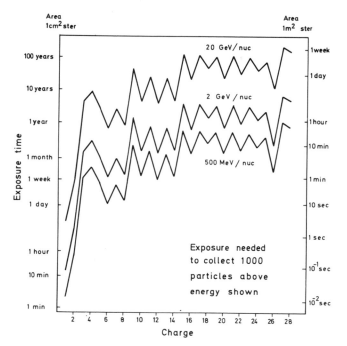

Fig. 7: The exposure time needed to collect 1000 particles of a given element.

be increased accordingly. It is also readily seen from the figure that if one is using a large instrument wherein a part either has an appreciable dead time or requires much power, one will have to discriminate against H and He due to their large flux (power is not in abundance in satellites or balloons).

These characteristics of the flux require that special attention is given to the long term stability of the instrument and to ensure that the quality of the data will not be greatly affected by variations in instrumental response as a function of incidence angle and impact point in the detectors. Two different approaches are used to reduce the effect of the geometrically dependent variations: One can design the instrument in such a way that the variations are minimized, or one can measure the incidence angle and the impact points and then correct for non-uniformities. Examples of both approaches will be given later.

It is convenient when discussing instruments to divide the energy ranges in which they operate into two groups: low energy where the particle can be slowed down and stopped before it is destroyed, i.e. where the particle range is shorter than the nuclear interaction mean free path, and high energy where the particle will interact before it can be stopped.

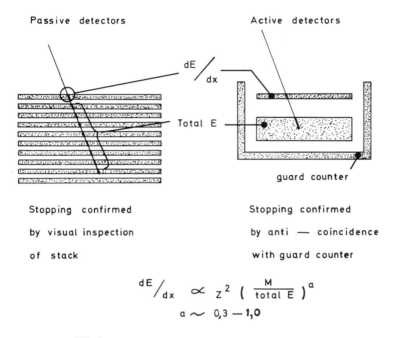

$$\frac{dE}{dx} \propto z^2 \left(\frac{M}{\text{total E}} \right)^a$$

$$a \sim 0{,}3 - 1{,}0$$

dE/dx vs total E method

Fig. 8: Principle of the dE/dx vs. total E method.

In the low energy range where nuclear interactions can be neglected most instruments are based on a variation of the so called dE/dx vs. Total E method. The principle of this method is shown in figure 8. The dE/dx signal, the energy loss rate, can be measured either in a single sheet from a stack of either nuclear emulsions or plastic sheets, or by the use of a very thin electronic detector. The total energy of the particle can be determined from the track length in the passive detector stack using range – energy relations or by measuring the output from a thick electronic counter wherein the particle is slowed down and comes to rest. It is of course important to be sure that the particle really gives up all its energy to the detector system. This can be confirmed either by observing that the track of the particle stops inside the passive counter stack or by demanding that there is no signal from a guard counter surrounding the active counters.

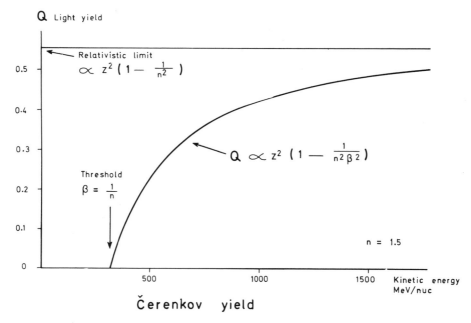

Fig. 9. Čerenkov counter light yield as a function of particle energy.

One type of active detector, the Čerenkov counter plays an important role in many instruments both at low and high energies. The principle of this detector is based on the fact that high energy particles moving with a velocity $\beta = \frac{v}{c}$ close to the velocity of light in vacuum c, will have a velocity greater than the velocity of light $c^1 = \frac{c}{n}$ in a medium of refractive index n, thereby producing a pulse of coherent light. An example of a light yield curve from a Čerenkov counter is shown in figure 9. It can be seen from the figure that the yield varies rapidly as a function of energy just above the threshold and the Čerenkov counter can give very precise measurements of particle energy in the range from threshold energy to around three times this energy. However, even at higher energies the Čerenkov signal being strictly proportional to Z^2 will provide a good measure of the charge of the incoming particle.

I will start the discussion of individual techniques used in the study of the isotopic composition by describing passive systems used at low energy, thereafter I will discuss active systems used at low energies, and finally I will discuss the possibilities of measuring the composition at high energies (>1 Gev/nuc).

The passive detector systems used are based on the analysis of particle tracks in stacks of thin plastic or nuclear emulsion sheets. Nuclear emulsions are used by the cosmic ray group in Lund. They base their method on accurate measurements, with nuclear track photometers, of the relation between mean track width (MTW) and residual range in the tracks. This relation is shown in figure 10.

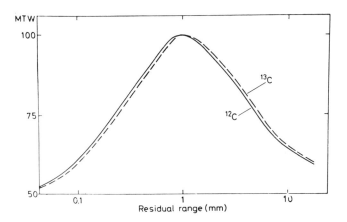

Fig. 10: The mean track width - range relations for C^{12} and C^{13}. The curves are arbitrarily normalized to 100 at the point where they reach their maximum value (Kristiansson 1971a)

From the measured MTW-range relation for the C^{12} isotope the corresponding curve for C^{13} has been calculated using the fact that the ionization and therefore also the mean track width only depend on the particle velocity for particles having the same nuclear charge. This gives as a straightforward consequence $R_1 : R_2 = M_1 : M_2$ for two particles with the masses M_1 and M_2, if the residual ranges R_1 and R_2 are measured from points in the tracks where the mean track width is the same (Kristiansson 1971a). The Lund group has found that in the charge region $3 \leq Z \leq 8$ the error in mass determination is smaller than ½ atomic mass unit for nuclei with a kinetic energy of a few hundred MeV/nucleon.

The analysis of thin plastic sheets has been performed by the groups in Kiel and Berkeley. The principle of the method used

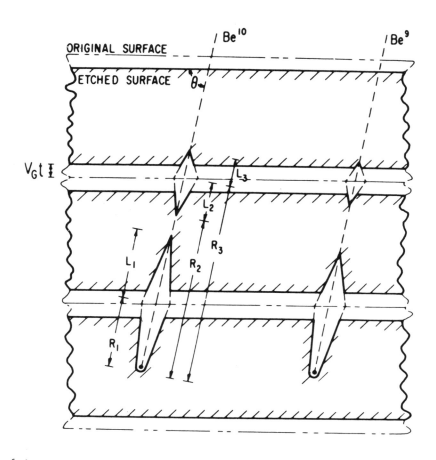

Fig. 11: Mass determination from etch cones in plastic. At a given ionization rate, the heavier isotope has a greater range, and the etching rate corresponding to that ionization rate will occur further from the end of the range of the heavy isotope than of the light isotope. (Price 1971).

is quite similar to the one used in the analysis of nuclear emulsions. After the stack has been exposed to the Cosmic Radiation the sheets are etched for several hours. The etching rate at the impact points of the particles is a monotonic function of the ionization rate and can therefore provide a measurement of dE/dx as shown for the case of Be in figure 11. For a given energy the measurement of the residual range from points with the same etch cone length will again provide information on the isotopic masses of the particles (a more detailed discussion can be found in Price and Fleischer 1971. Problems with variations in sensitivity from sheet to sheet in these stacks have so far prevented the groups from obtaining definitive results with this method (Beaujean and Enge 1973).

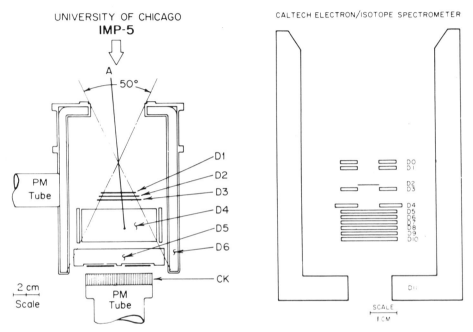

Fig. 12: Two small satellite instruments. IMP-5, University of Chicago, D1, D2 Li-drifted Si detectors (dE/dx counters), D4 Cs (Tl) crystal (total E counter) and D5 + D6 anticoincidence system (Garcia-Munoz et al. 1973). Caltech EIS, D0 to D10 fully depleted Si detectors (used as both dE/dx and total E counters) and D11 plastic-scintillator anti-coincidence counter (Hurford et al. 1973).

Among the active detector systems used at low energies many have been small satellite instruments with a typical collecting area of 1 cm^2 ster. These instruments have mainly been used for the study of the isotopic composition of Hydrogen and Helium, but instruments from the University of Chicago on IMP-5 and IMP-7 have been used to analyse isotopes up to Oxygen. Two examples of small satellite instruments are shown in figure 12. A common feature of these instruments is the use of several thin solid-state detectors that allow the experimenters to choose between different combinations of dE/dx - total E and multiple dE/dx - total E systems and thereby extend the useful energy range of the instruments. The use of some of the counters in anti-coincidence (e.g. the annular counters in the Caltech spectrometer) allows a reduction of the acceptance angle of the instrument making it possible to obtain data less affected by variations in response due to geometrical parameters.

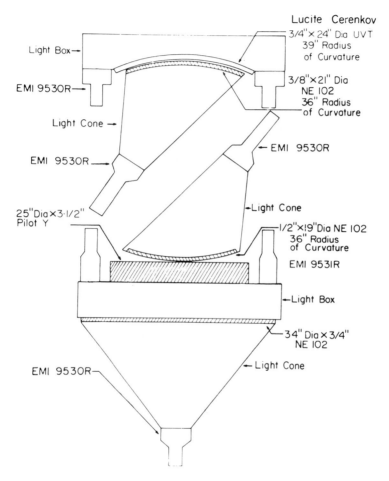

Fig. 13: Balloon borne instrument employing a combinaton of dE/dx total E and Čerenkov - total E techniques (Webber et al. 1973).

The balloon borne instruments are generally much larger than the satellite-borne ones and they can therefore collect data at a higher rate, so that the total amount of data collected in a balloon flight of ~30 hours duration is roughly equal to the amount of data collected in one year by the smaller satellite instruments. Figure 13 shows the latest instrument used by the group in New

Hampshire. This instrument is based on the conventional dE/dx – total E technique as well as simultaneous measurements using a new technique employing a Čerenkov – total E technique (Webber et al. 1973).

The instrument uses curved counters to minimize the effect of variations in path length with incidence angle. However, the variation in path length is still the dominant factor in resolution for particles with Z >8 and a position measuring system using proportional counters will be employed in later flights with this instrument. Based on detailed calculations of the resolution of the instrument, the group states that, at present, they are resolving adjacent isotopes up to a Z $\simeq$7 using the dE/dx total E technique and resolving isotopes separated by $\Delta A = 2$ up to a Z $\simeq$26 using the C x E technique, but that substantial future improvements in resolution can be made (Webber et al. 1973). The improved instrument should give mass resolutions <0.4 a.m.u. for Fe and lighter nuclei.

A new interesting method has been proposed by the group in Saclay for measuring the isotopic composition around 1 GeV/nucleon, a region intermediate between the low energy region where particles can be slowed down and the high energy region where they will interact.

This method, the Differential Slowing-Down method, is based on the fact that two particles having the same charge and energy/nucleon but different mass will lose a different amount of energy/nucleon due to ionization during passage of a thick absorber. By using Čerenkov counters to measure charge and velocity before and after passage of a thick lead absorber this method will make it possible to determine the isotopic composition of elements from Boron to Nickel over an energy range about 200 MeV/nucleon wide, located between .7 and 1.0 GeV/nucleon, depending on the charge of the incident nucleus. Figure 14 shows the experimental set-up and some results from a test carried out at the Berkeley Bevatron (Cassé et al. 1973). A balloon version of this experiment will be flown in 1975.

When the particle energy becomes greater than 1 GeV/nucleon it becomes important to use thin counters in order to keep the number of nuclear interactions low. As the nuclear interaction mean free path of a 5 GeV/nucleon iron nucleus is $\sim$20 g cm^{-2} in most counter materials and the corresponding ionization range is $\sim$200 g cm^{-2}, only 0.004% of the iron nuclei will come to rest before they collide and disintegrate, this then prevents the use of dE/dx – total E methods in the high energy/nucleon range. The active counters when used in this range permit the determination of charge and velocity of the particles, but one further measurement is needed in order to determine the mass of the particles.

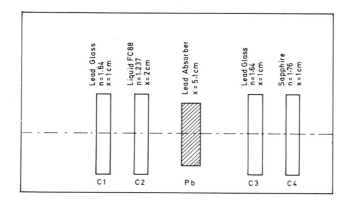

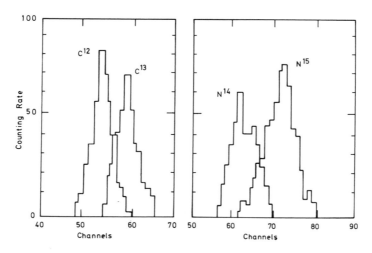

Fig. 14: Differential slowing-down method. The experimental set-up used in a test at the Berkeley Bevatron and the separation obtained of isotopes of Carbon and Nitrogen at 760 MeV/nucleon (Cassé et al. 1973).

For energies below ~3 GeV/nucleon one can measure the rigidity $(R = PM/Z)$ of the incoming particles by deflecting them in the magnetic field of a strong magnet. This approach is rather restricted in energy at present, mainly due to a lack of easy-to-use high precision devices needed to determine the particle position and direction before and after the deflection in the intense magnetic field. At present the group in Berkeley (Orth et al. 1974) is preparing an experiment to measure the abundance of Be^{10}, the most abundant "cosmic ray clock". This experiment uses a superconducting magnet and allows mass determination over a limited energy range around 2 GeV/nucleon.

In order to carry out direct mass determinations at still higher energies, say up to 20 GeV/nucleon, one would need track determining devices with a resolution an order of magnitude better than obtainable today, as the particles are only slightly deflected even in the very intense fields produced by superconducting magnets. The deflections can in fact be measured even at these energies, but not with the required precision $\frac{\Delta M}{M}$.

It might thus seem strange that it is possible to perform indirect measurement of the particle masses in the region 2-20 GeV/nucleon using the much weaker geomagnetic field (Lund etal. 1970, Lund et al. 1971). The reason for this is that the bending power of a magnetic field is determined by the line integral along the particle trajectory of the magnetic field strength over the distance where the field is acting on the particle. The weak but large-scale geomagnetic field ($HL \sim HR_{earth}$ ~1500 kilogauss meters) can thus bend the particles much more than the intense but small-scale field from a superconducting magnet ($HL \approx 4$ kilogauss meters).

The Danish-French, Lyngby-Saclay, collaboration is at present developing an instrument to be flown on one of the coming HEAO satellites. This instrument will determine the isotopic composition of the most abundant elements from Lithium to Nickel over the range 2-20 GeV/nucleon using the earth's magnetic field. The instrument is at present being tested in a series of balloon flights in order to develop new Čerenkov counter materials. One development model of the instrument is shown in figure 15.

The detectors are all Čerenkov counters, the two glass counters are mainly used to determine the charge and the liquid and gas counter measures the velocity at the low and high end of the energy range. Comparing the two signals from the glass counters one can detect whether a nuclear interaction has occurred during the traversal of the instrument.

The low index counters (liquid and gas counter) can only be used to measure velocity close to the counter threshold (refer to figure 8). New counters made from compressed silica powder are being developed (Linney and Peters 1972, Linney et al. 1973). This type of counter has the advantage that by varying the density of the compressed powder one can vary the refractive index over a wide range and thus make counters that are optimized in the energy range of interest. It is planned to use a powder counter in place of the liquid counter in the final version of the instrument and also to use a low-index silicagel counter under development at present (Cantin 1974) in place of the gas counter.

In order to increase the quality of the charge and velocity measurements the signals from the counters are corrected for fluctuations due to variations in incidence angle and impact point

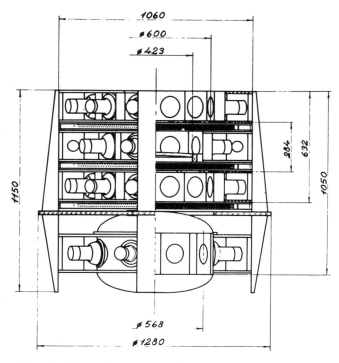

Fig. 15: Balloon-borne version of proposed HEAO instrument, Čerenkov counter materials from top to bottom: lead glass (n = 1.64), Fluorocarbon liquid (n = 1.24), lead glass, Freon gas (n = 1.015). Particle tracks are determined with neon flash tube arrays placed between the counters.

---

on the detectors. This is done by determining the particle track with a system of Neon Flash Tubes (Funch et al. 1973). The correction for geometrical effects is based on counter response maps obtained during pre-flight calibrations with ground level relativistic muons. It is important to note that by monitoring the computer response to the most abundant elements (C and O) one can update the counter maps and thereby correct for changes in the counter characteristics (dead phototubes etc.) during the flight. The instrument can thus determine charge and velocity of cosmic ray particles between 2 and 20 GeV/nucleon and by measuring these two parameters and the cosmic ray flux as a function of position in the satellite orbit one has sufficient information to determine the mean mass of an element relative to the mean mass of a reference element. The method is illustrated in figure 16.

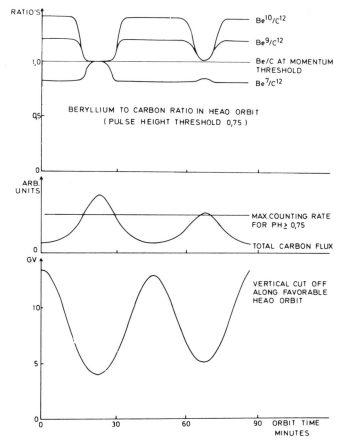

Fig. 16: Variations in the Be/C ratio (top), total C flux (center) and vertical rigidity cut-off (bottom) for a favourable HEAO orbit. Details in text.

A very simple picture of the effect of the earth's magnetic field on the cosmic ray particles is that only particles with a magnetic rigidity greater than the so called cut-off rigidity can penetrate sufficiently far down in the magnetic field to be detected by the instrument. As the earth's field to a good approximation is a dipole field, the effect of the field will be strongly dependent on the geomagnetic latitude of the instrument. The lower curve in figure 16 illustrates the variation of the cut-off rigidity during one orbit for a low altitude (~200 nautical miles) circular orbit inclined ~30° with respect to the equator as planned for HEAO.

As the cut-off rigidity becomes lower the flux of particles reaching the instrument will increase as shown in the middle curve in figure 16. By a suitable transformation of the measured Čerenkov pulse heights,

$$Q' = \frac{Q_{measured}}{z^2(1 - \frac{1}{n^2})} \quad \text{(suitably normalised)}$$

all pulse heights will fall in the range 0 to 1 and counting the number/sec of particles of a given element with pulse heights greater than .75, for example, corresponds to measuring the flux above a given velocity (or momentum/nucleon). Over part of the orbit the counting rate of particles above the velocity threshold will be constant and normalizing the ratio of the flux of two elements (say Be and C) to 1 over this part of the orbit, the flux ratio at other parts of the orbit will give a measure of the relative mean mass of Be to C. This effect is seen in the upper family of curves that shows the ratio of counting rates of Be and C above the momentum threshold. As the limiting factor changes from velocity to rigidity the relative flux of the isotopes will change in proportion to the mass/charge ratio i.e. the relative flux of the heavier isotopes will increase.

However, as the instrument cannot resolve isotopes directly the quantity that will be measured is not the ratio between fluxes of isotopes as shown on figure 16 but the ratio between fluxes of elements. The method will thus only give information on the relative mean mass of the elements, not on the isotopic composition. In the case where only two isotopes are present in the cosmic ray flux (e.g. He, Li, B, C, N) one can of course calculate the isotopic composition from the mean mass.

It is important to emphasize that the method does not require any detailed knowledge of the geomagnetic field but only the measurement of the flux ratio above a given velocity, above a given rigidity and the assumption that the flux of all the isotopes within an element has the same energy dependence. In theoretical calculations of the feasibility of the method a power law dependence is normally assumed (Lund et al. 1971) but when the experiment is carried out the energy dependence will be measured simultaneously.

What has been presented here is only the simplest possible analysis of the geomagnetic effects. A more careful study of the changes caused by the field, combined with the use of more detailed knowledge of the geomagnetic field, will make it possible to measure the isotopic composition directly (Peters 1974), although the analysis becomes very complicated and the effect of the instrument resolution has to be taken into account.

One of the most intriguing future possibilities of direct measurement of the isotopic composition in this energy region is the use of the socalled penumbral bands. Due to the fact that the earth's field is not a perfect dipole field, the effect of the field will not give rise to a sharp cut-off rigidity, above which the particles can reach the instrument, but to a range in rigidity, the penumbra, where narrow bands of allowed and forbidden rigidities alternate (Shea and Smart, 1967). It should be possible, when improved velocity resolution can be achieved (a factor 2 or 3 improvement over the present resolution is needed), to use the allowed penumbra bands as narrow slits and thereby obtain velocity separated isotopes.

IV RESULTS

The study of the isotopic composition of the arriving cosmic ray nuclei has just started to yield the first results in the last couple of years. As mentioned several times in the discussion of the instruments the main problem has been to obtain the instrumental resolution necessary to measure the masses to better than ~1 atomic mass unit.

Only the isotopes of Hydrogen and Helium have been clearly resolved for some time. I will not here discuss the numerous measurements already made of these isotopes but instead refer to two excellent reviews presented at the 12th and 13th international cosmic ray conferences (Simpson 1971, Stone 1973). Since the time of these reviews an important new result has been obtained by the group at the University of Chicago (Dwyer and Meyer 1974). This group is using the geomagnetic cut-off method to analyse the results from two balloon flights. The group has made a preliminary analysis of the data from the first of these flights and has found for the He^3/He^4 ratio, from 1.5 to 5 GeV/nucleon, a value of $(13 \pm 7)\%$. The new measurement extends the energy range over which this important ratio has been determined one order of magnitude towards higher energies. This is the first measurement of the Helium isotopes above the energy region wherein solar modulation plays a major role. When the final and more precise result of this analysis is available a new test of the different formulations of the transport equations used in propagation theory will be possible.

A common feature of the isotopic composition of the heavier elements ($Z \geq 3$) is that we expect to find many more abundant isotopes in the cosmic rays than we have found in solar system

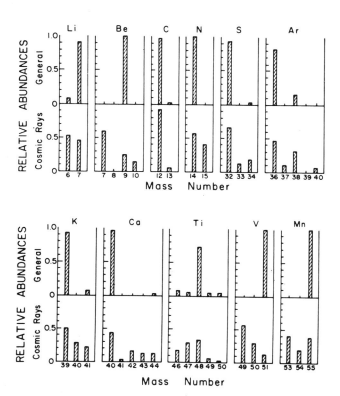

Fig. 17: Calculated relative isotopic abundances (Tsao et. al. 1973). Each element is normalized to unity.

material, as shown on figure 17. Besides the stable isotopes found in the solar system, long-lived radioactive isotopes and K-capture isotopes are also expected to be present in the cosmic ray flux. The K-capture isotopes, radioactive isotopes decaying by capturing one of the orbital electrons, are stable in the space environment as these energetic nuclei are fully ionized and the cross section for electron pick-up reactions is very small at energies above a few hundred MeV/nucleon. A table of the most important K-capture isotopes and their lifetimes (in the presence of orbital electrons) is given on the next page.

Most of the isotopes listed in Table I are produced in the spallation processes during propagation but if we can establish the presence of K-capture isotopes originated in the source, this would provide information on the time the particles have spent in the source region after ejection and before acceleration. This

Isotope	Halflife	
Be^7	53.2	days
Ar^{37}	36	days
Ca^{41}	8.10^4	years
Ti^{44}	48	years
V^{49}	330	days
Cr^{51}	27.8	days
Mn^{53}	$3.7.10^6$	years
Mn^{54}	303	days
Fe^{55}	2.6	years
Ni^{56}	6.1	days

Table I   K-capture Isotopes

---

problem is discussed in the contribution by H. Reeves.

The long-lived radioactive isotopes, listed in table II, are also mostly secondaries produced during the propagation.

Isotope	Halflife	
Be^{10}	$1.6.10^6$	years
Al^{26}	$7.4.10^5$	years
Cl^{36}	$3.1.10^5$	years
Mn^{54}	2.10^6	years
Fe^{60}	3.10^5	years
Ni^{56}	2.10^5	years

Table II   Long-lived Radioactive Isotopes

The isotopes Mn^{54} and Ni^{56} appear in both tables following a suggestion by M. Cassé (Cassé 1973a, 1973b), that although only K-capture decay is observed in the laboratory, positron decay is energetically possible and might occur when the nuclei are fully stripped.

In view of the fact that the present experimental data on the isotopic composition of the elements heavier than Helium are rather tentative and that many new results will be forthcoming shortly, I will restrict the following discussion to two important problems; the abundance of the "cosmic ray clocks" exemplified by Be^{10} and the abundances of the iron group.

The abundances of the radioactive isotopes will provide us with information on the cosmic ray propagation time as the following simple arguments show. As we know the total amount of interstellar material, $x(\sim 5 \text{ g cm}^{-2})$, traversed during propagation, a knowledge of the mean density of matter in the propagation region will allow us to determine the mean transit time from the equation

$$x = n c t$$

A measurement of the abundance of one of the radioactive isotopes produced during propagation gives an estimate of the mean density from the steady-state equation (production rate = decay rate).

$$\frac{dN}{dt} = N/\lambda$$

where $\frac{dN}{dt}$ is the production rate of the isotope (proportional to the mean density of matter), λ is the halflife of the isotope and N is the measured abundance of the isotope. We can thus determine a cosmic ray "age" from the measurement of the radioactive isotopes. It can also be seen from the last equation that at high energies, where the Lorentz time dilation is significant, the abundance of the radioactive isotopes will be higher. The Be^{10} energy spectrum for different assumed "ages" i.e. different mean densities is shown in figure 18.

The meaning of the "age" we can thus determine is clearly model dependent. In the standard leaky box model we assume that the sources are located within the galactic disk and that the disk is also the confinement region. Using the mean matter density of the disk, ~ 1 atom cm^{-3} one would expect a propagation time of the order of a million years. However, a measurement of an "age" of one million years cannot be used to exclude even the hypothesis of extragalactic origin of the cosmic rays. In the last case the "age" that we would determine, would not be the time since acceleration in another galaxy, but the time since arrival into our galaxy. The high density of matter in interstellar space compared to intergalactic space would result in a mean matter density which would be close to the disk value and the "age" determined would thus be close to the propagation time in our galaxy, although the time spent traversing the intergalactic space might be 1000 times longer. Similar problems would be the result of assuming that in the leaky box model the source region is surrounded by a dense cloud wherein a sizeable amount of the interactions take place. This could lead to the determination of

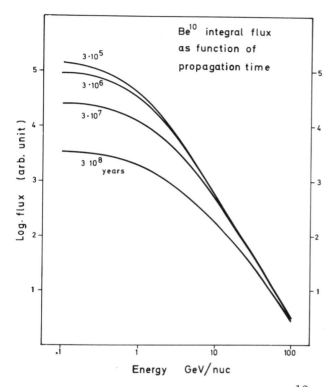

Fig. 18: The energy dependence of the Be^{10} flux as a function of mean density of the interstellar material in the confinement region.

an "age" greater than the actual time between acceleration and observation.

A precise determination of the energy spectra of the radioactive isotopes would give an indication of how well the mean density assumed represents the actual conditions during propagation. An increased production at the beginning of the propagation (as in a possible dense cloud) would lead to a relative over-abundance in the high energy region, an increase at the end of their life to a relative under-abundance. It probably will require 3 or 4 different instruments in order to cover the entire range of interest.

The experimental data obtained so far are shown in figure 19. Both mass spectra contain very few counts and the poor statistics make interpretation ambiguous. The resolution of the two mass spectra is rather similar and just about sufficient to resolve individual isotopes. The Chicago Group claims only to have

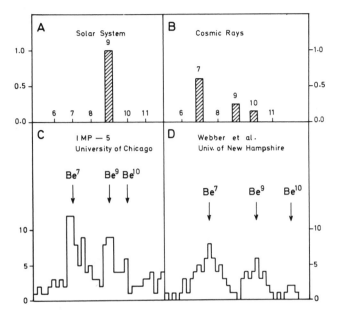

Fig. 19: Experimental mass distribution of Beryllium compared with the calculated distributions. A and B (Tsao et al. 1973). C (Garcia-Munoz et al. 1973) and D (Webber et al. 1973a).

resolved Be^7 from Be^{9+10}, whereas the New Hampshire Group claims to have resolved all three isotopes. Both Groups find that the Be^7 flux is half of the total Be flux. Based on a comparison between the observed number of Be^{10} nuclei (7) and the predicted number (18) the New Hampshire Group (Webber et al. 1973a) estimate a propagation age of $(3.4 \pm ^{3.4}_{1.3}) \times 10^6$ years.

The uncertainty quoted by the New Hampshire Group, though large, is due to statistical uncertainty on the observed number of Be^{10} only and thus does not represent the total uncertainty. The Group is here following a widespread but deplorable practice. It is useful to compare the more basic dataplots (like figure 20) with the end results (figure 19) in order to get an idea of the amount of data processing used in obtaining the results. The mass spectra shown in figure 19 are derived from two-dimensional plots like the one shown in figure 20 by summing the data along the calculated isotopic tracks.

The authors state that the positions of these tracks depends

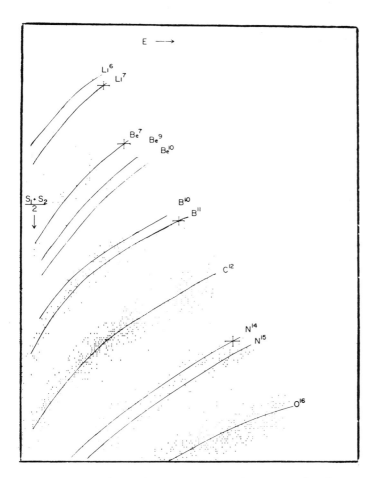

Fig. 20: Two-dimensional pulse height matrix for stopping particles (Webber et al. 1973). Solid lines give calculated response for various isotopes.

on the experimental data themselves. It is clear that in the case of inadequate statistics, as found in such experiments, the fitting of the parameters used in calculating these tracks plus the added uncertainty induced by background rejection procedures, will mean that the total uncertainty on any conclusion drawn from the experimental data will be considerably larger than the statistical one.

More definite results on the composition of Beryllium will probably be available in the near future as several new

experiments are under preparation. The possibility of using
heavy ions from high energy accelerators like the Berkeley
Bevatron will lead to improvements in the calibration of the
instruments, and thereby to more accurate results.

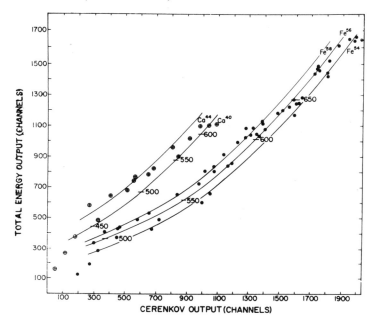

Fig. 21: Observed two-dimensional matrix of events for Ca and
Fe nuclei in C x E mode. Solid lines give calculated response
for various isotopes. Number of curves give calculated energies
in MeV/nucleon. (Webber et al. 1973).

It has not been possible so far to obtain mass spectra with good
resolution for the elements above oxygen and especially not for
elements in the iron group in any energy region. The Čerenkov-
total energy technique developed by the New Hampshire Group has
yielded the first results on the isotopic composition of iron
as shown on figure 21. It is clear from the figure that individual masses cannot yet be resolved. Again many groups are planning
new experiments and interesting results should also here be
available in the near future.

V   APPENDIX A   Composition at High Energies

Although the energy dependence of the charge composition in
the range 2 to 50 GeV/nucleon is not very well established at
present, a general picture is slowly emerging. The ratios

Fe/C+O and primary/secondary elements are increasing with energy
(a review of the experimental evidence can be found in the
rapporteur talks of Webber and Garcia-Munoz given at the Denver
Conference (Webber 1973, Garcia-Munoz 1973)). A comparison with
the changes in composition due to the propagation, shown in
figure 3, indicates that a possible interpretation of this energy
dependance is that the high energy cosmic ray particles are traversing less matter between the source regions and the solar
system. At high energies the composition of the arriving composition should thus approach the composition we can calculate at
lower energies by removing the effects of the spallation taking
place during propagation. The results of such a calculation are
shown in table III. It is of course a necessary assumption that
all the elemental abundances have the same dependence on energy
at the source if this calculated composition is to be valid at
higher energies. Using this additional assumption it is possible
to calculate the composition with respect to energy/nucleus, a
representation more useful at high energies than the standard
composition calculated in energy/nucleon. The result of such a
calculation is shown in Table IV. A comparison with a similar
table often used (p. 45 in Ginzburg and Syrovatsky 1964) shows
that the abundance of Helium is considerably lower now, and the
abundances of the elements heavier than Oxygen, correspondingly
higher.

It must be emphasised that the basis of these calculations
still is rather uncertain and that the picture soon will be
clarified as more experimental evidence becomes available.

H	$5 \cdot 10^4$	Si	20.5 ± 3
He	2600	P	$.2 \pm ^{.4}_{.2}$
C	100	S	$3 \pm .6$
N	11 ± 2	Ar	$.7 \pm .5$
O	109 ± 2	Ca	$2.2 \pm .8$
Ne	15 ± 2	Cr	$.3 \pm .3$
Na	$.8 \pm .4$	Mn	$.2 \pm ^{.4}_{.2}$
Mg	23 ± 2	Fe	22 ± 3
Aℓ	2 ± 1	Ni	$.8 \pm .2$

Table III: The Source Region Charge Composition (Shapiro and Silberberg 1974). C is normalised to 100.

Element	Mass	Relative content of nuclei %			
		energy/nucleon	energy/nucleus		
		$\gamma=2.5$	$\gamma=2.6$	$\gamma=2.7$	
H	1	94.5	49.5	43.1	36.7
He	4	4.9	20.6	20.6	20.2
C	12	.2	4.1	4.6	5.0
N	14	.02	.6	.6	.7
O	16	.2	6.9	7.9	8.9
Ne	20	.03	1.3	1.6	1.8
Na	23	.002	.09	.1	.1
Mg	24	.04	2.7	3.2	3.7
Al	27	.004	.3	.3	.4
Si	28	.04	3.0	3.7	4.3
P	31	.0004	.03	.04	.05
S	32	.006	.5	.7	.8
Ar	40	.001	.2	.2	.3
Ca	40	.004	.6	.7	.9
Cv	52	.0006	.1	.1	.2
Mn	55	.0004	.08	.1	.1
Fe	56	.04	9.1	11.9	15.1
Ni	58	.001	.3	.5	.6

Table IV: Composition at high energies. The flux is assumed to follow a power law in energy i.e. $N(E) \propto E^{-\gamma}$

ACKNOWLEDGEMENT

I would like to thank Professor B. Peters for many useful discussions during the preparation of this talk.

REFERENCES

Beaujean, R. and Enge, W., 1973. Proc. 13th Internat. Cosmic Ray Conf. 1, 111. Denver.

Cameron, A.G.W., 1968. In Origin and Distribution of the Elements, ed. L.H. Ahrens. Pergamon.

Cameron, A.G.W., 1973. In Explosive Nucleosynthesis, ed. D.N. Schramm and W.D. Arnett. Univ. Texas Press. Austin.

Cantin, M., 1974. Private communication.

Cassé, M., Goret, P., Koch, L., Maubras, Y., Mestreau, P., Meyer, J.P., Roussel, D., Soutoul, A., Valot, P. and Linney, A.D., 1973. Proc. 13th Internat. Cosmic Ray Conf. 4, 2901, Denver.

Cassé, M. and Goret, P., 1973. Proc. 13th Internat. Cosmic Ray Conf. 1, 584, Denver.

Cassé, M. 1973a. Astrophys. J., 180, 623.

Cassé, M. 1973b. Proc. 13th Internat. Cosmic Ray Conf. 1, 546, Denver.

Dwyer, R. and Meyer, P., 1974. Private communication.

Funch, O., Iversen, I.B., Lund, N., Rasmussen, I. Lundgaard and Rotenberg, M., 1973. Proc. 13th Internat. Cosmic Ray Conf. 4, 3023, Denver.

Garcia-Munoz, M., 1973. Invited and Rapporteur Papers, 13th Internat. Cosmic Ray Conf., Denver.

Garcia-Munoz, M., Mason, G.M., and Simpson, J.A., 1973. Proc. 13th Internat. Cosmic Ray Conf. 1, 100, Denver.

Havnes, O., 1973. Astron. and Astrophys. 24, 435.

Hayakawa, S., Ito, K. and Terashina, Y., 1958. Progr. Theor. Phys. Suppl. 6.1.

Hurford, G.J., Mewaldt, R.A., Stone, E.C. and Vogt, R.E., 1973. Proc. 13th Internat. Cosmic Ray Conf. 1, 93, Denver.

Jokipii, J.R. and Parker, E.N., 1969. Astrophys. J. 155, 799

Kristiansson, K., 1971. Astrophys. and Sp. Sci. 14, 485.

Kristiansson, K., 1971a. Symposium on Isotopic Composition of the Primary Cosmic Radiation, Lyngby, Denmark, ed. P.M. Dauber.

Kristiansson, K., 1972. Astrophys. and Sp. Sci. 16, 405.

Linney, A.D. and Peters, B., 1972. Nucl. Instr. & Meth. 100, 54.

Linney, A.D., Cantin, K., Koch, L., Maubras, Y., Mestreau, P., Roussel, D., Soutoul, A. and Valot, P., 1973. Proc. 13th Internat. Cosmic Ray Conf. 4, 2907, Denver.

Lund, N., Peters, B., Cowsik, R. and Pal, Y., 1970. Phys. Lett. 31B, 553.

Lund, N., Rasmussen, I. Lundgaard and Peters, B., 1971. Proc. 12th Internat. Cosmic Ray Conf. 1, 130, Hobart.

Mogro-Campero, A. and Simpson, J.A., 1972. Astrophys. J., 171, L5 and 177, L37.

Orth, C., Buffington, A. and Smoot, G., 1974. Private communication.

Peters, B., 1963. Pontificia Academiae Scientarium. Scripta Varia 25, 1.

Peters, B., 1974. To be published in Nucl. Instr. & Meth.

Price, P.B. and Fleischer, R.L., 1971. Ann. Rev. Sci. 21.

Price, P.B., 1971. Symposium on Isotopic Composition of the Primary Cosmic Radiation, Lyngby, Denmark, ed. P.M. Dauber.

Price, P.B., 1973. In Cosmochemistry ed. A.G.W. Cameron. Reidel, Dordrecht, Holland.

Raisbeck, G. and Yiou, F., 1973. Proc. 13th Internat. Cosmic Ray Conf. 1, 494, Denver.

Ramaty, R. and Lingenfelter, R.E., 1971. Symposium on Isotopic Composition of the Primary Cosmic Radiation, Lyngby, Denmark, ed. P.M. Dauber.

Schramm, D.N. and Arnett, W.D., 1973. Proc. 13th Internat. Cosmic Ray Conf. 1, 646, Denver.

Shapiro, M.M., Silberberg, R. and Tsao, C.H., 1973. Proc. 13th Internat. Cosmic Ray Conf. 1, 578, Denver.

Shapiro, M.M. and Silberberg, R., 1974. To be published in the Philosophical Transactions of the Royal Society of London, Series A.

Shea, M.A. and Smart, D.F., 1967. Journ. of Geophysical
   Research 72, 2021.

Silberberg, R. and Tsao, C.H., 1973a. Astrophys. J. Suppl. 25,
   315.

Silberberg, R. and Tsao, C.H., 1973b. Astrophys. J. Suppl. 25,
   335.

Simpson, J.A., 1971. Invited and Rapporteur Papers,
   12th Internat. Cosmic Ray Conf., Hobart.

Stone, E.C., 1973. Invited and Rapporteur Papers,
   13th Internat. Cosmic Ray Conf., Denver.

Tsao, C.H., Shapiro, M.M. and Silberberg, R., 1973. Proc. 13th
   Internat. Cosmic Ray Conf. 1, 107, Denver.

Webber, W.R., 1973. Invited and Rapporteur Papers, 13th Internat.
   Cosmic Ray Conf., Denver.

Webber, W.R., Lezniak, J.A. and Kish, J., 1973. Nucl. Instr.
   & Meth. 111, 301.

Webber, W.R., Lezniak, J.A., Kish, J. and Damle, S.V., 1973a.
   Astrophys. and Sp. Sci. 24, 17.

Webber, W.R., Lezniak, J.A. and Kish, J., 1973b. Astrophys.
   Journ. 183, L81.

NUCLEOSYNTHESIS AND GALACTIC COSMIC RAYS

Hubert Reeves

Centre d'Etudes Nucléaires de Saclay,
Institut d'Astrophysique de Paris

I   INTRODUCTION

　　The general theme of these lectures is the interaction between Nucleosynthesis and Galactic Cosmic Rays, the aim being twofold. Firstly: what is the nucleosynthesis that results from Galactic Cosmic Rays (G.C.R.). Secondly: from the nucleosynthesis, or more exactly, from a determination of the real abundances of the elements accelerated to be cosmic rays, can we say something about the origin of the G.C.R.

　　The stage may be illustrated as shown in Fig. 1. Our present "observational" position is illustrated in the middle of Fig. 1 by the observed flux and from the observed flux we can go two ways, forwards or backwards in time.

　　We can go backward in time, working our way, through the spallation effects, to the so-called G.C.R. sources. An important factor is the time delay between the moment the G.C.R. started to be fast and the moment that they reached us, this time denoted by T_{CR} in Fig. 1. From the G.C.R. source we can go backwards once more and introduce the time δt, which is defined in the following manner. Let us consider an iron nucleus in the G.C.R. δt is the time between the moment the iron nucleus we observe became an iron nucleus and the time at which it became fast, reaching the relativistic velocity with which we observe it. A determination of these time parameters could in principle be used to obtain information on the processes responsible for the G.C.R.

　　The last section of these lectures will consider what happens when we move forwards from the centre of the Fig. 1. The G.C.R.

move around in the Galaxy where their effect is two-fold. Firstly, they may be a source of ionisation of the interstellar medium, though, at the present time, their is some doubt of the importance of this process. Secondly, in induced spallation reactions, we have a possible source for the formation of light elements.

II   THE PHYSICS OF SPALLATION REACTIONS

What happens when you bombard a nucleus with a proton? How does it break up? What are the partial channels that it uses? Data of this nature is fundamental to any discussion of spallation reactions. Let us now consider the collision of a proton and a carbon nucleus.

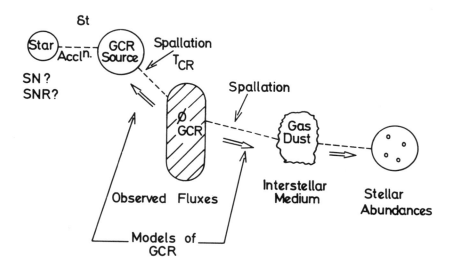

Fig. 1: The set up. We start from the "Observed fluxes" of G.C.R. and we move first backwards in time to discuss problems related to the origin of G.C.R. and next forwards to discuss the effect of the G.C.R. on the galactic gas and the nucleosynthesis of the light elements.

Curves showing the destruction cross-sections for the $^{12}C + p$ reaction are given in Fig. 2 as a function of energy. The total destruction cross-section σ_d, is shown on top. The lower energy region has resonances corresponding to compound nucleus formation. The lower curves show the cross-sections for the formation of the labelled products.

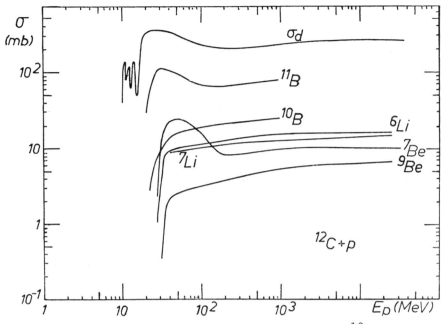

Fig. 2: Experimental data on the spallation of ^{12}C by protons as a function of incident energy. On top σ_d is the total destruction cross section which, after a few low energy resonances, shows a very smooth behaviour. Next the curve labelled ^{11}B shows the cross-section for forming the mass 11 products etc. ... Mass 11 is the most probable product, followed by mass 10 then by mass 7 (here the sum of ^{7}Li and ^{7}Be) then by mass 6 and then by mass 9. A similar situation is found for the case of spallation of ^{16}O.

One most important characteristic in these cross-sections is that all the reactions are endothermic, i.e. you need from 10 to 20 MeV and sometimes 50 MeV to induce the reaction. The cross-sections rise slowly to a platform and are essentially constant at energies >100 MeV. There is almost synchronisation between the value of the cross-section and the mass of the product. The most important product is that with mass 11, then that with mass 10, those with 7 and 6 and finally the product with mass 9. This gradation will be important when we discuss the cosmic ray data. There is also good data regarding the p + ^{16}O reaction. The cross-section curves show essentially the same features with the mass 11, 10, 7 and 6 together, 9, sequence. These elements have been studied carefully because they represent the most important contributions of the interstellar medium to the formation of light elements. Heavier targets are less important because they are rarer and because the thresholds for interaction are higher.

α-particles are also important in astrophysics, they number ~10% of the number of protons and data is now being obtained on the formation of light elements from α-particles. However, we do not expect these reactions to make a major contribution.

Experimentally, we have the result that the gradation of cross-section with mass is independent of the entrance channel and must therefore be related to some basic process of nuclear physics. It is possible to understand this result and reproduce it numerically (within a factor of 2 or so) using fairly simple ideas of nuclear physics. The first of these ideas, the Golden Rule of Fermi, tells us, that if you want to know the cross-section for some event, one way is to find out the number of ways in which the event you are studying can reproduce itself in nature, or, in terms of quantum mechanics, what is the phase space allowed to a given reaction. To know the ratio of two cross-sections you can compute the ratio of the phase space allowed to reach reaction.

This is the basis of the earliest model of nuclear physics reactions, the Compound Nucleus Model, which works well from threshold, to ~30 or 50 MeV. The model says that in the reaction, $p + {}^{12}C$ for instance, the proton is captured by the ^{12}C, giving ^{13}N in a very excited state. Now if you want to know how this is going to break up, you list all the possible ways e.g. you could have ^{7}Li + anything or ^{7}Be or ^{9}Be etc. To calculate the cross-sections, you just compute the phase space of each of the possible break up channels.

At higher energy (>50 MeV) we come into the region of spallation. The term spallation is not very well defined, but essentially it means that you get more particles out of the nucleus than you put in. The best model to explain the physics of spallation is the Serber "Two-Step" Model, which is essentially an extension of the Compound Nucleus Model.

Instead of being simply captured by the target nucleus, the incident proton knocks a certain number of nucleons in the nucleus and gets out. Some of the knocked nucleons also escape with some kinetic energy while other nucleons remain trapped there, bringing the residual nucleus to a certain excitation energy (usually described as a nuclear "temperature"). The temperature increases but to a lesser extent since the knock-on nucleons bring out an increasingly large fraction of the incident energy. Above a few hundred MeV the temperature reaches a saturation point: it hardly increases any more. This is the first step. In the second step the excited nucleus breaks into several pieces, and, as before, the probability or cross-section for any break-up is given by the corresponding volume of phase space. This model gives first physical reason why the cross-sections reach a "plateau" above a few hundred MeV: the nuclear temperature reaches a finite value and the break-up is thoroughly governed by the amount of excitation energy available.

The second thing that we can understand with our simple minded ideas of nuclear physics is the curves shown in Fig. 2, using some thermodynamics. Let us consider an example, the reaction p + ^{12}C, and see what the various resultant products are. A fundamental factor is the mass difference between the initial and final states because it is this which will govern the possibility of opening one channel or another.

If we consider the products ^{11}C + n + p, we find that this channel has a rather low threshold ~10 MeV. However, if we consider a channel where the product is ^{6}Li we find a much larger threshold ~30 MeV, the reason being that you have to remove more binding energy if you want to take out more particles. If we consider a 40 MeV proton incident on ^{12}C and consider the ^{11}C channel, we see that there will be ~30 MeV in kinetic energy for the products. There are many ways in which 30 MeV could be split amongst the products, hence a large phase space for this channel, hence a large cross-section. If we consider the same proton incident on the ^{12}C and consider the ^{6}Li channel, we find only ~10 MeV remaining in kinetic energy for the products, fewer ways of sharing it, hence, less phase space and a smaller cross-section.

This argument can be rephrased with some basic concepts of thermodynamics. Consider the effect of adding an amount ΔQ of kinetic energy to a nuclear system at temperature T. On the one hand, we increase the mass range ΔA that this nucleus may reach upon decaying; remembering that the average nuclear binding energy B is roughly independent of A and equal to ~8 MeV per nucleon we have:

$$\Delta Q = \frac{\Delta Q}{\Delta A} \Delta A = B \Delta A$$

On the other hand, the heat increase, ΔQ, corresponds to an increase in entropy ΔS or equivalently, in the logarithim of allowed volume of phase space Ω.

$$\Delta \log \Omega = \Delta S/k = \Delta Q/kT$$

We can then write the cross-section for a spallation reaction for losing ΔA nucleons as:

$$\sigma(\Delta A) = \Omega(\Delta A) = \exp\left[-B\Delta A/kT\right]$$

Since the sum of σ(ΔA) for all possible ΔA is simply the geometrical cross section we have:

$$\sigma(\Delta A) = \pi R^2 \, (B/kT) \exp\left[-B\Delta A/kT \right]$$

which is known as the "Rudstam" cross section. As discussed before, kT, the nuclear temperature is a slowly rising function of the incident proton energy. In fig. 3, a family of σ are plotted as a function of incident energy using an appropriate shape for kT. It is clear that the main qualitative features of spallation cross sections are well reproduced by these simple assumptions.

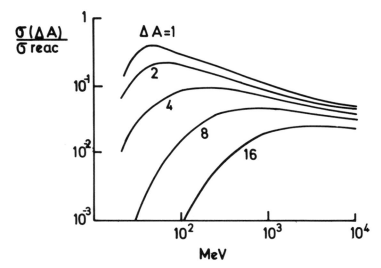

Fig. 3: Spallation cross-sections as calculated from the simple formula derived in the text. The curves ΔA are the cross sections for the removal of ΔA nucleons from a target of arbitrary mass by an incident proton of energy given in the abcissa. We note the ordering of the curves with increasing value of ΔA, and the absence of dependance on energy at high incident energy. Compare with figure 2.

The important characteristic to retain from this analysis of spallation is that spallation is very smooth, there are no sharp peaks in the cross-sections. We saw earlier that at low energy, you may expect strong effects but when you reach energies in excess of 100 MeV all the cross-sections gather into a small range.

We also see that small ΔA are more probable than large ΔA, because in general the Q value is smaller. If this were universally true, we would predict that the spallation cross-sections, when we look at p + ^{12}C for instance, would go in mass sequence 11, 10, 9, 7, 6 (there is no stable mass 8). However this is not observed. In Figure 2, we see 11 and 10 are in the correct place, 7 and 6 were quite similar and 9 was below. This behaviour is an example of the fact that when we are considering nuclei with a small number of nucleons it is imprudent to use statistical arguments. ^{9}Be is a rather special nucleus in that it just escapes being unstable. It has no excited state, and its isotopic brother, ^{9}B, is unstable to breaking into 2 alphas and 1 proton. This shows how the whole region of the periodic table is dominated by the stability of the α-particle. It is because of this near instability of ^{9}Be that the mass formulae used to calculate the cross-section are not valid and hence the production cross-section of ^{9}Be is well below the others.

We see, therefore, that these results on the spallation cross-sections can be understood in terms of these simplified models, though of course in any calculations using spallation cross-sections, we use the experimentally measured cross-sections which no theory can predict to the accuracy with which the cross-sections are often measured.

III CLOCKS FOR δt: ELECTRON CAPTURE ISOTOPES IN THE IRON PEAK

Most of the characterisitcs of nuclear forces result from a competition between a symmetry term favouring symmetric nuclei (as many neutrons as protons) and a Coulomb term which likes to have as few protons as possible. From this competition results two effects of great interest to us.

The first is the maximum in the binding energy/nucleon around iron. There is a simple law of calculus that states that when you have a maximum in a function, the first derivative of that function becomes small, so, with reference to the binding energy curve, we can say that the difference in the binding energy/nucleon becomes very small in the region around iron. The important thing for us is the effect that this has on the β-decay possibility of the nuclei in this region.

There are two ways in which a nucleus can β-decay. This can be illustrated by the following example. Let us consider the ^{13}N → ^{13}C decay: the ^{13}N being more massive than ^{13}C, it can eject part of its mass by forming an e^+ and a ν. This, of course, implies that the ^{13}N is quite more massive than the ^{13}C, since it has to furnish an extra electron mass of 0.5 MeV. If the difference in the binding energies is too small for an electron to be thrown out it is still possible for the nucleus to capture an electron from its electronic configuration and just emit a neutrino e.g. $^7Be + e^- \rightarrow {}^7Li + \nu$. This process is called electron capture β-decay. We note that it can take place only if the nucleus has electrons around it. In the lab. this is always so, but in space, where particles are moving with relativistic velocities, it is not so, hence for particles like ^7Be we have to talk about two lifetimes for β-decay, the lifetime in the lab (~53 days), and the lifetime in space which is ∞. In the region where the binding energy/nucleon goes to maximum we have several nuclei which are all in the position of having such small binding energy difference with their neighbours, that they cannot β-decay by the ordinary way and can only electron-capture, hence they are stable in space. These nuclei are $^{53}Mn, ^{54}Mn, ^{55}Fe, ^{56}Ni, ^{57}Co,$ and ^{59}Ni.

When we make a plot of the number of neutrons, N in the isotopes against the number of protons, Z, we find the second effect of the competition between symmetry terms and Coulomb terms: it moves the line of stability away from the N = Z line (Figure 4). In the low mass range the symmetry term wins and the stable nuclei have the same number of neutrons and protons. As you go to higher masses, the Coulomb term increases as Z^2 and the most stable configuration follows the hatched line. The nuclei mentioned above, those that β-decay by electron capture only, are shown by the dashed line.

When we have nucleosynthesis of matter in an explosive way (e.g. in supernovae) the nuclear reactions go extremely fast and, because the nuclear force is so much stronger than the β-decay force, there is no time for β-decay to take place. Therefore, the nuclei that you would make in explosive nucleosynthesis are always symmetric, i.e. have N = Z. After the explosive wave, these newly created nuclei then begin to migrate towards the line of stability, and, on their way, they have to go through the region of electron capture β-decay only. The question we have to answer is, when they were going through the region of electron capture β-decay (the dashed line in Figure 4) were the nuclei moving slowly or fast? If they were slow they would continue to move down to the stability line, but if they were fast they would not β-decay, as they would have no electron to capture, and would remain in the region of the dashed line.

This opens up the possibilities of having real clocks of the time, δt, between the generation of G.C.R. on the N = Z line, and their acceleration. If you accelerate nuclei at the same time that they are created they will stop on the dashed line of Figure 4 but if you want some time between nucleosynthesis and acceleration you will accelerate nuclei from the hatched region.

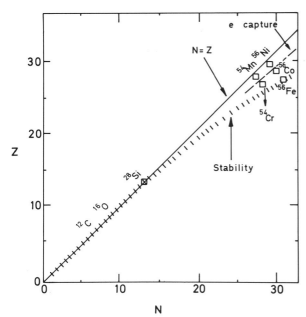

Fig. 4: The Z-N plane of nuclides. The hatched line is the region of nuclear stability where all the natural isotopes are found. The N = Z line is the region where the iron peak isotopes first appear when they are produced in explosive nucleosynthesis. In the region in between are located isotopes which are beta unstable by electron capture only, and hence are stable when moving at relativistic velocities (since they are stripped of orbital electrons). If the products of explosive nucleosynthesis are immediately ejected to fast speed, they will remain in the intermediate region.

Table I gives details of the properties of the various nuclei in the region of electron capture β-decay only.

The "Progenitors" are the isotopes as they are made in explosive nucleosynthesis, the "Intermediates" are those isotopes on the dashed line and the "Finals" are those in the hatched region of Figure 4.

How could we use this clock to find out something about δt? Let us consider the following example as an illustration.

In Figure 5 are plotted the universal abundances of the isotopes in the iron peak region. Those with the large circles are those which will be affected by electron capture β-decay. Now let us assume that this distribution of elements was

Progenitors		Intermediate	Final
^{53}Fe	9 min	^{53}Mn (lab 2×10^6 yr) (rel ∞)	^{53}Cr
^{54}Mn		(lab 313 d) (rel $\sim 3\times10^6$ yr)	^{54}Cr
^{55}Co	18 hr	^{55}Fe (lab 2.7 yr) (rel ∞)	^{55}Mn
^{56}Ni (lab 6.2d) (rel $\sim 2\times10^5$ yr)		^{56}Co (77 d)	^{56}Fe
^{57}Ni	36 hr	^{57}Co (lab 271d) (rel ∞)	^{57}Fe
^{59}Ni		(lab 8×10^4 yr) (rel $\sim 10^8$ yr)	^{59}Co

Table I. Clocks of δt in Iron Peak Nuclei (Cassé and Soutoul, 1974).

---

generated by a process of explosive nucleosynthesis. This implies that initially ^{56}Fe was not ^{56}Fe, but ^{56}Ni, ^{54}Cr was ^{54}Mn etc. As time went on, the decay took place from the left half circles to the right half-circles in the diagram, with the lifetimes indicated.

This indicates how the isotopic abundances in the iron-peak region in the G.C.R. could be used to tell us something about δt although of course we should not assume that the initial abundance of accelerated nuclei were as described in Figure 5.

Although we have as yet not isotopic ratios in G.C.R., the observed elemental Ni/Fe ~ 0.1 being in no serious disagreement with the universal abundance ratio (~0.05) may be taken as suggesting that we are actually observing elements with universal abundances and that $\delta t \gg 6$ days since otherwise ^{56}Fe would appear as ^{56}Ni and G.C.R. ratio would be Ni/Fe $\gg$ 0.1. Unfortunately ^{56}Ni is not stable even when relativistic; it has a small probability of ordinary β-decay with a poorly known lifetime of 2×10^5 yr, which, as discussed in the next paragraph, is similar to the G.C.R. travel time in space T_{CR}. However at

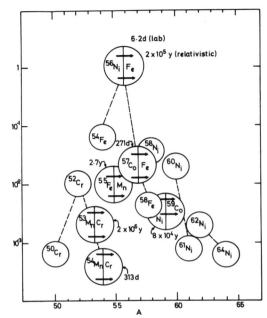

Fig. 5: Plot of universal abundances of isotopes in universal matter. The big circles correspond to the isotopes mainly decaying by pure electron capture. On the left semi-circle are the isotopes as they would appear soon after explosive nucleosynthesis, and on the right-hand side the isotopes remaining after all beta decay. For mass 56 for instance ^{56}Ni decays in 6.1 days in the laboratory but in ~2 x 10^{5}yr when relativistic (it has a small branching ratio for ordinary beta decay). (adapted from Cassé and Soutoul, 1974).

---

100 GeV/N the Lorentz time factor increases this lifetime by a factor one hundred. The fact that even at this high energy the "iron peak" has not become a "nickel peak" (Meyer private communication) confirms the assignment of $\delta t \gg 6$ days.

IV THE TRAVEL LIFETIME OF GALACTIC COSMIC RAYS

If you are to interpret the result obtained in the previous section, it is necessary to have some knowledge of the lifetime of the G.C.R. in the galaxy, since if it is much longer than the lifetime of ^{56}Ni you would expect a certain amount of decay of ^{56}Ni and hence a smaller Ni/Fe.

To estimate the travel lifetime of the G.C.R. we use two clocks, those of ^{36}Cl and ^{10}Be. ^{36}Cl is an isotope with a lifetime of 3×10^5yr and it decays into ^{36}Ar and ^{36}S. Figure 6 shows a plot of the expected ratio of Cl/Fe assuming that all the ^{36}Cl has decayed (lower line) and that none of the ^{36}Cl has decayed (upper line). Rasmussen has discussed the details of such a calculation in an earlier lecture. The experimental points appear to fall somewhere between the two extremes and it is probably not imprudent to say that some of the ^{36}Cl has decayed and that therefore the lifetime of G.C.R. is $> 10^6$yr (Cassé and Goret, 1974), if we take into account the fact that the measurements go up to a γ of about 3.

Figure 7 shows the results of a similar calculation using the ratio Be/B. Here the experimental points would seem to indicate that not all the ^{10}Be has decayed and hence the lifetime of the G.C.R. $T_{CR} \lesssim 8 \times 10^6$yr.

These limits are not very accurate but they are probably the best that we can do at the present time. From these limits you cannot say definitely what will happen to the ^{56}Ni which has a lifetime of the order of $2 \cdot 10^5$yr, with large uncertainties. But at 100 GeV the situation is clearer.

This is a new field of investigation which promises to tell us more about these two important parameters, δt, and T_{CR}.

V RELATION OF δt WITH CLASS OF MODELS OF THE ORIGIN OF G.C.R.

It is interesting to group all the various models for the origin of G.C.R. according to the range of δt implied by each model.

1. $\delta t = 0$

$\delta t = 0$ means that the acceleration of G.C.R. happens in hydro-dynamic explosion, i.e. the explosion is the source of both acceleration and nucleosynthesis. The result of this is that the chemical composition of the sources of the G.C.R., derived from the observed values by subtracting the spallation effects, should be identical with the ejecta of whatever source of explosive nuclear synthesis you are considering.

2. $0 \lesssim \delta t \lesssim 10^6$yr.

The reason for choosing this upper limit is that 10^6 yr is essentially the time it takes for a supernova to explode; go through the remnant stage and disappear, mixed with interstellar matter. In this type of model the acceleration takes place in an

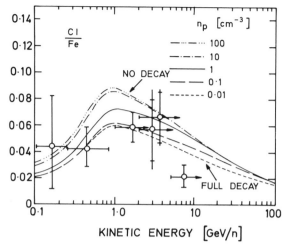

Fig. 6: Plot of the ratio Cl/Fe in GCR as a function of energy. The top curve represents the case where all the ^{36}Cl produced by spallation has not decayed (^{36}Cl has a life time of $\sim 3 \times 10^5 yr$). The lower curve corresponds to a full decay of ^{36}Cl. In spite of the large scatter of observational data, it appears that some ^{36}Cl is still present in the observed flux hence $T_{CR} > 10^6 yr$ (at 2 GeV we have a Lorentz factor of ~ 3).

---

expanding supernova remnant.

The composition of the G.C.R. "sources" will be given by two things. There is the component which is really coming from the supernova and there is a component of the interstellar matter due to the supernova remnant expanding and mixing. If you consider the Veil Nebula you have probably a remnant of $\sim 10^4 yr$ old and a ratio of supernova matter to interstellar matter of $\sim 10^4$. The matter that is the accelerated will be a mixture of these two components. The greater the value of δt the greater the proportion of the interstellar matter that will be mixed in (more exactly you will mix a small amount of atoms with $\delta t \ll 10^6 yr$ with a larger amount of atoms with $\delta t \gg 10^6 yr$ - interstellar gas cooked long ago).

3. $\delta t \gg 10^6 yr$

In models of this type there is no link between supernovae or supernovae remnants and G.C.R. The matter accelerated underwent nucleosynthesis long ago and its source is completely forgotten in the composition of the G.C.R. Models of this type are those of Fermi acceleration in the interstellar medium or origin

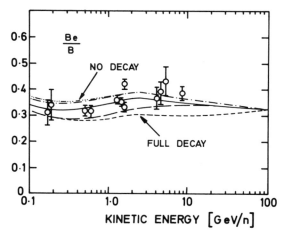

Fig. 7: Plot of Be/B in G.C.R. as a function of energy. The top curve represents the case where the ^{10}Be has not decayed (^{10}Be has a lifetime of 1.5×10^6yr). The lower curve corresponds to a full decay of ^{10}Be into ^{10}B. In spite of the large scatter of observational data it appears that some ^{10}Be has decayed. Hence $T_{CR} < 8 \times 10^6$yr. The various curves correspond to the same values of n_p as in fig. 6.

---

in White Dwarfs, etc. The chemical composition of the G.C.R. you would expect from a model of Fermi acceleration would be identical to that of universal matter.

VI   ANOTHER CLOCK FOR δt: THE "SOURCE" COMPOSITION OF G.C.R.

In a previous section we have seen how it is possible to get a measure of δt from the electron capture isotopes. In this section we will discuss another way by which we can get a handle on δt. From the chemical composition of the G.C.R. we search for the G.C.R. "sources", the real matter that is coming from whatever the source is. This search turns out to be more difficult than we first thought, because new obstacles appear at every step.

We first consider the G.C.R. composition as we observe it. Two things immediately strike you as you look at the data. There appears to be grouping into two different parts, the group of abundant elements, with a distribution which is quite similar to the universal abundance, i.e. H, He, C,N,O, etc. and the group of the rare elements, D, ^{3}He, Li, Be, B which are much more abundant than in universal matter. This may be interpreted by saying that these rare elements are not made in the source, but are made in space, and this tells us something about the amount of matter the G.C.R. have passed through. The situation may be schematically plotted as in Figure 8.

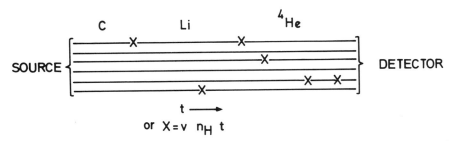

Fig. 8: Schematic view of the fate of high energy carbon nuclei moving through the galactic gas, with density n_H, describing the spallation of C into (for instance) Li and eventually He.

If we start with a flux of carbon nuclei at the sources and follow it through space where we have a hydrogen density n_H, the following will occur. One carbon may break into lithium which in turn may produce helium. Another may break into lithium but not into helium and others will not interact at all between the source and detector. One could then do a calculation and make a plot of the ratio of all the secondaries to the primaries as a function of the assumed path length, Figure 9. The ratio initially rises linearly with path length but eventually saturates, due to the destruction of the Li Be B, at a value of ~0.75. The experimental measurement of this ratio at 1 GeV gives a ratio of 0.25 which corresponds to a path length of ~6 g cm^{-2}. At 100 GeV the ratio falls to ~0.1 implying a smaller path length at this energy. The galaxy appears more and more transparent to the G.C.R. with increasing energy. We are also able to say that the Galaxy is leaking, or we would observe the saturation ratio of 0.75. A density n_H of 0.5 to 2 cm^{-3} gives an age of 3×10^6 yrs which is not inconsistent with the previous limits.

We are able to correct for the path length of the G.C.R. in the galaxy and can therefore calculate the chemical composition when they were accelerated (Westergaard et al. 1974). The composition at the "sources" is presented in Figure 10. In this plot the heavier elements are grouped together to improve the statistics. Figure 11 shows the universal abundance grouped in the same way. Before looking for differences, we should first be struck by the similarity between these curves. The relative abundance of the elements in each plot varies by ten orders of magnitude but, after correction for spallation, the difference between the two plots is never more than one order of magnitude.

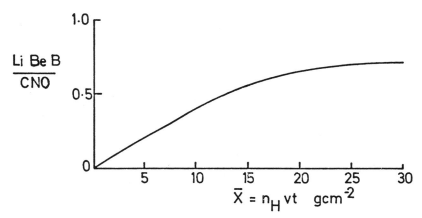

Fig. 9: Growth of the ratio of the secondary nuclides Li Be B to primary nuclides CNO as a function of the mean amount of matter traversed $\bar{X}$ for a model of G.C.R. propagation, involving a space distribution of sources (and resulting in an exponential distribution of path length with a mean value $\bar{X}$). (After Meneguzzi, 1974)

There are small differences, though we will now argue that they might not be real, they may just be due to preferential acceleration. We can define an overabundance ratio as, say, the ratio of iron to carbon in the G.C.R. divided by the ratio iron to carbon in the Universal abundance. These over-abundance ratios are plotted as a function of first ionization potential in Figure 12. We find that those elements with a low first ionization potential, ~5 or 6eV, are consistently "more over-abundant" than those with higher first ionization potential though the correlation is not perfect. This seems to suggest that differences between the two sets of abundances are due to preferential acceleration (Cassé and Goret, 1973 and Havnes, 1973).

Let us consider a simple model of how this could happen. Suppose we have a supernova remnant at a temperature of $\sim 10^4$. At this temperature those elements with a high first ionization potential will be mostly neutral, whereas those with a low first ionization potential will be completely ionized. If we assume some kind of electromagnetic acceleration, only the ionized fraction of particles will be accelerated. Relatively more of the elements with low first ionization potential will be accelerated simply because they are the more ionized species.

NUCLEOSYNTHESIS AND GALACTIC COSMIC RAYS

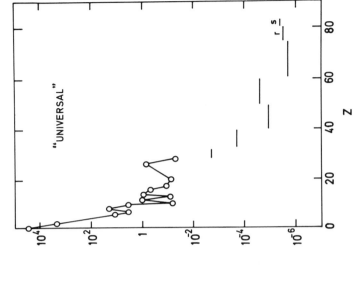

Fig. 11: Distribution of abundances in universal matter. For ease in comparisons the heavy elements have been grouped in the same way as in Figure 10.

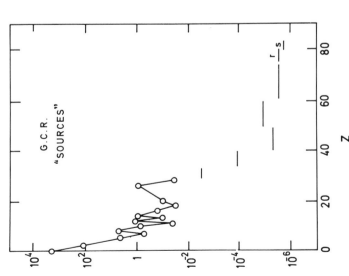

Fig. 10: Distribution of source abundances of G.C.R. fluxes, obtained after correcting the observed fluxes for the effects of spallation in space. The lines in the high mass range represent the sum of abundances of groups of nuclei in the corresponding mass range (because of the low counting rates, abundances of individual elements would not be statistically significant).

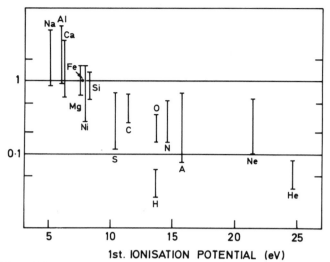

Fig. 12: Over-abundance factor in cosmic ray sources, as compared to universal matter, plotted as a function of the first ionization potential of each element (after Cassé and Goret, 1973).

We could test whether these differences are due to preferential acceleration or not by looking at isotopic abundances, since isotopes will not be affected by preferential acceleration. We could also make a test by looking at elements with similar ionization potentials. This is a study for the future.

With these new observations regarding the G.C.R. data, let us look again at the various classes of models discussed in section V and see how they best fit the data.

1. $\delta t = 0$

The heavy elements distribution that we expect to get in explosions of supernovae is likely to be quite "exotic" and therefore quite unlikely to present such a similarity with the universal abundance.

Let us look at a specific example, that of the r-process elements. If we compare the ratio of r/s elements in the G.C.R. sources we find that the ratio is apparently larger than the ratio in universal matter (but the uncertainties are large). From this data, people have concluded that this is a sign that the G.C.R. sources are supernovae, because the r-process means explosive nucleosynthesis and hence you relate to supernovae.

However, let us consider what happens when you have an r process. Consider a star with a number of heavy elements such as Fe in the centre. For an r process you require a lot of neutrons since you require that neutron capture rate be faster than decay rates (which can occur on the scale of seconds). The density of neutrons required would be ~ 1 to 10^3 g cm^{-3}. In such a situation a <u>large</u> fraction of all the iron will be transformed to heavier elements: an r/Fe ratio of the order of one would be expected.

The observed value of r/Fe in G.C.R. is $\sim 10^{-6}$, so the question is not whether you have a ratio of r/s that tells you that you have explosive nucleosynthesis but whether the r/Fe ratio really relates to prove explosive nucleosynthesis.

A further example is the ratio N/C+O. In nature nitrogen is not a primary element. To make nitrogen you first have to burn Hydrogen to make Helium and then burn Helium to make Carbon and Oxygen which is done in first generation stars. In second generation stars we have the CN cycle which forms Nitrogen. The observed N/C+O ratio in nature (~ 0.1) is the result of a long story involving many generations of stars throughout the life of the galaxy.

Due to spallation corrections, it is not easy to calculate the ratio N/C+O in the G.C.R., but interpretation of the best data gives a ratio not inconsistent with the Universal ratio.

Data from such ratios makes it very difficult to believe that we are just observing ejecta from any kind of explosive situation.

2. $\delta t \gg 10^6$ yr

Models in this category would predict that the composition of the G.C.R. should be exactly the same as the Universal composition. This would explain the great similarity between the two sets of abundances but the differences would all have to be explained by preferential acceleration. Isotope measurements, as previously discussed, should make the situation clearer. There is however one important point brought up recently by Catherine Cesarsky.

Measurements of the interstellar matter, by satellites such as Copernicus, have revealed that some elements are heavily depleted when compared to universal matter. For example the ratio of Calcium/H in the interstellar medium to Ca/H in the universal abundances is $\sim 10^{-3}$. Both U.V. and optical data give the same results.

Other elements appear to show the same behaviour, though the situation is not very clear due to complicated analysis of the data. However, it is thought that the abundance of Mg is ~ 0.1 that of universal matter, C and Si may be <0.3 that of universal matter. We have an explanation: space is full of grains, and some of these elements are very good at making stones, grains, and dust

These observations seem to have an interesting implication for the origin of G.C.R. in the range of $\delta t \gg 10^6 yr$ (acceleration in space). It is clear that you would not accelerate the dust in the same way as other particles (if you could accelerate it at all). You would certainly have some preferential acceleration effect.

In G.C.R., the abundance of Ca is small and when the spallation effects have been subtracted no real conclusions can be made as to whether Ca is present in the "sources". However, for Mg the situation is quite clear. The abundance of Mg in the sources appears to be quite <u>normal</u>, there is no reason to believe that there is any depletion.

This is an interesting point which works against the idea that G.C.R. are just accelerated from ordinary interstellar matter.

3. $0 \leqslant \delta t \lesssim 10^6 yr$

In these models there is a relation between supernovae and G.C.R., though not at the moment of explosion. We know that the Crab is accelerating electrons and because of their energy losses we know that they didn't originate at the moment of explosion.

As the supernova remnant expands, the ejecta of the explosion mixes with the interstellar medium and it is this mixture that is accelerated in the remnant. This accounts at least qualitatively for the quasi-similarity of the G.C.R. abundances and universal abundances. If the ratio of matter mixed to ejecta is large, you get quasi-similarity, but also small differences. To illustrate how this could work I have made a cook-book for G.C.R., Figure 13. It is pure imagination but shows, quantitatively, that it could work. The composition of the ejecta and the number of units of interstellar matter required have been invented to fit the G.C.R. data. However the real game is to go backwards from the G.C.R. data and the composition of the interstellar medium to find the composition of the ejecta.

The supernova remnant Cas A is an extremely strong source of radio emission. Cas A exploded ~200 yr ago. The idea was proposed

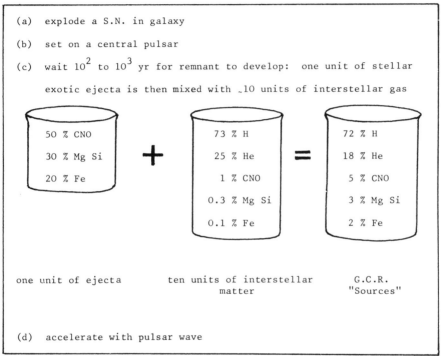

Fig. 13: Cook-book for G.C.R.

that if you could look at younger supernovae the radio flux would probably be much stronger. It was found, however, that the radio emission from young supernovae, ~1 or 2 yr old, in external galaxies is, allowing for distance effects, ~100 times weaker than that from Cas A. This would indicate that a supernova does not start to accelerate particles until 10^2 or 10^3 yr after it exploded. If you calculate the thermodynamical expansion of a supernova remnant this is about the time you need to mix ten units of interstellar matter with one unit of ejecta. (There is one important objection to this argument, which is: the reason that you don't see any radio emission is because it is all absorbsed by matter in the remnant. However, if the matter in the remnant is too dense to allow the radio emission to escape it will probably not allow the acceleration of particles to relativistic velocities).

This is all very illustrative and suggests that before we get to the real sources of the abundances of the G.C.R. we have to work much harder than we thought. As well as correcting the observed flux for spallation effects we have to consider the possibility of preferential acceleration and further we have to make correction for mixing of interstellar matter prior to acceleration. Only after making these three corrections can we claim that we know the composition of the real sources of G.C.R. and start to work out the physics of these sources.

In conclusion the data on the composition of the G.C.R. leads us to say that the most plausible class of model is one with δt in the range $0 \lesssim \delta_t \lesssim 10^6 \text{yr}$.

VII NUCLEOSYNTHESIS BY G.C.R. AND THE ORIGIN OF THE LIGHT ELEMENTS

In this last section, we are going to move forward, in Figure 1, from the observed fluxes to see what contribution the G.C.R. make to the nucleosynthesis of matter. If we look at the universal abundance curve we see that close to the very abundant elements, H, He, C, O etc. we find a group of extremely rare elements their ratio to the abundant group being 10^6 or 10^7. However we saw that when we looked at the observed flux of G.C.R., this rare group in the universal abundances, the Li, Be, B, are as abundant as iron. We have previously seen that these elements are produced in space as spallation products of C, N and O. The question we want to ask ourselves, is: could this process not only be the source of the fast component of the light elements, but could it be the source of <u>all</u> the light elements, Li, Be, B, <u>in nature</u>?

Before we go into any detail let us first make a simple-minded calculation to see that we are getting roughly the right numbers. We will then consider whether this process could be the origin of everything that is very rare.

We see that light element production by G.C.R. has two sources. There is the C, N, O in the G.C.R. striking the protons and α-particles in the interstellar matter, producing Li, Be and B at the same energy per nucleon (since we have a large particles hitting a small nucleus). There are three possible fates for these elements after they are formed in space.

Firstly, they could escape. We saw how the galaxy leaks
particles so a good fraction of them will leak out.

Secondly, some will be destroyed by nuclear reactions.

Thirdly, some will be decelerated by electronic
collisions and will mix in the interstellar gas.

We are interested in this third fraction. We have done
calculations to find out exactly what fraction of the particles
are decelerated. A good "rule of thumb" is the following:
Any CNO with energy >300 MeV will produce a Li, Be, B of the
same energy/nucleon which will rather, either be destroyed or
leak, than be decelerated. Any particle with energy <300 MeV will
be decelerated and join the interstellar medium.

There is a second component which is free of these difficulties, produced by the collision of a fast proton or α-particle with
a CNO from the interstellar medium. The products from such
reactions have energies ~5 or 6 MeV/nucleon, and all the Li, Be,
B are decelerated. In actual computation we find that the major
contribution at any one time to the formation of Li, Be, B is
this inverse reaction. As a result the galactic gas is
progressively enriched in light elements.

Let us compute the rate of formation of one isotope, say Be
as a monitor. The rate of formation today, which we multiply by
the age of the Galaxy, is given by the following:

$$T_G \times \frac{dn\left(^{Be}/H\right)}{dt} = \phi_p \ (E > 30 \text{ MeV}) \ n\left(\frac{CNO}{H}\right) \bar{\sigma} \ (CNO \to Be) \ T_G$$

where $\phi_p(E>30\text{MeV})$ is the flux of particles with energy >30 MeV
in space and is $10\text{cm}^{-2}\text{s}^{-1}$, (this is not well known due to solar
modulation effects).
$n\left(\frac{CNO}{H}\right)$ is the ratio of the density of CNO to H in the interstellar medium (everything is normalised to hydrogen to make
things simpler) and is ~10^{-3}. $\bar{\sigma}$ (CNO$\to$Be) is the mean cross-section and is ~5×10^{-27} cm^2. T_G is the age of the galaxy ~10^{10}yr.
The calculation gives a value for Be/H ~2×10^{-11}. The average
value for Be/H from stars is in the range 10^{-11} to $4 \ 10^{-11}$ so we
seem to be in the right ball park.

We continue by saying that we have made a number of
assumptions, which are not evident by themselves. One of these
assumptions is that the present rate multiplied by the galactic
age gives you the total time-integrated rate, which means that
we have assumed that the cosmic ray density has remained constant

and also that the relative density of CNO has remained constant. From the study of radioactive isotopes in meteorites we can infer that the flux of cosmic rays hasn't changed by very much over the last 10^9yr or so but we do not have any information about earlier times. In fact you can probably argue that the flux was larger in the beginning if you believe that the G.C.R. are related to supernovae. Arguments based on the statistics of stars with various metal abundances indicate that the process of star formation and element build-up were stronger in the first few $\times 10^9$ years of the galaxy. However the CNO was presumably less abundant in same period so that there might be some compensation.

As well as questioning the constancy of ϕ and the constancy of n(CNO) we could also include a factor which allows for the destruction of these light elements. There is such a process in the galaxy, namely "astration", the formation of stars. Any light element that is caught up in stars will be immediately burned and hence will be lost for the galactic gas. The best estimate for the lifetime of the gas for being caught by astration is $\tau^* \sim 5 \; 10^9$yr, which is long enough compared to the age of the galaxy so that there is probably no major destruction of light elements. Again we do not know whether the value of τ^* was different in the past or not.

Having seen that we have got the correct order of magnitude for Be, we want to become a little more sophisticated, and want to look in detail at what happens to the other elements. The next step is to work out a propagation model in which you try to account for all the observables of the G.C.R. (Westergaard et al. 1974). Using Be as a monitor we can calculate an average value of ϕ and n(CNO) and using these we are able to calculate the relative abundances of the other light elements from this spallation mechanism. Ratios of the various light elements will be independent of any variations in ϕ, n or τ^*, as all the elements will be affected in the same way.

Before proceeding it is necessary to make a more detailed review of the observations of the abundances of these light elements in Nature.

A summary of the observations regarding Li, which is the best known of the light elements, is shown in Table II.

Lithium has been observed in more than one hundred stars. The abundances do vary with a number of parameters but are understandable in the following plausible way. Plot the abundances in a cluster, e.g. Pleiades, as a function of stellar type (the region of type FGK, is where the line is best observed). If you take very young stars, e.g. T Tauri stars, which have not yet started their nuclear burning, you have a plot which is about a straight line with a value ~5 10^{-10}. If you have a cluster

Table II: Observations of Li

"Initial"	Hyades	Praesepe	Pleiades	NGC 2264	T Tauri Stars
$\frac{Li}{H}$	~ 8 10^{-10}	~1.6 10^{-9}	~10^{-9}	~10^{-9}	~5 10^{-10}

CCI	Solar	Red Giants	Interstellar
1.5 10^{-9}	10^{-11}	up to 10^{-7}	~3 to 6 10^{-10}

$^7Li/Li^6$ = 12.5 ± 0.5 (solar system)

that is a little older, the line shows that the older stars with the longest nuclear burning times have less Li. The values given above are the "initial" values interpolated from the measurements. We find that we have nuclear destruction of Li in stars as a function of time. For fixed spectral class, the older the cluster, the less the Li. For a fixed age, the later type stars in a cluster have less Li than the early type stars. The lithium abundance from stars fits in well, within a factor of 2, with an initial abundance of 10^{-9}.

The meteoritic value taken in the Carbonaceous Chondrites of type I (CCI) also fits in well. The surface of the sun gives a much smaller value, but this we expect because the sun is a G star and has decreased its abundance of Li. There are a small number of red giant stars, i.e. evolved stars, which do show much higher values up to 10^{-7}, suggesting that later in the life cycle of a star, there is a process which generates fresh lithium. We are not presently interested in this later process, but only in the initial "universal" abundance.

We know very little about the ratio of $^7Li/^6Li$, in stars except that whenever it has been looked for, no ^6Li has been found, indicating that the ratio is >10. We have very good measurements from the solar system which are all in good agreement, giving the ratio $^7Li/^6Li$ = 12.5 ± 0.5.

The situation regarding the observational data on Be is shown in Table III.

The best result that we have at the present time is the star Vega, which gives a value 10^{-11}. CCI give $\sim 2\ 10^{-11}$. The solar value which has been accepted for many years is 10^{-11}.

Interstellar	Solar	Stars	Vega	CCI
$<7\ 10^{-11}$	10^{-11}	from $4\ 10^{-11}$ to 10^{-12}	10^{-11}	$2\ 10^{-11}$
Ratio B/Be	Solar	Vega	CCI	
	$\lesssim 10$	~ 10	150	

Table III: Observations of Be

The interstellar value is an upper limit of 7×10^{-11}. A value of 2×10^{11} would seem to fit most of the data and is taken as "universal".

The situation regarding Boron, shown in Table IV, is rather confused.

Interstellar	Solar	Vega	CCI
$\lesssim 7\ 10^{-11}$	$\lesssim 10^{-10}$	$\sim 10^{-10}$	$3\ 10^{-9}$

Ratio $^{11}B/^{10}B = 4 \pm 0.4$

Table IV: Observations of Boron

There are really two sets of data, those on the left of Table IV giving a value $\sim 10^{-10}$, and the CCI data, which is usually but not always a very good indicator of solar system abundances. In my view a star is more likely to give you a cosmic abundance than the solar system because of all the history of chemical fractionation in the CCI.

The ratio $^{11}B/^{10}B$ again comes from meteoritic data; values from the moon, earth and meteorites, agree well with a ratio of 4 ± 0.4.

How do all these observed values compare with what one would obtain by computation based on spallation of the G.C.R. Table V is a summary of the results of such computations and the above observations.

	$^7Li/^6Li$	$^{11}B/^{10}B$	$^6Li/Be$	B/Be	$^{50}V/^{56}Fe$
Calc	~2	2 to 3	~4	10 to 20	~0.2 10^{-6}
Obs	12.5	4	~5	10	0.8 10^{-6}

Table V. Computed and Observed Ratios of Light Elements

Let us start with lithium: the observed and the calculated ratio of $^7Li/^6Li$ do not fit. One should remember that in calculations of these ratios the dominant factor is not the propagation model or the astrophysics, but the cross-sections which are well known. We should now ask which is guilty: is the ^7Li overabundant or is the ^6Li under-abundant. To answer this we look at the ratio $^6Li/Be$ which appears to fit quite well with the calculation, hence we conclude that the formation of ^7Li by this process is not sufficient to explain the observed $^7Li/^6Li$ ratio. The ratio B^{11}/B^{10} looks as though it is well explained by this spallation model. The B/Be ratio agrees well if one takes the observed value from Vega as being the cosmic abundance. Hence we conclude that the formation of ^6Li, ^9Be, ^{10}B and ^{11}B are satisfactorily explained by spallation in G.C.R. but that the ^7Li is not. Can we explain any other rare elements? We find that the abundance of the He and D from spallation is too small by a factor of 10^3 to account for the cosmic abundances. There is one further very rare isotope, ^{50}V, which could possibly be produced by the spallation of Fe. The observed ratio $^{50}V/Fe$ is $0.8 \; 10^{-6}$ and the calculated value is $0.2 \; 10^{-6}$, so we conclude that a significant part if not all of the ^{50}V could be produced by spallation of Fe in the same way that ^6Li, Be, B are the offspring of the CNO.

Let us consider a possible origin for the ^7Li. We have done some recent work on the low energy cosmic rays (E < 50 MeV) to see what would happen if they were more copious at low energy than we believe at present (Menegussi and Reeves, 1974).

There is a measured amount of heating and ionisation in the interstellar medium. There have been several possible mechanisms brought forward to account for these observations, one, by Spitzer, being the low energy GCR. With the results from the Copernicus satellite on the abundance of the elements and more especially on their ionized fraction, this theory ran into some difficulty. CI and CII were detected by Copernicus but the ratio of CIII/(CI + CII) was very small. This was held to prove that G.C.R. could not ionize the interstellar medium to any great extent because a particle of a few MeV should provide a distribution of all ionized states. The fact that no CIII was found could put a limit on the flux of low energy cosmic rays. However the situation has recently changed since workers in the field of Atomic Physics of excitation and recombination mechanisms found that there are a number of processes which could quench high energy excitation states much faster than was previously believed. Hence, on these grounds, the number of low energy cosmic rays could be raised significantly. We cannot rule out the possibility that these low energy G.C.R. could generate the missing ^{7}Li, although energetically speaking, this hypothesis seems rather unlikely.

You could also argue that you had a lot of low energy particles in a supernova remnant before it expanded largely. In this case, as in the previous one, most of the lithium would not come from the spallation of CNO, but, because of the cross-sections, would come from the He + He reaction. It is not impossible that a major part of the ^{7}Li comes from such a process. However, if we had a remnant with enough low energy particles to form the ^{7}Li we would also expect reactions of the type p + ^{12}C → ^{12}C* + p which would decay giving a 4.4 MeV γ-ray. In a few more years we could test this hypothesis by looking for localised sources of nuclear γ-rays.

Personally my favourite origin for the ^{7}Li is that it is left over from the Big Bang. Figure 14 shows a computation by Wagoner (1973) of the yield of particles from the Big Bang as a function of the present universal density of matter. We do not know the density very well but we cannot set limits on it, the lower limit approximates to the density of visible matter, and the upper limit corresponds to a density of ~10^{-29} g cm^{-3}, where we could see the deceleration of expanding galaxies.

The production of ^{4}He in the Big Bang is more or less independent of density. The only other elements formed in the Big Bang besides ^{4}Helium are D, ^{3}He and ^{7}Li, which are found in quantities close to explaining the observed universal abundances. It is also interesting to note that these elements are just the ones that you do not get from G.C.R.

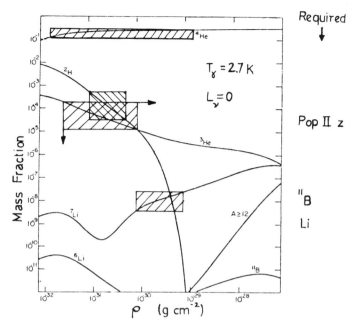

Fig. 14: Yield of light elements in Big Bang Nucleosynthesis as a function of the assumed value of the present universal density of matter (adapted from Wagoner 1973). The height of the various boxes represents the uncertainties on the initial galactic value of the abundance of the isotopes D, ^{3}He, ^{4}He and ^{7}Li, and their width the corresponding uncertainty in the universal density required to fit the abundance. It appears that in the range $5\text{-}7 \times 10^{-31}$ g cm^{-3}, one could account for all four isotopes.

So a plausible and pleasant solution, which may not be true, but is simple, is that the light elements ^{4}He, ^{3}He, D and ^{7}Li were formed in the Big Bang, while the other elements ^{6}Li, ^{9}Be, ^{10}B, ^{11}B and ^{50}V are the products of G.C.R.

REFERENCES

Boesgaard-Merchant, A., Praderie, F., Hack, M., 1974, in preparation.

Cassé, M. and Soutoul, A., 1974, Durham Conference on Isotopes in Cosmic Rays.

Cassé, M.,and Goret, P., 1973, Proc. 13th Int. Cosmic Ray Conf., Denver, Vol. 1, p.584. "Atomic properties of the elements and their acceleration to cosmic ray energies".

Cesarsky, C., private communication, to appear in Westergaard et al., 1974.

Goret, P. and Cassé, M., 1974, Durham Conference on Isotopes in Cosmic Rays.

Gunn, J.E. and Ostriker, J.P., 1971. Ap. J. Letters, <u>164</u>, L95.

Havnes, O., "On Cosmic Rays and Magnetic Stars" Astron. Astrophys., 1973, <u>24</u>, 435.

Meneguzzi, M., Reeves, H. and Audouze J., 1971. Astron. Ap. <u>15</u>, 337.

Meneguzzi, M. and Reeves, H., 1974, in preparation.

Reeves, H., Audouze, J., Fowler, W.A. and Schramm, D.N., 1973. Ap. J., <u>179</u>, 909.

Reeves, H., 1973. "Nucleosynthesis and the Origin of the Galactic Cosmic Rays".

Reeves, H., 1974. "On the Origin of the Light Elements, Ann. Rev. Astron. Astrophys. Vol. 12.

Shapiro, M., Silberberg, R. and Tsao, CH., "Composition of relativistic cosmic rays near the earth and at the source" Space Research XII Academic-Verlag Berlin, 1972.

Wagoner, R.V., Fowler, W.A. and Hoyle, F., 1967, Ap. J., <u>148</u>, 3.

Wagoner, R.V., 1973 Ap. J., <u>179</u>, 343.

Westergaard, N. et al., 1974, in preparation.

GALACTIC STRUCTURE, MAGNETIC FIELDS AND COSMIC RAY CONTAINMENT

K.O. Thielheim

Institut fur Reine und Angewandte Kernphysik
University of Kiel, W. Germany

I   INTRODUCTION

Cosmic ray research has become predominantly a branch of astrophysics. Under this aspect, the origin and transfer of cosmic ray particles through interstellar space are the objects of main interest. In trying to understand these phenomena one soon is lead to the conclusion that the structure and dynamics of cosmic magnetic fields have to be studied in advance. Specifically the structure of the galactic fields has to be known. In working on the structure of magnetic fields one eventually finds out that it is necessary to have basic information on the other constituents of the galaxy as well. This is due to the fact that the magnetic fields are coupled to the interstellar plasma which in turn through gravitational forces is linked to the distribution of stellar matter.

Following this line of thought I will refer in the first part of my lecture to some empirical results and theoretical interpretations on the structure of the galactic disk and on the spiral structure. In the second part of my lecture I will give a short resumé on observational methods and results pertaining to the structure of the magnetic fields in our galaxy. These will be followed by some theoretical arguments on the possible origin and topography of these fields. A preliminary model of the global magnetic field in the galactic disk ("Version I") will be the basis for a first attempt to reproduce empirical data. The third part of these lectures will be devoted to some short notes on cosmic ray particle response to different types of field configuration.

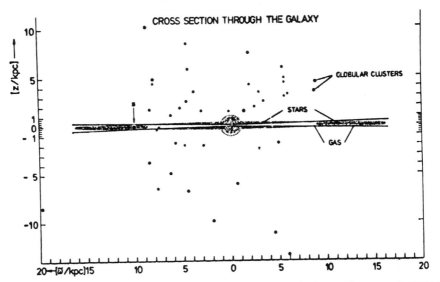

Fig. 1: Cross section through the galaxy (after Mezger (1973)).

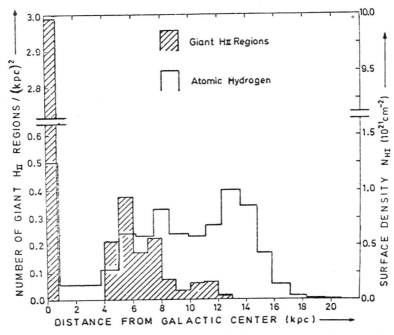

Fig. 2: Radial distribution of interstellar hydrogen in the galactic disk (after Mezger(1973)).

GALACTIC STRUCTURE, MAGNETIC FIELDS AND COSMIC RAY CONTAINMENT 167

It should be emphasized that rather controversial ideas still exist on most of these items. Especially, suggestions referring to the origin of cosmic ray particles as well as of galactic magnetic fields still remain far from being unanimous.

Since this is intended to be a lecture rather than a conference report, no attempt is made to give complete references of people who have contributed to the results presented here.

II GALACTIC STRUCTURE

1. Basic Information on the Structure of our Galaxy

The total mass of our galaxy is known to be about $1.8 \cdot 10^{11}$ $M_{\odot}$, 90% of which are contributed by stars while about 10% represent the interstellar gas and only about 1% are due to the interstellar dust. The chemical composition of matter incorporated into the interstellar gas is represented by about 70% hydrogen, 28% helium and about 2% of other elements. The greater part of galactic matter is concentrated within a very flat disk illustrated schematically by figure 1 taken from a recent article by Mezger (1973). The projected mass density as a function of distance from the galactic centre exhibits a sharp increase towards the latter. This is demonstrated for atomic hydrogen in figure 2.

The thickness of the galactic disk defined with respect to the density distribution of interstellar hydrogen increases with increasing distance from the galactic centre as is illustrated by a diagram of Lequeux (1973) in figure 3. Opposed to this global structure, spiral arm features manifest themselves within the local region through the density distribution of different types of arm tracers as is shown in figure 4.

Global features of the spiral structure are derived from the distribution of interstellar hydrogen incorporated in H 1 regions, as is obvious, for example, from the well known results of Kerr and Weaver cited by Simonson (1970) and reproduced in figure 5.

2. Theoretical Arguments Referring to the Structure of the Galactic Disk

The most prominent feature of galactic structure of course is that almost its entire mass is collected in the form of a flat disk. This may be understood theoretically within a first approach in terms of first order momentum equations of Vlasov's equation for its main constituents namely stars, plasma and cosmic radiation which, for reasons of simplicity, may be formulated under the assumption of translational invariance with respect to two

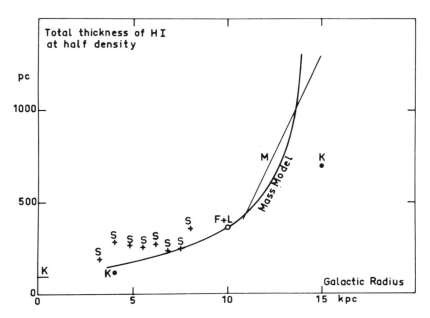

Fig. 3: Thickness of neutral hydrogen layer as a function of distance from galactic centre (after Lequeux (1973)).

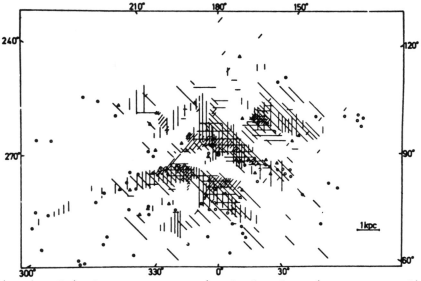

Fig. 4: Spiral arm structure in the local region corresponding to different arm tracers (after Lequeux (1973)).

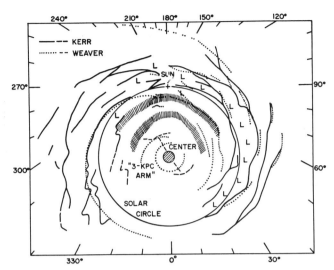

Fig. 5: Neutral hydrogen distribution in the galactic disk as interpreted by Kerr and Weaver (after Simonson (1970)).

---

coordinates say the x- and y-coordinates. The following deductions are essentially based on works by Oort (1959), Camm (1950), Parker (1966) and Kellman (1972) and finally result in a generalization of Camm's relation for the distribution of the various galactic constituents with respect to the z-coordinate given by Fuchs and Thielheim (1974).

Hydrostatic equilibrium in the thermal plasma is described by

$$\frac{d}{dz} P_{pl} = \frac{1}{c} [j_{pl}, \mathcal{B}]_z + g\rho_{pl} \qquad (2.1)$$

where P_{pl}, j_{pl} and ρ_{pl} are the pressure, electric current and mass density of the thermal plasma. $\mathcal{B}$ is the magnetic field vector and g is the total gravitational acceleration. All these quantities are understood as functions of z according to the symmetry adopted. Analogously hydrostatic equilibrium in the cosmic ray gas corresponds to the relation

$$\frac{d}{dz} P_{cr} = \frac{1}{c} [j_{cr}, \mathcal{B}]_z \qquad (2.2)$$

involving the pressure P_{cr} and the electric current j_{cr} in this component.

Addition of (2.1) and (2.2) and application of Maxwell's equation leads to

$$\frac{d}{dz}\left\{P_{pl} + P_{cr} + \frac{B^2}{8\pi}\right\} = g\rho$$

The plasma is suggested to be isothermally distributed

$$P_{pl} = \langle v_{th}^2 \rangle \rho_{pl}$$

The ratio of magnetic field energy to other types of energy present in the galactic disk, is generally assumed to be governed by the following relation

$$P_{pl} + P_{cr} + \frac{B^2}{8\pi} = Q^2 \rho_{pl} \quad ; \quad Q^2 = \text{const.}$$

which is rather an ad hoc assumption and therefore subject to criticism.

Eventually hydrostatic equilibrium in the stellar component, which also is assumed to be distributed isotropically, is described by

$$\langle v_{sz}^2 \rangle \frac{d}{dz} \rho_s = g\rho_s$$

Finally, Poisson's equation

$$-\frac{d}{dz} g = 4\pi G \{\rho_s + \rho_{pl}\}$$

is applied, where ρ_s is the density of stellar matter. Substituting ρ_{pl}, ρ_{cr} and g results in a non-linear differential equation for $\rho_s(z)$,

$$\rho_s'' - \frac{1}{\rho_s}(\rho_s')^2 + \frac{4\pi G}{\langle v_{sz}^2 \rangle}\{\rho_s^2 + \rho_{plo}\rho_s (\rho_s/\rho_{so})^{\langle v_{sz}^2 \rangle / Q^2}\} = 0$$

which by means of $(\rho_s')^2 = V(\rho_s)$ is easily reduced to a first order linear differential equation. Its solution gives the inverse form

$$z = \int_{\rho_{so}}^{\rho_s} \frac{d\rho_s}{\rho_s} \left[C - \frac{8\pi G}{\langle v_{sz}^2 \rangle}\left\{\rho_s - \frac{Q^2}{\langle v_{sz}^2 \rangle} \rho_{plo}(\rho_s/\rho_{so})^{\langle v_{sz}^2 \rangle / Q^2}\right\} \right]^{-\frac{1}{2}}$$

where C is an arbitrary constant.

An approximation is performed by means of the Taylor series expansion

$$\rho_s^{<v_{sz}^2>/Q^2} \simeq \rho_s^{<v_{sz}^2>/Q^2} + \{\rho_s - \rho_{so}\} \frac{<v_{sz}^2>}{Q^2} \rho_{so}^{<v_{sz}^2>/Q^2 - 1}$$

from which finally the density distribution of stellar matter is found in the form

$$\rho_s = \rho_{so} \cdot \cosh^{-2}\left\{\frac{\pi G \sigma}{<v_{sz}^2>} z\right\}$$

where

$$\sigma = \frac{<v_{sz}^2>}{2\pi G}\left[C + \frac{8\pi G}{<v_{sz}^2>}\left(1 - \frac{Q^2}{<v_{sz}^2>}\right)\rho_{plo}\right]^{\frac{1}{2}}$$

is the mass density projected onto the galactic plane.

Similarly the gravitational potential is found in the form

$$g = -2\pi G \sigma \tgh\left\{\frac{\pi G \sigma}{<v_{sz}^2>} z\right\} \qquad (2.3)$$

while the mass density of the thermal plasma is given by

$$\rho_{pl} = \rho_{plo} \cosh^{-2<V_{sz}^2>/Q^2}\left\{\frac{\pi G \sigma}{<v_{sz}^2>} z\right\} \qquad (2.4)$$

Parameters σ and $<V_{sz}^2>$ may be determined by a best fit of (2.3) to values of g observed by Hill for $|z| < 488$ pc as is shown in figure 6.

One finds $\sigma = 1.6 \times 10^{-2}$ g cm^{-2} and $<V_{sz}^2>^{\frac{1}{2}} = 17$ km s^{-1} which is a rough agreement with data published by other people. If Q^2 is determined by a best fit of (2.4) to Schmidt's observational data giving $Q = 102$ kms^{-1}, the distribution of stellar and plasma density respectively result in the form shown in figures 7 and 8.

Relation (2.4) may be helpful for the construction of simple models of the magnetic fields in our galaxy.

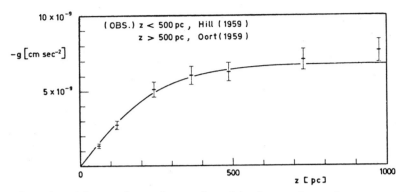

Fig. 6: Adaptation of gravitational acceleration near the galactic plane (after Fuchs and Thielheim, 1974).

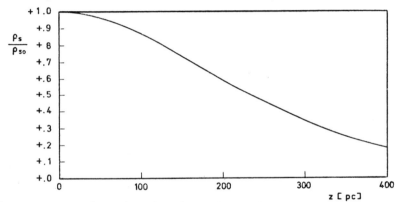

Fig. 7: Relative distribution of stellar matter near the galactic plane

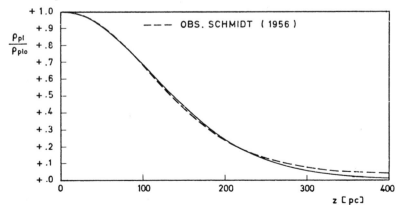

Fig. 8: Relative distribution of thermal plasma near the galactic plane.

III GALACTIC MAGNETIC FIELDS

1. Methods and Results

The degree of ionization of interstellar hydrogen is estimated to be about 100% in H II-regions and of the order of 10% in H I-regions. Consequently the electric conductivity in both regions is so high, that the magnetic field lines of force are "frozen" into the interstellar plasma.

Observation of the spatial distribution of hydrogen present in H I- and H II-regions, therefore, also delivers information on the large scale pattern of magnetic field. Specifically the concentration of hydrogen along the spiral arms suggests an alignment of magnetic field lines following the spiral structure. Further support for this suggestion is lent by the fact that in other spiral galaxies as in M51 the spiral structure observable in the brightness temperature distribution of synchrotron radiation closely follows the optical spiral pattern.

Reversing this argument the magnetic field lines may be looked upon as tracers for the kinematics of interstellar hydrogen.

(a) <u>Optical polarization</u>. The optical light of stars in many cases is found to be linearly polarized. For illustration a compilation by Selck and Thielheim (1974) of data referring to distances between a 100 and 200 pc is reproduced in figure 9. Analogous data for distances between 600 and 1000 pc are shown in figure 10.

Polarization is believed to be produced through scattering by elongated dust grains oriented with their larger axis preferentially in a direction vertical to the field vector in a statistical equilibrium under the combined influence of paramagnetic absorption and thermal motion (Davis and Greenstein (1951)). Both the orientation of the electric vector and the magnitude of polarization incorporates information on the magnetic field along the line of sight. But the interpretation of these data involves the problem that the spatial distribution of dust grains and of interstellar hydrogen has to be well known as well. (As will be discussed later the interpretation of the magnitude of polarization also involves physical properties of dust grains). Application of this method is restricted to the local region, i.e. to distances less than about 4 kpc.

The magnitude of polarization typically is of the order of several percent. Large scale correlations in the magnitude of polarization as well as in the alignment of the electric vector

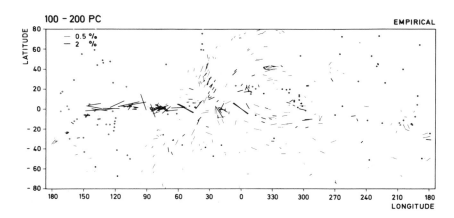

Fig. 9: Polarization of optical light from stars between 100 and 200 pc (data collected from different authors, quoted by Selck and Thielheim (1974)).

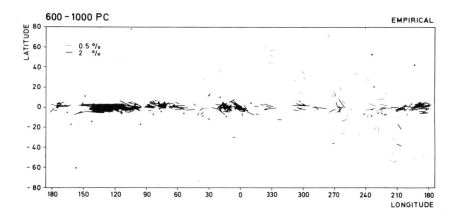

Fig. 10: Polarization of optical light from stars between 600 and 1000 pc (data collected from different authors, quoted by Selck and Thielheim (1974)).

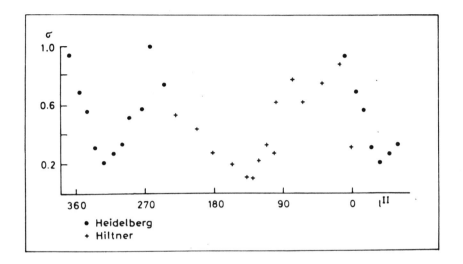

Fig. 11: Deviations from the mean direction of polarization as a function of galactic longitude (after Elsasser (1971)).

are suggested by these data. Deviations from the mean direction of polarization as a function of galactic longitude (Elsasser (1971)) shown in figure 11 demonstrate a systematic pattern in the scatter of the alignment of the electric vector. In most parts of the equatorial region there is a tendency for the electric vector to be aligned parallel to the galactic equator which is interpreted in the sense that there is a large scale ordered field configuration parallel to the local arm.

Part of the spurs and ridges observed in the celestial distribution of the direction of the electric vector may be produced by supernova remnants.

Correlations present in the magnitude of polarization reflect this spatial distribution of density and/or of properties of dust grains rather than the spatial distribution of magnetic field energy. This is suggested by maps showing the celestial distribution of the visual absorption A_V represented in figure 12 and figure 13 for distances of 100 to 200 pc and 600 to 1000 pc

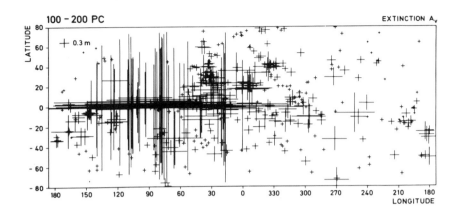

Fig. 12: Celestial distribution of visual extinction of stars in the region 100-200 pc (data collected from different authors, quoted by Selck and Thielheim (1974)).

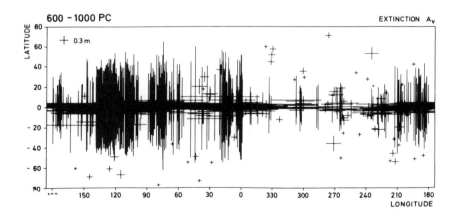

Fig. 13: Celestial distribution of visual extinction of stars in the region 600-1000 pc (data collected from different authors, quoted by Selck and Thielheim (1974)).

respectively. In comparison with the celestial distribution of the magnitude of optical polarization shown before, regions of strong absorption generally correspond to regions of high polarization values.

(b) Faraday rotation. Radio frequency radiation received from extragalactic sources often is linearly polarized. The angle through which the orientation of the electric vector is rotated is found to be proportional to the square of wavelength, which of course is typical for Faraday rotation in a magnetized plasma. The coefficient of proportionality, namely, the rotation measure is reproduced in figure 14 for 139 radio galaxies and quasars, the data having been collected from papers of different authors including Mitton (1972) by Nissen and Thielheim (1974). Analogously, rotation measures of 38 pulsars from various authors including Manchester (1974) are shown in figure 15.

These data which incorporate information on the line-of-sight component of the magnetic field vector are of special interest since it is only through the observation of Faraday rotation that conclusions may be drawn on the sign of the field vector.

Problems arise from the fact that, especially in the case of extragalactic sources, fields inherent to these objects contribute to observed rotation measures. Further contributions may arise from metagalactic or extragalactic fields. It should be noted that a correlation has been claimed between rotation measures and red shifts for very distant objects (Sofue et al. 1968). The spatial distribution of thermal electrons along the line of sight of course is involved in the interpretation of these data on which, in the case of pulsars, additional information is derived from observed dispersion measures of radio signals. Application of this method is restricted essentially to the local region due to solid angle arguments as far as extragalactic sources are concerned. The observation of pulsars is even more restricted to distances of the order of 1 kpc.

Possible correlations in the celestial distribution of rotation measures have been studied mostly by looking at diagrams of the form presented in figures 15 and 16 suggesting the prevalence of positive or negative rotation angles respectively in certain regions of galactic coordinates. For a more rigorous analysis one has to look at the (normalized) auto-correlation function.

$$<RM.RM\ (\Delta \theta\ )>_\theta\ /<RM.RM\ (0)> \quad (3.1)$$

defined with respect to the angular separation on the celestial sphere. The mean value in the numerator is calculated by summation over all pairs of sources within given ranges $\Delta \theta = 10°$ for

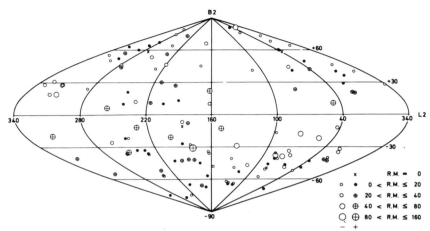

Fig. 14: Rotation measures of 139 radio galaxies and quasars (data collected from different authors, quoted by Nissen and Thielheim (1974)).

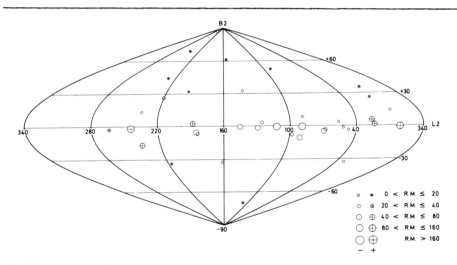

Fig. 15: Rotation measures of 38 pulsars (data collected from different authors, quoted by Nissen and Thielheim (1974)).

extragalactic sources or $\Delta \theta = 20°$ for pulsars. The denominator is calculated by summation over all sources.

The form of the autocorrelation function defined in (3.1) may be discussed for some idealized field configurations. Thus in the case of a longitudinal field the autocorrelation function would have the form represented qualitatively in figure 16a, whereas in the case of a quasi-longitudinal field the autocorrelation function would correspond to a diagram shown in figure 16b. As is obvious, the form of the autocorrelation

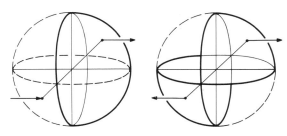

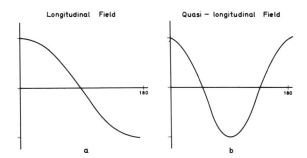

Fig. 16: Autocorrelation function of rotation measures predicted for different types of field configuration.

---

function is independent from the alignment of the field.

The autocorrelation function (Nissen and Thielheim (1974)) computed for the 139 extragalactic sources incorporated in figure 14 are shown in figure 17. We did not make use of data referring to unidentified radio sources since the nature of these sources is not well known. The existence of correlations (Michel and Yahil (1973)) within a range of $40°$ is confirmed but there is also some evidence for an anticorrelation at about 90 degrees and a correlation at about 180 degrees suggesting a quasi-longitudinal field configuration.

The autocorrelation function computed for 38 pulsars incorporated in figure 15 is shown in figure 18. Again a correlation within about 40 degrees is evident and there is some

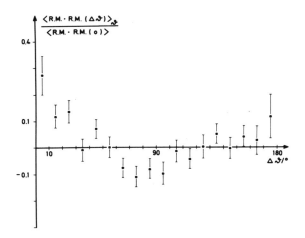

Fig. 17: Autocorrelation of rotation measures of 139 radio galaxies and quasars (after Nissen and Thielheim (1974)).

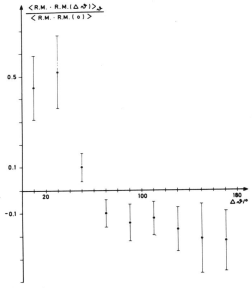

Fig. 18: Autocorrelation of rotation measures of 38 pulsars and quasars (after Nissen and Thielheim (1974)).

indication for an anticorrelation at larger angular separations of pulsars. The autocorrelation function therefore resembles the one predicted for a longitudinal configuration in agreement with what has been found by Reinhardt (1972). The striking differences appearing in the autocorrelation function of these two categories of sources will be discussed later.

(c) **Synchrotron radiation.** The celestial distribution of radio frequency radiation at a frequency of 150 MHz due to Landecker and Wielebinski (1971) is reproduced in figure 2 . Intensity values attached to the different curves are given in units of brightness temperature. Problems arise from the necessity to separate discrete sources, of which some 104 have been catalogued, from galactic contributions. The general features of the celestial distribution of galactic synchrotron radiation are characterized by an intensity maximum at the galactic centre and by regions of high intensity around the galactic equator.

In some regions the background radiation is found to be linearly polarized. Galactic radio radiation in this frequency range is identified through its spectral index as synchrotron radiation emitted by relativistic electrons gyrating in interstellar fields and may therefore be interpreted in terms of the spatial distribution of magnetic field energy and cosmic electron density along the line of sight. For electron densities of the order of 10^{-2} cm^{-1} s^{-1} GeV^{-1} at 1 GeV. The magnitude of magnetic fields strength that results is about 10^{-5} G.

In principal, the application of this method to investigate galactic fields is not restricted to the local region. Local disturbances may partly be due to supernova remnants as is suggested by the diagram of Berkhuijsen et al. (1971) represented in figure 20, which may be compared with analogous features appearing in the celestial distribution of stellar polarization.

(d) **Gamma radiation.** The celestial distribution of cosmic gamma radiation has recently been measured in the satellite experiments. Early results representing the intensity distribution along the galactic equator (Kraushaar et al. (1972) are reproduced by the histogram in figure 21. This distribution is characterized by a broad maximum at the position of the galactic centre and some smaller maxima additionally. The observed distribution is believed to be composed of contributions from discrete sources and from the galactic disk. The latter has been identified as being due to the decay of π^o mesons produced in inelastic encounters of cosmic ray particles with nuclei of the interstellar gas (Stecker (1970), Cavallo and Gould (1971), Fichtel et al. (1972) and Stecker (1973)).

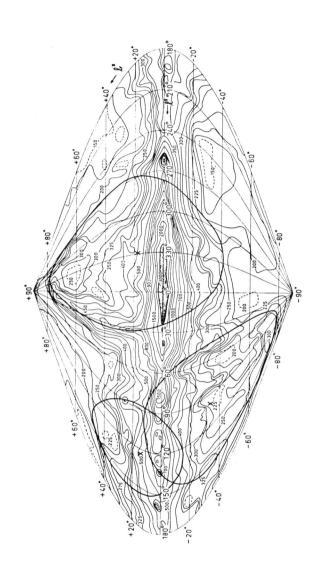

Fig. 19: Celestial distribution of brightness temperature at 150 MHz (after T.L. Landecker and R. Wielebinski. Aust. J. Phys. Supp. 1970 quoted by Berkhuijsen et al. (1971)).

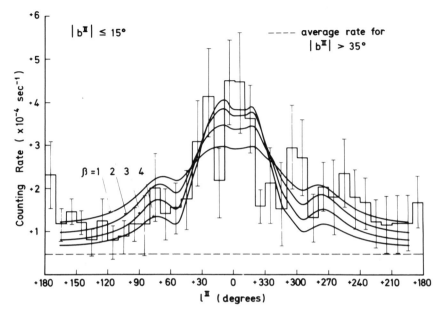

Fig. 20: Empirical counting rate of galactic gamma radiation (after Kraushaar et al. (1972)). Curves represent theoretical predictions for different values of the parameter β. (Schlickeiser and Thielheim (1974a)).

According to this interpretation the observed distribution involves the spatial distribution of cosmic ray particles in the 1-10 GeV region as well as the distribution of hydrogen density along the line of sight. Both quantities are believed to be coupled to the magnetic field. The latter therefore may be discussed in terms of observed gamma ray distribution. The application of this method for the investigation of interstellar fields is not restricted to the local region.

Further information on galactic fields are obtained from observations of Zeeman splitting particularly on the 21 cm line of neutral hydrogen in dense clouds. In principle this method is of great interest since it is the only one delivering direct information on the magnetic field strength but unfortunately its application is restricted to regions which may not be typical for the overall conditions on the dynamics of interstellar matter, cosmic ray transfer and from the analogy of our galaxy to other spiral galaxies of similar type.

Eventually one is lead to the conclusion that in the galactic plane there is a rather uniform distribution of field energy. At least in the local region the field strength is of the order of 3 µG and there is a tendency for an alignment of the field lines parallel to the local arm.

2. Conjectures on the Origin and Structure of Fields.

Cosmic magnetic fields are found to be associated with a wide scale of astrophysical objects on which they may be the source of valuable information. The best known example of course is the terrestrial field, observations on which lead to conclusions about the conditions existing inside the earth as well as in its surroundings. Manifestations of the solar magnetic field give information on the processes connected with the outer layers of the sun. Other stars are known to posses magnetic fields as well, some of which are considerably stronger than the solar field. Extremely strong magnetic fields are connected with pulsars.

The magnetic fields associated with the galaxy are coupled to the interstellar plasma. The magnetic field lines of force may thus be looked upon as tracers of the kinematics of the interstellar medium. Magnetic fields are also known to exist in many other galaxies some of which are similar to our own galaxy. Although observed from larger distances in some cases the internal structure of these objects is easier to recognise than that of our galaxy. Some objects on the galactic scale are believed to have magnetic fields stronger than the local field by some orders of magnitude. Magnetic fields have also been found on an even larger scale in galactic clusters as has been demonstrated by Burbidge in his lectures and analogously the existence of the metagalactic field is quite conceivable. It should be noted that the hypothesis of a primordial field associated with the early universe has been put forward.

Eventually we arrive at the conjectures put forward to explain the origin of the magnetic fields on such a variety of scales as is known to exist. Two extreme points of view with respect to the origin of cosmic fields have been proposed. One of them is that in each system the magnetic fields have their origin in the preceding larger systems, the ultimate origin of all fields being the primordial field. According to this line of thought the galactic field was produced by compression and winding up of field lines during the condensation of the galaxy from the gaseous matter distributed in the early Universe. Almost the same picture could be applied to the condensation of a star from the magnetised interstellar plasma. The other extreme point of view is that the magnetic fields of our galaxy have come into existence by some not yet understood dynamo mechanism immediately before, or in the first stages after, the gravitational collapse through which the disk was formed. In this case the original form of the magnetic field belonging to the pre-galaxy may be thought to have resembled the structure of a dipole field sheared by the differential rotation of matter.

If the galactic fields originated from a larger field the configuration inside the galaxy would primarily be expected to be open in the sense that field lines would run from regions inside the galaxy to regions outside. If, alternatively, the magnetic field originated from a dynamo mechanism associated with the galaxy the configuration inside the disk would primarily be expected to be closed, but these features may have changed in time through reconnection of field lines. The notion of a closed configuration implies the existence of a polarity for the galaxy defined through the orientation of the dipole contribution with respect to the angular momentum of the galaxy.

Dynamo mechanisms may be thought of according to which this polarity is different for galaxies made of matter, from those made of antimatter (Nissen and Thielheim 1974).

3. Model Calculations

(a) Model version I. It appears to be meaningful to discuss the various forms of empirical and theoretical information available on the structure of galactic fields in terms of field models which of course in any case are strongly idealized. By comparison of data calculated by means of such models with those derived from observations one may finally be lead to improved versions of field models. To begin with we wish to consider a simple field model version I which has been published by us (Thielheim and Langhoff 1968) in connection with the calculation of trajectories of high energy cosmic ray particles in our galaxy.

We wish to adopt the notion that the field is mainly concentrated in a flat disk of half-thickness z_0 and radius R_0. Therefore, we start to formulate the component vertical to the z-axis, which is considered to give the main contribution, by writing down a factor $\exp(-z^2/z_0^2 - R^2/R_0^2)$. Reversal of the field direction on both sides of the galactomagnetic equator is achieved by a factor z. Clearly, this pancake shaped object has to be modulated by a spiral structure. We wish the field to be stronger (by a coefficient k) in the galactic arms than in the inter-arm regions. The location of the spiral arms may be described in polar coordinates R and ϕ defined in the galactic plane $z = 0$. A possible choice for this spiral function providing two arms equidistantly wound for higher values of R is $\phi_1(R) = (aR/b)\,\text{arctg}(R/b)$ with adequate coefficients a and b. This spiral structure is easily impressed on our model by a factor $1 + k^2 \cos^2(\phi - \phi_1(R))$. After introduction of a constant factor c for the adjustment of the mean field strength and an additional factor $1 - \exp(-R^2/R_1^2)$ suppressing the field in the immediate vicinity $R \simeq R_1$ of the origin for technical reasons we end with the formulation

$$H_{a_o} = cz \cdot \exp\{-z^2/z_o^2 - R^2/R_o^2\}\,\{1-\exp(-R^2/R_1^2)\}\,\{1+k^2\cos^2(\phi-\phi_1(R))\} \tag{3.2}$$

The direction of this field component may be specified by a unit vector $\vec{a}_o$ orthogonal to the unit vector $\vec{z}_o$. Following the conjectures of the quasilongitudinal model, $\vec{a}_o$ is parallel or antiparallel to the tangent vector to the spiral function $\phi = \phi_1(R)$ where the former is defined. For the example given above the angle between the tangent vector and the radius vector $\vec{R}_o$ is $\text{arctg}((aR/b)\text{arctg}(R/b) + R^2/(b^2 + R^2))$. In the regions between the two branches of the spiral function the unit vector $\vec{a}_o$ easily interpolated by taking the angle between $\vec{a}_o$ and $\vec{R}_o$ to be independent of ϕ and equal to ε,

$$\vec{a}_o = \vec{R}_o \cos\varepsilon + [\vec{z}_o, \vec{R}_o]\sin\varepsilon$$

The remaining field component H_z parallel to the z-axis is uniquely determined through $\text{div}\,\vec{H} = 0$ and $H_z \to 0$ for $|z| \to \infty$ resulting in the somewhat unhandy expression

$$H_z = 2cz_o^2\{1+k^2\cos^2(\phi-\phi_1(R))\} \cdot \exp\left(\frac{z^2}{z_o^2} - \frac{R^2}{R_o^2}\right) \cdot \cos\varepsilon$$

$$\{(1-\exp(-R^2/R_1^2))(1-2ab^2R^2\text{tg}\varepsilon/(b^2+R^2)^2)\cos^2\varepsilon - 2R^2/R_o^2\}/R$$

$$+ (2R/R_1^2)\exp(-R^2/R_1^2).$$

The adjustment of this model to the geometry of our galaxy is performed with parameter values $z_o = 0.175$ kpc, $R_o = 10$ kpc, $R_1 = 2$ kpc, $a = 1$, $b = 1.5$ kpc. It should be well understood that at the best this formulation can do no more than give a formal description of some of the typical features of the global field structure.

For investigations on local field disturbances due to supernova shells, (Selck and Thielheim (1974) unpublished) the expression given for the equatorial component of the magnetic field may be reduced to the form $H_{ao} = C \cdot z \cdot \exp(-z^2/z_o^2)$. Considering a supernova shell with an outer radius $r = R_{II}$ and an inner Radius $r = R_I$ the magnetic field is described in polar coordinates $r, \underline{\theta}, \underline{\phi}$, defined with respect to the centre of the supernova, where the polar axis has the direction of the tangent to the local arm, by the form

$$\underline{H} = C \; (r.\sin\underline{\theta}\sin\underline{\phi} . (f^{1/2} + z_{\hat{N}})$$

$$.\exp\left[-(r \sin\underline{\theta} \sin\underline{\phi} (f)^{1/2} + z_{\hat{N}})^2/z_o^2 \right] . \left[f \cos\underline{\theta} . \vec{\underline{r}}_o \right.$$

$$\left. - (\tfrac{r}{2} df/dr + f) . \sin\underline{\theta} . \underline{\phi}_o \right]$$

$z_{\hat{N}}$ is the position of the centre of the supernova in the coordinates used in (3.2). $f = f(r)$ is an arbitrary, continuous and part by part differentiable function describing the structure of the shock front. $f(r)$ is given by $f(r) = (r-R_I)^2/(R_{II}-R_I)^2$ for $R_I \leqslant r \leqslant R_{II}$ which approximately fits earlier results while $f(r) = 0$ for $r < R_I$ and $f(r) = 1$ for $r > R_{II}$. Obviously $H = C \cdot z \cdot \exp(-z^2/z_o^2)$ for $r > R_{II}$ and $H = 0$ for $r < R_I$ and div $\underline{H} = 0$ everywhere.

It should be noted that the component of $\vec{H}$ tangential to the sphere $r = R_I$ and $r = R_{II}$ is not necessarily continuous, which means that currents may be present on the two spheres. Application of this procedure to the supernova remnant which is believed to be responsible for the north galactic spur lead to a compression of field lines as indicated in figure 22.

(b) Reproduction of stellar polarization. The relative intensity of polarized light (referring to unit total intensity) is, according to the theory of Davis and Greenstein, given by

$$I_{\pi} = C_s \cdot C_g \cdot A_v \; \frac{\int_o^D N_g(R) dR \; \frac{(H_{\perp}(R),Ro)^2}{n_H(R)}}{\int_o^D N_g(R) \; dR} \qquad (3.3)$$

where

$$C_s \propto \frac{\sigma_A/\sigma_T - 1}{\sigma_A/\sigma_T + 2}$$

is a coefficient depending on particle properties relevant to the

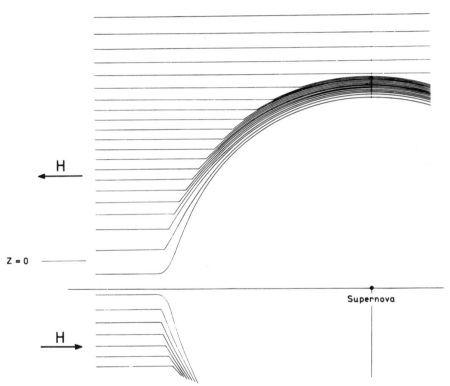

Fig. 22: Model of magnetic field compressed by a supernova shell (Selck and Thielheim (1974) unpublished).

scattering of light, σ_A and σ_T are extinction cross-sections referring to parallel or anti-parallel orientation of the larger body axis with respect to the electric vector. The coefficient $C_g \propto \chi''/a\, T^2\, \omega$ is a parameter depending on particle properties relevant to the orientation of these grains. χ'' is the imaginary part of the magnetic susceptability for grains rotating at an angular velocity ω, a is the size of the grains and T is the temperature of the interstellar gas. A_V is the visual absorption, D is the distance of the star from the observer, and n_g (R) and

$n_H(R)$ are the number densities of grains and hydrogen atoms respectively. $\bar{H}(R)$ is the magnetic field vector and $\bar{R}o$ is the unit vector pointing from the observer to the star under consideration.

Strictly, this theory is applicable only for $F = C_g H^2/n_H(R) << \frac{1}{3}$. In the case of saturation $F >> \frac{1}{3}$ a cut off has to be introduced $H^2 = n_H(R)/3C_g$.

The polarization matrix of the polarized contribution to relative intensity I_π is given by

$$\underline{\underline{P}}_\pi \propto C_s \cdot C_g \cdot A_v \cdot \frac{\int_o^D dR\, n_g(R) \frac{(\bar{H}(R),\bar{R}_o)^2}{n_H(R)} \begin{pmatrix} \sin^2\theta, & \sin\theta\cos\theta \\ \sin\theta\cos\theta, & \cos^2\theta \end{pmatrix}}{\int_o^D dr\, n_g(R)}$$

from which the degree of polarization is obtained in the form

$$P_\pi = 1 - 4 \frac{\text{Det}(\underline{\underline{P}}_\pi)}{(\text{Tr}(\underline{\underline{P}}_\pi))^2} \tag{3.4}$$

The total degree of polarization may now be calculated with the help of (3.3) and (3.4) from

$$p = I_\pi P_\pi$$

Eventually the **angle** θ between the electric vector and the northern direction may be calculated from

$$\sin^2\theta = \frac{1}{P} \left\{ \frac{\int_o^D dR\, n_g(R) \left[\frac{(\bar{H}(R),\bar{R}_o)^2}{n_H(R)}\right] \sin^2\theta(R)}{\int_o^D dR\, n_g(R) \left[\frac{(\bar{H}(R),\bar{R}_o)^2}{n_H(R)}\right]} - \frac{1-P_\pi}{2} \right\}$$

The degree of polarization p as well as the angle θ specifying the direction of polarization has been calculated for several thousand stars (a) using version I of the model and (b) using a superposition of version I and a model of supernova remnants incorporating a detailed description of the shock front.

The factor $k = C_s \cdot C_g$ has been adapted to observed data through a best fit procedure. k, which is specified by the properties of the grains e.g. whether they are para-, dia- or ferromagnetic, is found to be almost independent of the distance if all stars are taken into account, suggesting model version I gives a fair mean description of what is really existing (figure 23). But k may strongly depend on the distance if stars are taken into account only from smaller parts of the celestial sphere, as is obvious, for example, in the afore-mentioned region where a supernova remnant is suspected to exist.

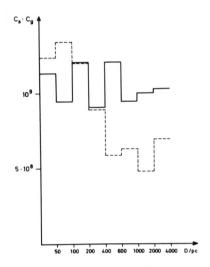

Fig. 22: Adaptation of calculated to observed degree of polarization as a function of distance from the sun. Calculations were performed with model version I for the whole celestial sphere (full line), and for the region on the celestial sphere incorporating the supernova under consideration (broken line).

Maps prepared from calculated polarization vectors may be directly compared with those representing observed data (figures 9 and 10). Theoretical results based on version I of the field model reproduce a good deal of the general features in empirical data, especially at larger distances from the sun. For example it produces the alignment of polarization vectors in most parts of the equatorial plane as well as convergence of polarization vectors in the direction tangential to the local arm (figures 23 and 25). But of course, there is no statistical scatter in the theoretical data.

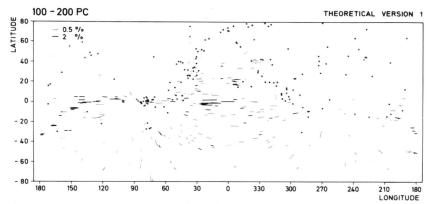

Fig. 23: Calculated polarization of optical light from stars between 100 and 200 pc (M. Selck and K.O. Thielheim, unpublished).

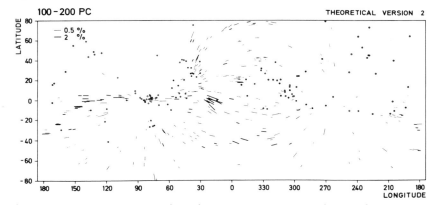

Fig. 24: As above, taking into account the field disturbance produced by the supernova under consideration.

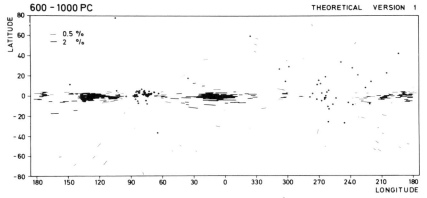

Fig. 25: As in figure 23, for stars between 600 and 1000 pc.

Theoretical results especially those for distances between 100 and 200 pc based on the field model incorporating the supernova remnant reproduce some of the general features observed within the region of the North Galactic Spur and the Cetus Arc. (figure 24).

(c) Reproduction of Faraday Rotation. Theoretical autocorrelation functions have been computed by Nissen and Thielheim (1974) for radio galaxies and quasars as well as for pulsars by means of the well known relation

$$R.M. = 8.1 \times 10^5 \cdot \int_0^R N_e \cdot B_L \, dL$$

The theoretical autocorrelation function has been computed for the ensemble of 139 radio galaxies and quasars mentioned before using model version I. These results are shown in figure . The absence of irregularities in the abstract formulation of the model is obvious. This form of the autocorrelation function is typical for a quasi-longitudinal field structure.

Adoption of theoretical to observed rotation measure leads to a positive value of local electron density. This means that the polarity suggested by model version I is realistic in the sense that the present field configuration may have developed from an original dipole-like structure, the magnetic momentum vector of which was antiparallel to the angular momentum vector of the galaxy.

The local density of relativistic electrons found by the best fit procedure is in reasonable agreement with results from other types of argument.

The autocorrelation function for 38 pulsars is shown in figure 27. This form of the autocorrelation function supports a longitudinal field configuration.

The different forms of the autocorrelation function for extragalactic sources and for pulsars may be interpreted in the sense that the solar system is placed at a certain distance from where the field reversal takes place similar to that anticipated in model version I. The model itself certainly is not very realistic, the correlation coefficients between observed and rotation measures being 0.436 for radio galaxies and quasars and 0.692 for pulsars respectively. This is also obvious from the representation of celestial rotation measures shown in figures 28 and 29.

(d) Reproduction of galactic gamma radiation. The expected counting rate of galactic gamma rays from a narrow band along the

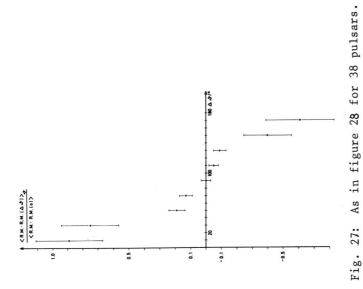

Fig. 27: As in figure 28 for 38 pulsars.

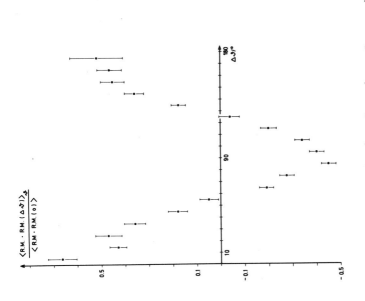

Fig. 26: Autocorrelation function of rotation measures of 139 radio galaxies and quasars calculated by means of an idealized field model version I (after Nissen and Thielheim (1974)).

galactic equator has been calculated for comparison with the
OSO-III data (Schlickeiser and Thielheim 1974a). Adopting
the specific production rate given by Cavallo and Gould the
integral spectral flux per steradian is determined by the integral
of the product of hydrogen density n_H and cosmic ray density N

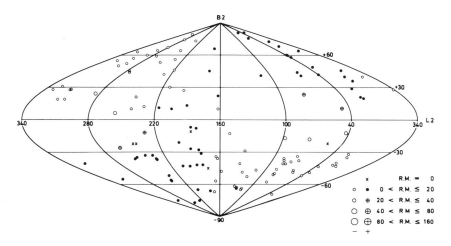

Fig. 28: Rotation measures of 139 radio galaxies and quasars
calculated by means of an idealized field model version I
(after Nissen and Thielheim (1974)).

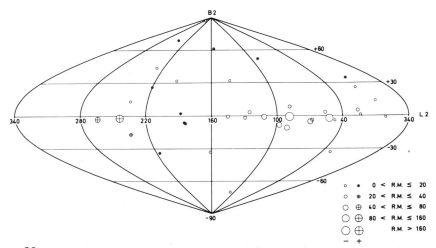

Fig. 29: Rotation measures of 38 radio galaxies and pulsars
calculated by means of an idealized field model version I
(after Nissen and Thielheim (1974)).

divided by the demodulated cosmic ray density N_s at the position of the solar system which is performed along the line of sight:

$$F(> 100 \text{ MeV}; l^{II}, b^{II}) =$$

$$\frac{1.8 \times 10^{-25}}{4\pi} \int_0^\infty dl^{II} \frac{n_H \cdot N}{N_s} \text{ cm}^{-2} \text{ s}^{-1} \text{ sterad}^{-1}$$

by application of the acceptance function

$$3.5 \exp\left[-\{(L^{II} - l^{II})^2 + b^{II\,2}\}/225\right]$$

we arrive at the expected counting rate:

$$R(>100 \text{ MeV}; l^{II}) = \frac{3.5 \times 1.8 \times 10^{-25}}{4\pi}$$

$$\int_{l^{II}-15}^{l^{II}+15} dL^{II} \exp(-(L^{II} - l^{II})^2/225)$$

$$\int_{-15}^{+15} db^{II} \cos b^{II} \exp(-b^{II\,2}/225)$$

$$\int_0^\infty dl \frac{n_H \cdot N}{N_s} \text{ s}^{-1} \tag{3.5}$$

The relation between the energy density of the magnetic field and the density of interstellar hydrogen obviously results through the evolutionary process which was sketched at the beginning of this lecture and which finally led to the following field configuration.

Within the working hypothesis allowing the evaluation of (3.5) a relation of the form

$$n_H \cdot N \propto (\text{abs}(H_{ao}/z))^\beta \tag{3.6}$$

may be considered in which the parameter β has to be discussed in connection with observational data.

The expected counting rate (3.5) calculated for $\beta = 1, 2, 3$ and 4 normalised by a best fit to the empirical results of Kraushaar et al. (1972) is shown in figure 20. The smallest mean square deviations from the empirical results are found for β within the range 2-3. The factor of proportionality in relation (3.6) is given by $n_{H_s}/(abs(H_{ao_s}/z_s))^\beta$.

According to the model used here the ratio of low energy cosmic ray density within the 4 kpc-arm to the local arm density results to be about 6 in agreement with what has been found by Stecker et al. (1974). This of course will have to be predicted by a theory of the dynamic equilibrium existing within the galactic disc.

As may be seen from figure 20 the existence of a maximum of gamma ray intensity at the galactic centre may be understood entirely on the basis of a spiral field configuration. There is no necessity for postulating an extra gamma ray source at the galactic centre. The ratio of the central to the tangential gamma ray intensity (after subtraction of an extragalactic isotropic gamma ray component) is essentially influenced by the relation between hydrogen and cosmic ray gas densities and the magnetic field strength.

As is obvious, from figure 34 a high resolution gamma ray survey (aperture of the order of 1°) would yield detailed information on the structure of the spiral function of our galaxy. Empirical data quoted here are not yet of this quality. Clearly, such data would be extremely helpful for the construction of more realistic versions of the large scale magnetic field model and thereby yield information on the details of galactic evolution.

Similarly, the intensities

$$I(>100 \text{ MeV}; \vec{1}) = \frac{1.8 \cdot 10^{-25} n_{H_s}}{4\pi (abs(H_{a_{o_s}}/Z_s))^\beta}$$

$$\int_{-10}^{+10} db^{II} \cos b^{II} \int_0^\infty dr \, (abs(H_a/Z))^\beta \text{ cm}^{-2} \text{ rad}^{-1} \text{ s}^{-1}$$

have been calculated for comparison with SAS-II data

(Schlickeiser and Thielheim 1974b). β between 2 and 3 is found to give the best fit. These results are reproduced in figures

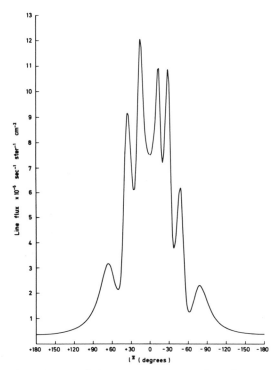

Fig. 30: High resolution galactic gamma ray line flux calculated on the basis of an idealized version I (after Schlickeiser and Thielheim (1974a)).

31 and 32. Aside from the central increase, which is fairly reproduced, there are two increases from the local arm in the calculated data.

(e) Conclusions. Comparisons of predicted with observed data presented above indicates that while some basic features of the empirical results (existence of a global field structure, alignment of field energy concentration with the local spiral arm) are fairly reproduced there still is considerable discrepancy with respect to detailed features of the spiral structure as well as of local disturbances although part of the latter may be reproduced by models of supernova remnants. Improvements of the formal description of galactic fields should have to take into account a more accurate conjecture of the structure of the

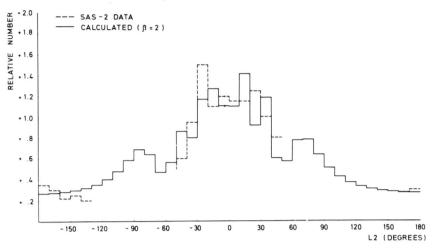

Fig. 31: Predicted and empirical distribution of high energy gamma rays along the galactic plane in the region $-10° < b^{II} < 10°$ for parameter $\beta = 2$ (after schlickeiser and Thielheim, 1974b).

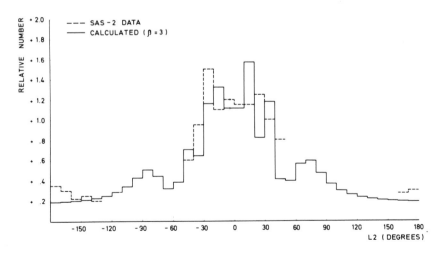

Fig. 32: Predicted and empirical distribution of high energy gamma rays along the galactic plane in the region $-10° < b^{II} < 10°$ for parameter $\beta = 3$.

galactic disk based for example on relations as those presented in section II. Moreover the spiral configuration given by the model field should be brought into agreement with recent H I results as well as with observed distribution of galactic radiation. In any case even an improved model may correspond to the mean distribution of magnetic fields rather than to the actual distribution which, of course, might be studied more closely by models incorporating the kinematics of interstellar hydrogen within the local region.

III COSMIC RAY RESPONSE TO DIFFERENT TYPES OF FIELD CONFIGURATION

Discussing the cosmic ray response to different types of field structure one may adopt the naive notion of particles following the magnetic field lines of force, provided that there is not too much turbulence within a certain range of magnetic rigidity and that diffusion phenomena considered by Osborne in his lectures, might be neglected. According to this conjecture if the field has a closed configuration cosmic ray particles are expected to be of galactic origin, in this case the problem arises of explaining the high degree of isotropy found.

However, if the field configuration is an open one and particles are of galactic origin the problem of containment arises. If, on the other hand, cosmic rays are of extra-galactic origin the problem is to deal with the background radiation which should exist as well as the anisotropy.

In reality there may well be efficient mechanisms allowing the transfer of cosmic ray particles from one closed configuration to another one. For example, such mechanisms seem to exist in interplanetary space permitting the transfer of low energy cosmic ray particles from the sun to the earth although there are probably no field lines running from the sun directly to the earth. Similarly the exchange of cosmic ray particles between a closed galactic field configuration and a probably weaker extragalactic field may be rather effective although a formal theory describing this process does not exist. In any case, particles above about 10^{18} eV should originate from regions outside the galaxy, for example, from those giant radio sources which are being considered by Burbidge in his lectures.

As a working hypothesis we may consider the idea that different types of galactic and extragalactic astrophysical objects may contribute to the observed particle intensity, the essential point being that the acceleration mechanisms in all these sources are based on analogous electrodynamic processes resulting in the same shape of the energy spectrum. Diffusion phenomena, of course, due to galactic containment should in this case manifest themselves in the detailed features of the primary spectrum, as is being discussed by Wolfendale in his lectures. The situation obviously is different for electromagnetic radiation where different types of processes such as thermal radiation, synchrotron and plasma oscillations exist and manifest themselves by different features within the distribution of frequencies.

REFERENCES

Berkhuijsen, E.M. et al., Astron. & Astrophys, 14, 252, 1971.

Camm, G.L., MNRAS, 110, 305, 1950.

Cavallo, G., Gould, R.J., Nuovo Cimento, 2B, 77, 1971.

Davis, L., Jr., Greenstein, J.L., Ap. J., 114, 206, 1951.

Elsasser, H., in: "Structure and Evolution of the Galaxy" ed. by L.N. Mavridis (D. Reidel) 1971.

Fichtel, C.E., Hartmann, R.C., Kniffen, D.A., Sommer, M., Ap. J. 171, 31, 1972.

Fuchs, B. and Thielheim, K.O., to be published, 1974.

Kellman, S.A., Ap. J., 175, 353, 1972.

Kniffen, D.A., Hartmann, R.C., Thompson, D.J., Fichtel, C.E., Ap. J., 186, L105, 1973.

Kraushaar, W.L., et al., Ap. J., 177, 341, 1972.

Landecker, T.L. and Wielebinski, R., Aust. J. Phys. Supp. 1970 (quoted by E.M. Berkhuijsen et al., 1971).

Lequeux, J., in: "The Interstellar Medium", ed. by K. Pinkau (D. Reidel) 1973.

Manchester, R.N., Ap. J., 188, 637, 1974.

Mezger, P.G., in: "The Interstellar Medium", ed. by K. Pinkau (D. Reidel) 1973.

Michel, F.C. and Yahil, A., Ap. J., 179, 771, 1973.

Mitton, S., Mon. Not. R. Astr. Soc. 155, 373, 1972.

Nissen, D. and Thielheim, K.O., Astrophys. & Sp. Sci. (in the press) 1974.

Oort, J.H., BAN, 15, 45, 1959.

Parker, E.N., Ap. J., 145, 811, 1966.

Reinhardt, M., Astron. & Astrophys, 19, 104, 1972.

Schlickeiser, R. and Thielheim, K.O., Astron. & Astrophys, 34, 167, 1974a.

Schlickeiser, R. and Thielheim, K.O., to be published, 1974b.

Selck, M. and Thielheim, K.O., unpublished 1974 (prepared from results of various authors including Mathewson, D.S., Ford, V.L., Mem. R. Astro. Soc., 74, 139, 1970).

Simonson, S.C., Astron. & Astrophys., 9, 163, 1970.

Sofue, Y., Fujimoto, M., Kawabata, K., Publ. Astron. Soc. Jap., 20, 388, 1968.

Stecker, F.W., Astrophys. Sp. Sci. 6, 377, 1970.

Stecker, F.W., Ap. J., 185, 499, 1973.

Stecker, F.W., Puget, J.L., Strong, A.W. and Bredekamp, J.H., Ap. J., 188, L59, 1974.

Thielheim, K.O. and Langhoff, W., J. Phys. A. (Proc. Phys. Soc.), Ser. 2, 1, 694, 1968.

GALACTIC PROPAGATION OF COSMIC RAYS BELOW 10^{14} eV

J.L. Osborne

Department of Physics
University of Durham, U.K.

I   INTRODUCTION

   This chapter is concerned with the question of whether the observed anisotropy and lifetime of cosmic rays can be reconciled with the hypothesis of galactic origin in discrete sources such as supernovae or supernova remnants. The propagation of the electron component of the cosmic ray flux is discussed in the article by Meyer. Here we consider the nuclear component only. In the energy region above 10^{14} eV the energy density of the cosmic rays and the pressure exerted by them are so low that they must have a negligible effect on the magnetic fields through which they propagate. At the lower energies considered here the mutual interaction between the particles and the field has to be taken into account.

II   OBSERVATIONAL DATA

   In this section a brief summary is given of the observed properties of the cosmic ray flux that bear on the propagation in the Galaxy.

1.   Anisotropy

   An anisotropy of the primary cosmic ray flux with respect to the Galaxy would appear to an observer on earth as a sidereal variation of the secondary particle intensity. The observed magnitude of the anisotropy, δ, is the fractional amplitude of a first harmonic sinusoidal variation of intensity with sidereal

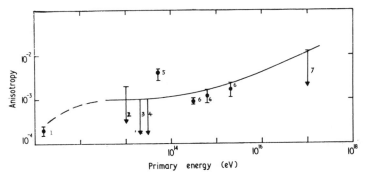

Fig. 1: Measurements of cosmic ray anisotropy: 1, Elliot et al. (1970); 2, Sherman (1953); 3, Cachon (1962); 4, Kolomeets et al. (1969); 5, Delvaille et al. (1962); 6, Daudin et al. (1956); 7, Lapikens et al. (1971).

---

time that best fits the data. Figure 1 shows the values of δ from a number of experiments plotted at the estimated median energy of the primary cosmic rays producing the signal. The error and upper limits indicate one standard deviation. The curve is an estimate of the upper limit to the anisotropy. The relatively large value of Delvaille et al. (1962) has not been confirmed by more recent observations. The median primary energy of 1.5×10^{11} eV attributed to the point of Elliot et al. (1970), may not be high enough to avoid dilution of a galactic anisotropy by the effects of the interplanetary field. The curve is therefore shown as a dashed line in this region. A preliminary result of the Peak Musala experiment, Gombosi et al. (1974), may be added to these data. An anisotropy of 1.3×10^{-3}, significant at a confidence level 0.7% has been observed for air showers corresponding to a mean primary energy of 6×10^{13} eV. One concludes that in the range of primary energy from 10^{11} eV to 10^{14} eV δ is in the region 10^{-4} to $\sim 10^{-3}$. This means that the streaming velocity of cosmic rays is between 10 and 100 km s^{-1}.

2. Amount of Matter Traversed

From the proportion of secondary Li, Be and B in the cosmic ray flux, produced by the spallation mainly of C, N and O on the interstellar gas, and the proportion of nuclei with 17<z<25 produced by spallation of Fe the total amount of matter traversed by the cosmic rays between their source and the earth can be deduced. The observed composition at about 10^9 eV/nucleon requires an approximately exponential distribution of this 'grammage' with a mean of 6 g cm^{-2}. A number of recent experiments have extended the measurements to higher energies;

Ramaty et al. (1973) have given a summary. All indicate that the proportion of secondary nuclei is decreasing with energy. At the highest energies observed ($4 \; 10^{10}$ eV/nucleon) the grammage has decreased to about 2.5 ± 1.5 g cm^{-2}. This corresponds to a travel time of $1.5 \; 10^6$ yr if all of the grammage is traversed in an interstellar medium of effective mean density 1 atom cm^{-3}, assuming, as is conventional, that a negligible amount of material is traversed in the source. If the cosmic rays propagate mainly in the intercloud regions the travel time could be 10 times longer. It is important to note that there is at present no overlap in the energy ranges for which the grammage and the anisotropy are known.

3. Constancy of the Cosmic Ray Flux

The measured abundances of radioactive isotopes formed in meteorites by the interactions of cosmic rays give some indication of the long term time variations in the cosmic ray flux. They appear to show that the present day value of the flux is within a factor of two of the flux averaged over the last 10^5 to 10^7 yr (Geiss 1963). The energy range of the cosmic rays producing the transmutations is 10^9 to 10^{10} eV.

III DIFFUSION OF COSMIC RAYS

To reconcile the above observations with a galactic origin of the bulk of the cosmic rays it is apparent that the galactic magnetic field must regulate the propagation of the particles and their escape from the Galaxy. Ones direct knowledge of the detailed structure of this magnetic field is poor, however, and one is more likely to get information on it from the constraints imposed by the properties of the cosmic rays than the converse. The widely used approach, introduced by Fermi has been to postulate that particles random walk in some fashion through an irregular magnetic field so that the cosmic ray density satisfies a diffusion equation. The effective diffusion mean free path is, to some extent, a free parameter. Any value that is deduced from the properties of the cosmic rays must not, of course, be inconsistent with our knowledge of the field or the interstellar medium. Even for the interplanetary field, of which there is much more information, no complete theory yet exists for accurately determining diffusion coefficients from the power spectrum of field fluctuations. A survey of the various approaches to this problem has been given by Fisk et al. (1974).

1. 3-dimensional Diffusion

It can be argued that cosmic ray propagation approximates to 3-dimensional diffusion (see for example Ginzburg and Syrovatskii, 1964). A natural, but not unique, choice for the diffusion mean free path is 10 to 30 pc, the characteristic scale of the observed magnetic field inhomogeneities. The general equation for the concentration of cosmic rays N ($\underline{r}$, t, E) as a function of space, time and energy is then

$$\frac{\partial N}{\partial t} - \nabla \cdot (D\nabla N) + \frac{\partial}{\partial E}\left(\frac{\partial E}{\partial t} N\right) + \frac{N}{T} = Q(\underline{r}, t, E)$$

The diffusion coefficient D may be a function of direction, position, and energy. For 3-dimensional isotropic diffusion $D = \lambda c/3$ where the diffusion mean free path λ may vary with energy. The third term on the left hand side represents energy loss of the particles during propagation. This is important for electrons but may be neglected for nuclei. In the fourth term $1/T = 1/T_c + 1/T_e$. T_c is the nuclear collision loss time to be used in the calculation of the production of secondary nuclei by spallation. When considering the proton component this term can be neglected if one assumes that the 'grammage', deduced from the heavier nuclear composition, applies to the protons also. T_e is an escape time for cosmic rays from the Galaxy. This has sometimes been used instead of taking a spatial boundary to the diffusion region. The approximation may be valid when the object of the calculation is to obtain a distribution of spallation products. It cannot be used, for instance, when the object is to obtain a value for the anisotropy of cosmic rays at a given point in the Galaxy. The term Q ($\underline{r}$, t, E) is the source function. It may be a δ-function in space and time, corresponding to production of cosmic rays in discrete events such as supernova explosions.

The solution of the 3-dimensional diffusion equation for various boundary conditions and source regions was discussed by Ginzburg and Syrovatskii (1964). At that time it was generally accepted that radio observations indicated the existence of spherical halo of the Galaxy with radius R $\sim$ 15 kpc filled with relativistic electrons. The proton component would occupy the same region. For cosmic ray production in the disc and free escape from the halo boundaries, the escape time of cosmic rays would be $\tau \approx R^2/2\lambda c$. For $\lambda = 10$pc this gives $\tau \approx 3 \times 10^7$ yr. The mean effective gas density in the halo and disc together would be lower than the local interstellar density by the ratio of the volume of the disc to that of the halo so that this escape time is quite consistent with the observed grammage. The sun is sufficiently close to the plane of symmetry of this system to account for the low anisotropy of cosmic rays. More detailed observations have, however, led to a reinterpretation of the radio data in terms of a large proportion of the high latitude synchrotron emission coming from relatively nearby, very much expanded, supernova remnants (see the article by Thielheim). There is also a lack of confirming evidence for radio halos around other spiral galaxies. Nevertheless calculations on the electron spectrum and radio background radiation reported by Bulanov and Dogel (1974) indicate a spheroidal halo extending ~10 kpc from the disc. A firm choice between disc and halo confinement regions must await the determination of the proportion of Be^{10} (half life 1.6×10^6 yr) surviving in the cosmic ray spallation products. The current status of these measurements is given in the article by Rasmussen.

For a disc confinement region extending a distance h on either side of the galactic plane and free escape from the boundaries the escape time would be $\tau \approx h^2/\lambda c$. With $\lambda = 10$pc as before and $h \approx 100$pc the escape time is only ~3000 yr. To obtain the observed grammage a diffusion mean free path $\lambda < 0.1$pc is required. (One should bear in mind that the grammage is related to the age of particles reaching the earth rather than the mean escape time from the Galaxy. However, unless one is considering the effects of nearby discrete sources the two times are not very different). Also, with $\lambda = 10$pc, unless the earth were very symmetrically positioned with respect to the source distribution the anisotropy would be too large.

If cosmic rays were produced by random discrete sources in the Galaxy such as supernovae the anisotropy at a given point would vary with time. Sometimes, by chance, it would be lower than the average. Ramaty et al. (1970) calculated the probability distribution of anisotropies for given values of λ assuming one source every hundred years randomly situated in a disc 200pc thick. An upper limit to the anisotropy, δ, of 10^{-3} would be observed for half the time if $\lambda = 0.1$pc while it would be expected for 2% of the time if $\lambda = 4$pc. If the upper limit was $\delta < 2 \times 10^{-4}$ then $\lambda = 0.7$ pc for it to be achieved 2% of the time. For $\lambda = 10$ pc the chance of observing $\delta < 10^{-3}$ would be very small.

2. 1-dimensional Diffusion

The gyroradius of a proton of energy E(GeV) in a field of $B(\mu G)$ is $(E/B) \times 10^{-6}$pc. Thus in the galactic magnetic field of a few μG the gyroradii of protons below 10^{14}eV are very much smaller than the scale of any observed irregularities in the field. The cosmic rays reaching the earth will then be, to some extent, bound to the galactic magnetic field lines passing within one gyroradius of the earth. The effect of this on the flow of cosmic rays must be considered. The concept of 3-dimensional diffusion may have to be abandoned in favour of 1-dimensional diffusion along field lines unless the rate of separation of adjacent field lines is high.

The limits on the streaming velocity of cosmic rays and their lifetime when traversing interstellar gas in the disc indicate that they must travel a net distance of only a few hundred pc before leaving a disc confinement region. Jokipii and Parker (1969) have considered the problem of transporting particles many gyroradii perpendicular to the regular component of the galactic magnetic field in such a short distance. They contend that although the field lines run, on average, parallel to the plane of the disc the individual lines will be expected to random walk to the 'surface' due to turbulent motion of the interstellar gas. At this surface, at a height of ~130pc from the plane, the gas density is low enough that the pressure of the cosmic rays can cause a bubble instability in the magnetic field and the cosmic rays can freely escape from the 'end' of the field line. The average length of the field lines between such points, deduced from the observed turbulent motion of the gas and data on polarization of starlight, agrees with the streaming distance of cosmic rays.

The configuration of the field line intersecting the earth is not known but a probability distribution for its length has been derived by Jones (1971). He goes on to find the probability distribution of cosmic ray anisotropy for a model of one-dimensional diffusion where cosmic rays are injected continuously and uniformly along the field line. The anisotropy and mean age of the cosmic ray flux as seen by an observer at a given point on the line is then constant in time and dependent solely upon the distances to the two ends. With diffusion coefficient D and source strength S the diffusion equation is

$$D \frac{\partial^2 N}{\partial x^2} + S = 0$$ where x is measured along the field line.

For free escape at x = o and x = h the concentration of cosmic rays is

$$N(x) = S(hx-x^2)/2D$$

and the anisotropy is

$$\delta = - \frac{\lambda}{N} \frac{dN}{dn} = - \frac{\lambda (h-2x)}{x (h-x)}$$

Folding in a field line length distribution one obtains a probability distribution of anisotropy which peaks at zero corresponding to the earth being at the mid-point of its line. The observed upper limit to the anisotropy determines the range of positions about the exact mid-point of the line in which the earth must be. A very low anisotropy is, of course, always possible but from the size of this range one may make a judgement of the probability of the earth being in the required position. Table I shows values obtained by Dickinson and Osborne (1974). The field line length distribution was generated assuming that the line is randomly displaced up or down by 45pc for each 100pc along the line until it reaches 135pc from the galactic plane.

δ	λ = 1pc	λ = 3pc	λ = 10pc
$<10^{-3}$	0.30	0.16	0.049
$<4 \times 10^{-4}$	0.18	0.065	0.020
$<10^{-4}$	0.049	0.016	0.005

Table I: Probability of observing an anisotropy, δ, less than the given value as a function of diffusion mean free path, λ. A continuous and uniform injection of particles on to the field line is assumed.

This approach may be developed by taking a more physically likely source distribution: instantaneous sources randomly distributed in space and time. At a fixed point on the line the properties of the cosmic ray flux (concentration, mean age and anisotropy) will vary with time. For given values of the diffusion parameters one may then calculate, for instance, the fraction of the time during which the anisotropy is not greater than the observed upper limit. From the size of this fraction one may make a judgement of how reasonable is the choice of propagation parameters. A value has to be adopted for the average time interval between sources on a given field line. To convert the rate in the galaxy ($1/26$ yr^{-1}) to rates on a line one must know the size of the supernova remnant at which cosmic rays escape on to the galactic field lines. If one estimates that this would occur between radii of 10 and 30pc then (time interval) x (field line length) = 1.6×10^6 to 1.4×10^7 yr kpc. It should be noted that, if tight binding of the particles to field lines were true, the entire disc could be populated with cosmic rays only if the size of the source regions were at least of this order.

An example of the variation of concentration mean age and anisotropy with time at a given point on a field line is shown in figure 2. The correlation between the variations in anisotropy and concentration should be noted. If we are at a time when the anisotropy is low the cosmic ray concentration will automatically be close to its time average value in agreement with

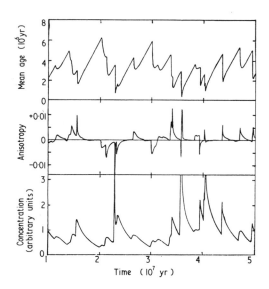

Fig. 2: An example of the time variation of concentration, mean age and anisotropy of cosmic rays at the mid point of a field line of length 3 kpc in the disc.

---

observation. One can calculate the average age of cosmic rays and the fraction of time that δ is less than any given value for all positions on lines of various lengths. Again, to find the overall probability of observing δ the field line length distribution is folded in. Some results are given in Table II. The probabilities are within a factor of two of those for a uniform source distribution. They increase for smaller values

δ	λ = 1pc	λ = 3pc	λ = 10pc
$<10^{-3}$	0.43	0.16	0.046
$<4 \times 10^{-4}$	0.24	0.084	0.018
$<10^{-4}$	0.091	0.038	0.004

Table II. Probability of observing an anisotropy, δ, as in Table I but for random discrete sources. A mean time and space interval between sources of 4×10^6 yr kpc is assumed.

---

of λ but a constraint on reducing λ is imposed by the average age of the cosmic rays. It increases from 2×10^6 to 7×10^6 yr as λ decreases from 10pc to 1pc. One concludes that, on the basis of this model of propagation, although the observed properties of the cosmic ray flux could be the result of our being at a particular time in the earth's cosmic ray history the probabilities are rather small particularly if δ is indeed $<10^{-4}$.

3. Compound Diffusion

For either 3-dimensional or 1-dimensional diffusion in a disc confinement region a mean free path $\lambda < 1$ pc is required to account for the low streaming velocity. Lingenfelter et al. (1971) introduced the concept of compound diffusion in order to obtain these low velocities with values of λ consistent with the observed field irregularities. The cosmic rays are taken to remain on their field line where they propagate by 1-dimensional diffusion, with mean free path λ_p, due to scattering from minor irregularities in the field while the field lines experience 3-dimensional random walk with a step size λ_m. The probability density of particle displacement, s, along the field line in time t is then

$$f_1(s,t) \propto (\lambda_p t)^{-\frac{1}{2}} \exp(-3s^2/4\lambda_p ct)$$

while the probability density for a three dimensional displacement r of a field line of length s is

$$f_2(r,s) \propto (\lambda_m s)^{-3/2} \exp(-3r^2/4\lambda_m s)$$

The compound probability density for a net displacement r in time t is thus

$$f(r,t) = \int_r^{ct} f_1(s,t) f_2(r,s) ds$$

The result is that the net displacement of particles is proportional to $t^{\frac{1}{4}}$ rather than $t^{\frac{1}{2}}$ as in simple diffusion and anisotropies $\delta \approx 10^{-4}$ can be obtained with $\lambda_p \approx \lambda_m \approx 30$ pc. There are strong objections to this concept, however. The anisotropy is low in spite of the long mean free path because it is effectively the 'macro-anisotropy' resulting from the averaging of 'micro-anisotropies' over many field lines. Allan (1972) has emphasised a point made by Ginzburg and Syrovatskii that, if 1-dimensional diffusion is to hold, adjacent field lines must have the geometry of flux tubes at least one gyroradius wide. Although the sun is moving with respect to the interstellar medium its velocity is such that, even in several years, it will not move across the magnetic field a distance corresponding to the gyroradius of 10^{11}eV particles. Thus the anisotropy observed at the earth would be the larger micro-anisotropy on the field line passing through the earth as discussed in section 2.

4. Separation of Field Lines

The necessary short mean free path appears to show that the cosmic rays are scattered not only by large scale irregularities in the field but also by hydromagnetic waves in the interstellar medium. The reversal of the particle's direction due to these will be the net result of many small pitch angle scatterings, and in reversing its direction the particle will move, on average, one gyroradius across the field (Skilling 1970). Sideways drift of the particle due to field gradient and curvature is also of this order in one diffusion mean free path. If, after the sideways step the new field line remains parallel to the old one then effectively one has 1-dimensional diffusion. If field lines, initially one gyroradius apart, separate rapidly one must abandon this concept.

The rate of separation of field lines, caused by the background spectrum of interstellar turbulence, has been considered by Skilling et al.(1974). For a Kolmogorov spectrum of wave intensity versus wave number, $I(k)dk \propto k^{-5/3}dk$, extending from $k_1 = (L_1)^{-1}$ onwards, the lines separate with distance, r, along the line in proportion to r^3. A separation of L_1 is reached in $\sim L_1$ along the lines. The length L_1 is identified with the observed large scale irregularities in the field (i.e. $L_1 \approx 30$ pc). This separation in distance L_1 is not very sensitive to the form of the turbulence spectrum provided it is no steeper than $\propto k^{-2}$. Skilling (private communication) has also considered the effect of a sharp cut off to the turbulence spectrum at $k_2 = (L_2)^{-1}$ due to the dissipation of turbulent energy by viscosity. While the separation of the lines is $<L_2$ the separation increases exponentially with distance along the line. With $L_2 \sim 610^{-3}$ pc the separation doubles every 0.3 pc. If the initial separation is equal to the gyroradius of a 10^{12} eV proton a distance of 31 pc leads to a final separation of ~ 30 pc: if the initial separation corresponds to the gyroradius of a 10^9eV proton only an additional 3 pc distance is required.

These results imply that at the earth we may, after all, be observing particles reaching us along a number of essentially independent field lines. However, the concept of 1-dimensional diffusion along these lines also breaks down and compound diffusion still does not apply.

IV ENERGY DEPENDENT PROPAGATION

1. Self-Confinement of Cosmic Rays

We have seen that the low streaming velocity of cosmic rays implies the existence of hydromagnetic waves in the insterstellar medium with an appropriate energy density. For particles of a given energy, resonant scattering will occur on waves having wavelengths a few times the gyroradius. The mean free path for reversal of the direction of the particle will be approximately equal to the gyroradius multiplied by the ratio of the energy density in the ambient galactic magnetic field to the energy density in the waves causing the scattering. Wentzel (1968) and Kulsrud and Pearce (1969) have shown how the required waves may originate. When cosmic ray particles stream down their density gradient with a streaming velocity V_s greater than the Alfven velocity V_A an instability exists which generates waves. Thus the cosmic rays themselves produce the waves which scatter them. The growth rate of the waves that resonate with particles of energy E is

$$\Gamma_G = \frac{eB}{2\pi^{\frac{1}{2}} M_H c} \frac{N(>E)}{n_i} \left[\frac{V_s}{V_A} - \frac{\gamma+2}{3} \right] \qquad (4.1)$$

where the differential energy spectrum of cosmic rays is taken as a power law of index $-\gamma$, B is the ambient magnetic field, $N(>E)$ is the number density of cosmic rays of energy $>E$ and n_i is the number density of ions in the interstellar medium. The Alfven velocity $V_A = B/(4\pi \rho_i)^{\frac{1}{2}}$ where ρ_i is the ionised gas density. In the galactic plane $V_A \approx 70$ km s^{-1}. The authors considered damping of the waves due only to collisions between charged particles moving with the waves and neutral atoms. The damping rate is then

$$\Gamma_D = G\, n_H \qquad (4.2)$$

where n_H is the density of the neutral gas and the constant G depends on its temperature. The equilibrium streaming velocity is obtained by equating (4.1) and (4.2). For the bulk of the cosmic rays, i.e. those with energy of a few GeV, their number density is such that they would stream little faster than $1.5 \times V_A$.

For the particles of several hundred GeV, of which the low streaming velocity is actually observed, there are problems with this self-confinement mechanism. Kulsrud and Cesarsky (1971) concluded that, because of the sharply falling energy spectrum, these high energy cosmic rays do not have a number density sufficient to produce a strong instability and they can therefore stream very much faster than the Alfven velocity contrary to observation. This is certainly true in the galactic plane but Skilling (1971) and Holmes (1974a) have pointed out that, because n_i and n_H decrease with height above the plane, the growth rate of the waves increases while their damping rate decreases. Cosmic rays of energy greater than a few GeV need only travel a certain distance away from the galactic plane before they reach a region of strong scattering. Skilling has labelled the region close to the galactic plane where there is essentially no scattering the 'free zone' and the region of strong scattering the 'wave zone'. At the boundary only a fraction $\sim V_A/c$ of the incident particles can be transmitted. Thus it is proposed that there exists reflecting boundaries on either side of the galactic plane at a height which increases with energy. Holmes has shown how this not only saves the self-confinement mechanism but naturally leads to a grammage that decreases with energy. A simplified derivation of the energy dependence, based on that of Holmes (1974b) is given below.

Let us assume that B is independent of distance from the galactic plane, z and that the neutral and ionised gas densities vary respectively as

$$n_H \propto \exp - (z/z_o) \text{ and } n_i \propto \exp - (az/z_o) \tag{4.3}$$

The factor, a, is observationally uncertain but would be expected to be ~ 0.5. Substituting these expressions in equations (4.1) and (4.2) and equating them, the streaming speed is found to be

$$V_s = K_1 \exp\left(\frac{az}{2z_o}\right) + K_2 E^{\gamma-1} \exp - \left[\frac{z}{z_o}\left(1 + \frac{a}{2}\right)\right] \tag{4.4}$$

where K_1 and K_2 are independent of z and E. The streaming velocity out of the galaxy of particles of a given energy is determined by the minimum streaming speed. The height at which this occurs is the boundary, z_B. Then from (4.4)

$$z_B(E) = z_o(\gamma-1)(1+a)^{-1} \ln(E/E_o)$$

where

$$E_o = \left[\frac{a}{2+a}\frac{K_1}{K_2}\right]^{\frac{1}{\gamma-1}}$$

is the highest energy for which the galactic plane, $z = 0$, is in the wave zone. When a particle encounters the boundary its transmission probability is $\sim V_A(z_B)/c$ and the residence time in the free zone

$$\tau(E) \propto z_B/V_A(z_B) \propto \ln\left(\frac{E}{E_o}\right) \times \left(\frac{E_o}{E}\right)^{\frac{a(\gamma-1)}{2(1+a)}}$$

The logarithmic factor shows the increase in boundary height with energy while the other factor is the decrease in reflection probability. The grammage in the free zone is $X = \tau c \bar{\rho}$ where $\bar{\rho}$ is the mean density of the interstellar gas between $z = 0$ and $z = z_B$. Thus

$$X \propto \left[1 - \left(\frac{E_o}{E}\right)^{\frac{\gamma+1}{1+a}}\right] \left(\frac{E_o}{E}\right)^{\frac{a(\gamma-1)}{2(1+a)}} \tag{4.5}$$

For $\gamma = 2.5$ and $a = 0.5$ and a continuous uniform production of cosmic rays near to the galactic plane an observer there would see
$$X \propto E^{-0.25} \text{ for } E > E_o.$$

Similar energy dependences can be obtained if the z-distributions of interstellar matter are assumed to be gaussion rather than exponential. The uncertainties in the properties of the interstellar medium are sufficient that the predicted energy dependence can always be made consistent with observation. A break in the observed energy spectrum is expected at E_o. Some of the arbitrariness can be removed if such a break can be identified. Holmes suggests $E_o = 8$ GeV but the position of a break in this energy region may be confused by the effects of solar modulation.

A complication of the picture of self-confinement that should be taken into account is the additional non-linear damping of the waves due to wave-wave interactions. This process by which Alfven waves are degraded to longer wavelengths and into sound waves is discussed by Wentzel (1974). The damping does not depend on the gas density and its effect, to a first approximation is not to alter the boundary height but to increase the streaming velocity in the wave zone. This in turn increases the transmission probability and reduces the grammage. Holmes (1974b) shows that agreement with the observed grammage can still be obtained. There is some doubt, however, whether the streaming velocity from the galaxy can be kept sufficiently low beyond 100 GeV.

2. Self-Confinement with Discrete Sources

It is interesting to apply the model of self-confinement to the case of random, discrete sources. Dickinson (private communication) has made some preliminary calculations. In obtaining the results presented here the following assumptions were made. The ionised and neutral gas were taken to have identical distributions, $n \propto \exp -(z/z_0)^2$ with $z_0 = 165$ pc. The transmission coefficient was $2 V_A(z_B)/c$. Within the 'free zone' the particles move freely along field lines. Scattering by large scale irregularities and the rate of separation of adjacent lines are assumed to have the effect of 3-dimensional diffusion with a mean free path alternatively 10 pc or 30 pc. It is then possible to calculate the contribution to the cosmic ray flux of any instantaneous point source, after a time τ, which occurred at a distance R from the sun. Figure 3 shows a sample of the results from the solution of the diffusion equation with a mean free path of 10 pc and a reflecting boundary appropriate to 10^3 GeV particles. It is apparent that sources contribute significantly to the total flux up to ages of about 5×10^6 yr and to distances of about 5 kpc.

If one assumes that supernova explosions are the sources of cosmic rays one may calculate contributions to the flux using the observed positions and estimated ages of supernova remnants. However the total contribution from the known sources, which date back to only a few times 10^5 yr, will be a small fraction of the total flux. In order to calculate the anisotropy and grammage using this model it is necessary to use a randomly generated source distribution. Initially 2×10^4 sources were chosen within 5 kpc of the sun and within 100pc of the galactic plane. For an assumed rate for the whole galaxy of one supernova in 50 yr the oldest sources were 2.25×10^7 yr. The concentration, anisotropy age and grammage were calculated, initially, for these sources, and then at intervals of 5.4×10^3 yr over a period of 10^5 yr adding new sources at the appropriate rate to simulate the time variation. The results are shown in figure 4. It can be seen that the results for a 10 pc and a 30 pc mean free path are not significantly different. For graphs (a), (b) and (c) the points show the time-average values while the error bars indicate the full range of fluctuation of the quantities over the 10^5 yr period. The concentration decreases as a power of the energy showing the way in which the propagation modifies the source spectrum. The grammage decreases with energy in a manner consistent with observations. By far the largest fluctuations are in the anisotropy which varies by 2 orders of magnitude. The minimum values are within a factor of 2 or so of the experimental upper limits but the fraction of the time they lie in this range is small. A number of refinements needs to be applied to these calculations. The boundary height is obtained neglecting the

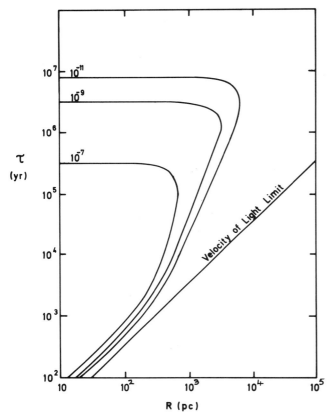

Fig. 3: Contours showing the flux, in arbitrary units, from a cosmic ray source as a function of its age, τ, and distance from the sun, R.

---

time-variation of the cosmic ray intensity. If the intensity near the boundary increases, the height of the boundary will decrease. The thickness of the gaseous disc and the rate of supernova explosions depend quite strongly upon distance from the galactic centre. The incorporation of these effects is not likely to reduce the anisotropies, however.

V CONCLUSIONS

Self-generated waves may not be sufficient to confine the cosmic rays beyond ~100 GeV. It is possible that a spectrum of waves may be generated by the large scale turbulent motion of the

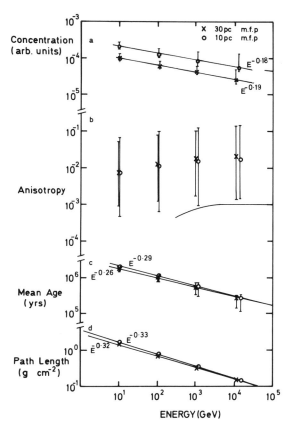

Fig. 4: Variation with energy, assuming random discrete sources, of the following observables: (a) the concentration, in arbitrary units, (b) the anisotropy (full-line shows experimental upper limit from figure 1), (c) the mean age, (d) the grammage (i.e. path length in g cm^{-2}).

---

gas. Those waves with wavelengths equal to the gyroradii of particles at the top of the energy range being considered here, 10^{13} to 10^{14}eV, will almost certainly be present with sufficient amplitude. Jokipii (1971) has argued, on grounds of the energy input required, that waves generated by some kind of external stirring cannot confine particles down to the lowest energies. If the interstellar turbulence is isotropic the cosmic rays must experience a degree of Fermi acceleration on being scattered. The time for doubling the energy of the cosmic rays is proportional to the diffusion mean free path, λ and for $\lambda \simeq 1$ pc it approximates to the age of cosmic rays. For all cosmic rays, down to the lowest energies, to be confined by externally generated turbulence an unknown, and unreasonably large, source of turbulent energy is required. If only those particles with energies beyond ~100 GeV are confined in this way the energy input needed is naturally much reduced. Diffusive confinement thus involves either transfer of energy from the turbulent motion of the interstellar gas to the cosmic rays or, in the case of self-confinement, from the cosmic rays to the gas. Although it has not been taken into account in the discussion in section IV the self-confined cosmic rays may be expected to give up in the order of half of their energy in escaping from the Galaxy. This energy heats the gas and at the boundary of the disc would give it bulk motion outwards.

It appears that if cosmic rays are diffusively confined to the galactic disc the mechanism may change from self-confinement to that due to externally generated waves in the region of 100 to 1000 GeV. This is the region, at present below which we have information on the grammage only and above which there is information on the drift velocity only. It may well be wrong to extrapolate the energy dependences of these quantities respectively to higher and lower values and it is thus of great interest to close the energy gap experimentally.

REFERENCES

Allan, H.R., 1972. Astrophys. Lett. 12, 237.

Bulanov, S.V. and Dogel, V.A., 1974. Astrophys. and Space Sci. 29, 305.

Cachon, A., 1962. Proc. 6th Interam. Sem. on Cosmic Rays, La Paz, Vol. 2 (University of San Andreas) p. 39.

Daudin, J., et al. 1956. Nuovo Cim., 3, 1017.

Delvaille, J., Kendziorski, F. and Greisen, K., 1962. J. Phys. Soc. Japan 17, Suppl. 3,76.

Dickinson, G.J. and Osborne, J.L., 1974. J. Phys. A., 7, 728.

Elliot, H., Thambyahpillai, T. and Peacock, D.S., 1970. Acta. Phys. Acad. Sci. Hung. 29, Suppl. 3, 491.

Fisk, L.A., Goldstein, M.L., Klimas, A.J. and Sandri, 1974. Astrophys. J., 190, 417.

Geiss, J., 1963, Proc. 7th Int. Conf. on Cosmic Rays, Jaipur, Vol. 3 (Bombay: TIFR) p. 434.

Ginzburg, V.L. and Syrovatskii, S.I., 1964. The Origin of Cosmic Rays (Oxford: Pergamon).

Gombosi, T., et al., 1974. Report at the 4th European Symposium on Cosmic Rays, Lodz.

Holmes, J.A., 1974a. Mon. Not. R. Astron. Soc., 166, 155.
1974b. Ph.D. Thesis, University of Oxford.

Jokipii, J.R., 1971. Proc. 12th Int. Conf. on Cosmic Rays, Hobart, Vol. 1 (Hobart: University of Tasmania), pp. 401-6.

Jokipii, J.R. and Parker, E.N., 1969. Astrophys. J., 155, 799-806.

Jones, F.C., 1971. Proc. 12th Int. Conf. on Cosmic Rays, Hobart, Vol. 1 (Hobart: University of Tasmania), 396-400.

Kolomeets, E.V., Nenolochnov, A.N. and Zusmanovich, A.E., 1970. Acta. Phys. Acad. Sci. Hung. 29, Suppl. 1, 513.

Kulsrud, R. and Pearce, W.P., 1969. Astrophys. J., 156, 445.

Kulsrud, R.M. and Cesarsky, C.J., 1971. Astrophys. Lett. 8, 189.

Lapikens, J., et al. 1971. Proc. 12th Int. Conf. on Cosmic Rays, Hobart, Vol. 1 (Hobart: University of Tasmania) p. 316.

Lingenfelter, R.E., Ramaty, R. and Fisk, L.A., 1971. Astrophys. Lett. 8, 93-7.

Ramaty, R., Reames, D.V. and Lingenfelter, R.E., 1970. Phys. Rev. Lett. 24, 913.

Ramaty, R., Balasubrahmanyan, V.K. and Ormes, J.F., 1973. Science 180, 731.

Sherman, N., 1953, Phys. Rev. 89, 25.

Skilling, J., 1970. Mon. Nat. R. Astron. Soc. 147, 1.

Skilling, J., 1971. Astrophys. 170, 265.

Skilling, J., McIvor, J. and Holmes, J.A., 1974. Mon. Nat. R. Astron. Soc. 167, 87.

Wentzel, D., 1968. Astrophys. J., 152, 987.

Wentzel, D., 1974. Ann. Rev. Astron. and Astrophys. Vol. 12.

POSSIBLE EXPLANATIONS OF THE SPECTRAL SHAPE

A.W. Wolfendale

Physics Department,
University of Durham, U.K.

I   INTRODUCTION

A satisfactory model for the origin of cosmic rays must explain the many facts which are known about these particles and quanta. In the present work the nucleonic component is the main concern, particularly those particles above 10^{12}eV. The form of the primary spectrum at these energies is briefly examined and then the relative merits of a number of origin models are studied. The paper does not aim to be comprehensive; instead attention is confined to those models with which the author has been concerned.

II   THE PRIMARY SPECTRUM

The problems involved in measuring the energy spectrum of nucleons and nuclei above 10^{12}eV are well known. Just above the lower limit there are some measurements with balloon − and satellite-borne ionization calorimeters and some indirect results have come from measurements of nucleons and muons in the atmosphere but the bulk of the data have been derived from extensive air shower studies. Figures 1 and 2 review the situation. Figure 1, taken from the analysis by Kempa et al. (1974), is a composite of measurements above 2.10^{13}eV from EAS and below 5.10^{13}eV from an extrapolation of measurements on nuclear active particles at various depths in the atmosphere (there is a small overlap). The nuclear active particle data give the spectrum in terms of energy per nucleon; conversion to energy per nucleus (an increase by a factor 1.7) has been made assuming that the mass composition is unchanged from its measured value at 10^{10}eV/nucleon.

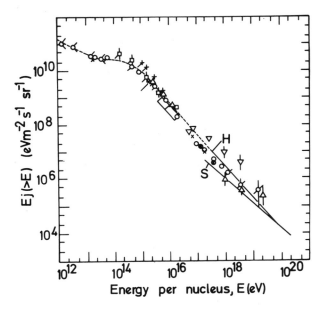

Fig. 1: The integral spectrum of primary cosmic ray nuclei (Kempa et al., 1974). The sources of the data are given in that work.

The most striking feature of the spectrum is the clear difference in slope below and above about 10^{15}eV. Below this value, the integral exponent is in the region of -1.6 to -1.7 and above it the exponent is -2.0 to -2.2.

An interesting feature is the apparent 'bump' in the region of 10^{14}-10^{15}eV (note that the ordinate is energy multiplied by integral intensity; the intensity is falling rapidly with increasing energy everywhere, including the 'bump' region). The 'high' intensities are largely those from experiments carried out at mountain altitudes where the comparatively low energy showers are near their maximum development. A good feature is that different experiments in the same energy region give very similar intensities but a disquieting feature is that these showers appear to reach their maximum of development higher in the atmosphere than expected (see the work of Wdowczyk and Wolfendale, 1973). The reason could be that the primary nuclei are more massive than has been assumed here or there could be a drastic change in the Nuclear Physics of the particle interactions in this region. If the latter were true a downward trend in the intensities could result but it is difficult to see that they could go very far. Complete removal of the bump appears to be rather difficult.

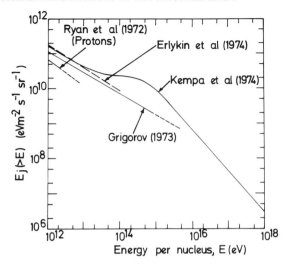

Fig. 2: The integral spectra of primary nuclei, nucleons and protons from various analyses. The line marked Kempa et al. (1974) is the best line through the points of Fig. 1. 'Erlykin et al. (1974)' relates to primary nuclei and is derived from sea level muon data. The line denoted 'Grigorov (1973)' refers to the spectrum of all nuclei measured by that author.

Figure 2 shows the best line from Figure 1 together with the results of other measurements. The line denoted by Erlykin et al. (1974) is derived from the measured single muon spectrum at ground level, and below, taken together with an assumed model for high energy interactions. The assumption is also made that the mass composition is unchanged from its value at 10^{10}eV/nucleon. It can be seen that there is good agreement with the 'best' line up to 10^{13}eV but that there is no support for the idea of a spectral upturn (i.e. the 'bump'). However, the Nuclear Physics is by no means certain in this region and the muon data cannot be regarded as evidence against the bump. The 'Grigorov' spectrum of Figure 2 also gives no support but these measurements suffer from the well known disadvantage of appearing to indicate a rapid fall off in primary proton intensity above 10^{12}eV which is not supported by other experiments; the lack of agreement with the spectrum of Kempa et al. is thus not necessarily too serious.

In conclusion, there is strong evidence for an increase in slope of the primary spectrum, at above 10^{15}eV, and there may well be a bump in the spectrum at an energy a little below this value.

III. GALACTIC ORIGIN

The view advanced by many authors is that the bulk of cosmic rays originate in Galactic sources and that the spectral shape is a consequence of an energy-dependent diffusion coefficient. On general arguments a gradual change of primary exponent would be expected and as was remarked in the last section there is experimental evidence for this (a superimposed 'bump' could possibly also be accomodated).

Interest centres on an examination of the relevant Astronomical quantities which indicate the particle energy, E_c, at which the change of exponent should commence and the numerical magnitude of the change ($\Delta\gamma$). Bell et al. (1974) have recently examined the characteristics of the magnetised 'clouds' of neutral hydrogen which are commonly thought of as providing the scattering centres and which thereby cause the diffusive motion. Using the data of Heiles (1967) and Ames and Heiles (1970) on cloud diameters and that of Verschuur (1970) on magnetic fields they have derived the expected diffusion coefficient as a function of proton energy. The analysis is clearly approximate in that the astronomical data only refer to a limited region of space and it has been necessary to assume that this region is representative and, furthermore, simplifying assumptions have been made about the nature of the clouds and the manner in which they scatter charged particles. However, the analysis is probably the best that can be done at the present time. Assuming that the particles can escape freely from the surface, or the ends, of the local inter-arm spur, and, following experiment, that there is a mean longitudinal magnetic field of strength 3-4 μ gauss, the analysis gives $E_c \simeq 4 \times 10^{15}$ eV and $\Delta\gamma \simeq 0.9$. Furthermore, the mean lifetime of protons below E_c is predicted to be $\tau \sim 2 \times 10^5$ years. Comparison can be made with the experimental values; experimentally $E_c \sim 10^{15}$ eV and $\Delta\gamma \simeq 0.6$, not far from prediction and τ is not inconsistent with expectation (no values of τ are available at such high energies but the evidence is that τ is $\sim 2 \times 10^6$ years at $E \simeq 10^{10}$ eV and is falling with increasing energy).

There is thus some circumstantial evidence in favour of the galactic diffusion model, as far as about 10^{17} eV; the reason for the energy restriction is that at higher energies the predicted anisotropy of particle arrival directions becomes rapidly greater than observation. Above 10^{17} eV another model appears imperative. One possibility is to allow an increasing fraction of heavy nuclei in the primary beam and this displaces the upper limit of validity; however, lack of any measured anisotropy of the highest energy primaries still cannot be accounted for.

IV. EXTRAGALACTIC ORIGIN

POSSIBLE EXPLANATIONS OF THE SPECTRAL SHAPE

1. Energy density

   In some ways the simplest assumption to make about cosmic ray origin is to assume that all the particles are of extragalactic origin and that they fill the whole of the Universe, although an immediate problem that arises, is the large energy density, $\sim 1\text{eV cm}^{-3}$; if this were Universal it would be exceeded only by the energy density of matter itself. Recent discoveries of very energetic radio sources of large linear dimension make the energy density problem rather less severe however, and furthermore it is always possible to hypothesise that the 'low' energy primaries, say below 10^{12}eV, are of Galactic origin, in which case the Universal energy density falls to about $2 \times 10^{-2}\text{eV cm}^{-3}$.

   An interesting model which uses the change of slope of the primary spectrum at $E_c \sim 10^{15}\text{eV}$ is that first mentioned by Hillas (1968). Here, the primaries are assumed to have been produced predominantly at very early epochs ($z \sim 15$) when the Universal blackbody temperature was about 40K. For this temperature, the threshold energy for electron pair production in p-γ collisions is E_c and by suitable choice of z-dependence of the primary intensity, the measured value of $\Delta\gamma$ can be reproduced. So far, of course, the model is ad hoc and designed to fit the measured spectral shape but Strong et al. (1973, 1974a) and Strong (1974) have devised a way of testing the model in an independent fashion. The test involves following through the energy released into electron pairs in the p-γ collisions. The electrons interact with the black body photons by the Inverse Compton effect which, together with photon-starlight photon interactions, causes an extensive electron-photon 'shower' in the Universe (interactions with intergalactic matter are ignored – a valid procedure in view of the lack of evidence so far for significant intergalactic densities). The bulk of the energy is carried by γ-rays and there is immediately the possibility of explaining the experimentally measured diffuse γ-ray background in this way.

   Before examining the spectral shapes the energy densities can be considered. The energy density available in the primary proton spectrum (which is converted into γ-rays) is that contained between an onward extrapolation of the spectrum below 10^{15}eV and the measured spectrum above this limit. The best estimate is $\sim 7 \times 10^{-5}\text{eV cm}^{-3}$ and this can be compared with the actual energy density in the measured diffuse γ-background (above 1 MeV) of about $3.5 \times 10^{-5}\text{eV cm}^{-3}$. The near-agreement of these energy densities is encouraging. Figure 3 gives a comparison of the observed and predicted spectra. It will be seen that there is rough agreement as to the overall spectral shape although the highest points, those above about 30 MeV, appear to be falling below the predicted line. When measurements are extended above 100 MeV it should be possible to pronounce rather firmly either

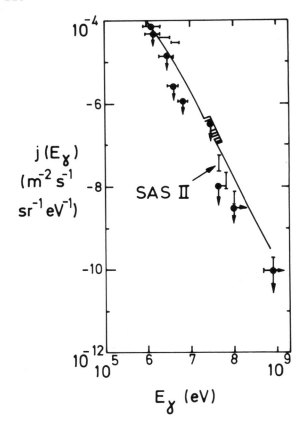

Fig. 3: The diffuse γ-ray spectrum. The experimental points are a representative set taken from the summary by Strong et al. (1974a); a more recent and comprehensive set is given in the article in the present volume by Stecker (1975).

The line is the prediction by Strong (1974) for the model in which the change of slope of the primary proton spectrum arises from interactions with the black body radiation.

for or against this origin theory.

Further aspects of extragalactic origin are considered in the next section.

V. MIXED GALACTIC AND EXTRAGALACTIC ORIGIN

The possible existence of a bump in the spectrum and the fact that an extrapolation of the directly measured spectrum below 10^{12}eV when extrapolated onwards meets the measured intensities at $\sim 10^{18}$eV led Strong et al. (1974b) to propose a composite model. The model has some interesting features and is worthy of description.

A contender for the spectrum bump is pulsar-accelerated particles. Although the details of the acceleration mechanism are not well understood (see the article in this Volume by Pacini) a number of calculations have been made with various assumptions. Ostriker and Gunn (1969) have given specific predictions for spectral intensities and Karakula et al. (1974) have used these data together with experimental information on the density of pulsars in the Galaxy and the residence time of cosmic rays to predict an absolute primary spectrum. Figure 4, taken from Strong et al. (1974b), gives the resulting spectrum together with other relevant spectra. The agreement with observation in the region of 10^{15}eV is interesting, and perhaps significant.

Turning to the extragalactic component, the assumption here is that the particles are produced with a constant spectral index, mainly at low values of z, and interactions with the radiation

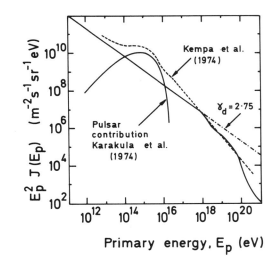

Fig. 4: Composite primary spectrum, after Strong et al. (1974b). The component with differential exponent at production of γ_d = 2.75 is extragalactic.

The line marked Kempa et al. (1974) is the best fit line of Figure 1.

field at 2.7K are operative. Figure 4 shows the resulting spectrum; the onset, successively, of electron pair production ($\sim 10^{18}$eV) and pion production ($\sim 6 \; 10^{19}$eV) are clearly visible. Although there can be no question of an exact agreement between the summed contributions (Pulsar + E.G.) and experiment, the intensities are sufficiently close to warrant further attention.

Probably the most interesting region is that above 10^{17}eV and attention will be confined here. This is because whatever the source of the particles at lower energies it seems very likely that the more energetic ones are extragalactic. The most serious problem is the lack of observation of the sharp cut-off in intensity above $\sim 6 \times 10^{19}$eV due to the pion threshold in p-γ interactions (Greisen, 1966; Zatsepin and Kuzmin, 1966). Figure 4 demonstrates this difficulty (the divergence of the 'best' line and the theoretical curve at the highest energies).

A number of attempts have been made to overcome the problem. For example, Brecher and Burbidge (1972) have suggested a model in which there is a measure of containment of particles within galactic clusters. Such containment would arise from the existence of magnetic fields which may be present in the space occupied by the cluster. Provided that the fields are not so strong as to cause trapping lives comparable with the mean time between pγ collisions in which pion production occurs, the rapid attenuation would not occur. However, it then becomes difficult to account for the whole spectral shape in a straightforward way (the lifetime will be energy dependent).

An alternative explanation which is able to explain the experimental data available at present has been put forward by Strong et al. (1974c). These authors have drawn attention to the fact that even on a truly Universal origin model, in which energetic cosmic rays are generated in a representative fraction of all galaxies (but not including our own) then, due to our galaxy being in a denser than average aggregate - the 'Supercluster' - the black body cut-off will be reduced. Strong et al. use the data of Allen (1973) and de Vaucouleurs (1953) to show that the average density of galaxies in the supercluster is ~ 0.5 galaxies mpc^{-3} compared with a Universal value of ~ 0.02 galaxies mpc^{-3}. Thus the 'local' density of galaxies is some 25 times higher than the Universal average and this fact, taken with the values for the Supercluster radius and the Hubble radius indicates that in the absence of blackbody attenuation some 10% of the extragalactic cosmic rays would have come from the Supercluster. When the black body attenuation is now allowed for it is apparent that, due to the smaller distances involved, its effect on the Supercluster particles will be small and in turn there will be a smaller high energy 'cut-off'. Figure 5 summarises the situation for two values for the Supercluster fraction: 13% and 6%; it is likely

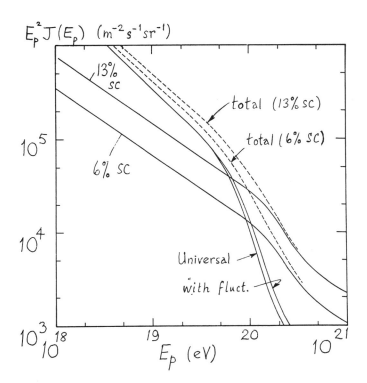

Fig. 5: Predicted primary spectra at very high energies for models in which the particles arise from extragalactic sources. In the 'Universal' model the average density of sources is constant over the Universe. The dotted lines relate to the situation where there is enhanced density in the Supercluster. The percentages refer to the ratio of the Supercluster contribution to that from the rest of the Universe. 'With fluct.' relates to inclusion of the effect of fluctuations in interaction point in the Universe and fluctuations of inelasticity in the p-γ collisions; although the fluctuation correction is not large it is significant.

---

that the true fraction lies in this range. The smaller cut-off compared with the uniform Universal origin model can be clearly seen.

Finally, comparison is made with experiment in Figure 6. Only two experiments have given adequate statistical precision in the measured intensities to afford a meaningful comparison and these have been used in the Figure. It should be noted that although there are no energy uncertainties indicated, they are undoubtedly

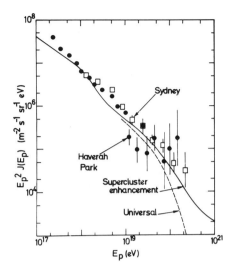

Fig. 6: Comparison of observed and predicted primary intensities above 10^{17}eV. The Haverah Park data are those of Edge et al. (1973). The Sydney points come from the work of Bell et al. (1974); they differ from the line marked S (Sydney) in Figure 1 because in the present comparison the nuclear physical model which fits the Haverah Park data has been used to convert from measured muon size to primary energy rather than the preferred model of Bell et al.

The Supercluster enhancement corresponds to the 10% value (interpolated from Figure 5). The 'Universal' curve is the spectrum expected for a constant average density of sources throughout the Universe.

---

present, largely because of lack of knowledge of the high energy physics which governs the transformation from measured shower parameter to primary energy, at the very high energies in question. Neglecting these uncertainties for the moment Figure 6 indicates that there is as yet no evidence against the Supercluster enhancement model but that the Universal model is probably ruled out. If the uncertainties are included, however, then the Universal origin model is probably also not ruled out, either. A distinction between the models must await further experimentation and improvements in energy estimates; more data would also aid the search for anisotropies in arrival directions; the Supercluster enhancement model would predict significant anisotropies above about 10^{20}eV.

REFERENCES

Allen, C.W., Astrophysical Quantities, 3rd Edition (Univ. of London, Athlone Press), 1973.

Ames, S. and Heiles, C., Ap. J., 160, 59, 1970.

Bell, M.C., Kota, J. and Wolfendale, A.W., J. Phys. A., 7, 420, 1974.

Bell, C.J. et al., J. Phys. A., 7, 990, 1974.

Brecher, K. and Burbidge, G.R., Astrophys. J., 174, 253, 1972.

Edge, D.M. et al., J. Phys. A., 6, 1612, 1973.

Erlykin, A.D., Ng, L.K. and Wolfendale, A.W., J. Phys. A., 1974, (in the press).

Greisen, K., Phys. Rev. Lett., 16, 748, 1966.

Grigorov, N.L., private communication, 1973.

Heiles, C., Ap. J. Suppl., 15, 97, 1974.

Hillas, A.M., Canad. J. Phys., 46, S623, 1968.

Karakula, S., Osborne, J.L. and Wdowczyk, J., J. Phys. A., 7, 437, 1974.

Kempa, J., Wdowczyk, J. and Wolfendale, A.W., J. Phys. A., 7, 1213, 1974.

Ostriker, J.P. and Gunn, J.E., Astrophys. J., 157, 1395, 1969.

Strong, A.W., Wdowczyk, J. and Wolfendale, A.W., Nature, Lond. 241, 1973.
    J. Phys. A., 7, 120, 1974a.
    J. Phys. A., 7, 1489, 1974b.
    J. Phys. A., 1974c(in the press).

Strong, A.W., private communication, 1974.

Vaucouleurs, G. de, Astrophys. J., 58, 130, 1953.

Verschuur, G.L., I.A.U. Symposium No. 39 (D. Reidl Publ. Co., Dordrecht-Holland), 150, 1970.

Wdowczyk, J. and Wolfendale, A.W., J. Phys. A., 6, 1594, 1973.

Zatsepin, G.T. and Kuzmin, V.A., Sov. Phys., - JETP, 4, 114, 1966.

THE COSMIC RAY ELECTRON COMPONENT

P. Meyer

Enrico Fermi Institute and Department of Physics
University of Chicago

I   INTRODUCTION

   Certainty about the presence of an electron component in the
primary cosmic radiations was established by 1950 through new
observations in radio astronomy. It was discovered at the time
that our galaxy is emitting a continuous spectrum of radio waves
over a fairly wide band that is definitely not thermal in nature.
The origin of this radio emission was soon interpreted as synchrotron radiation produced by highly relativistic electrons spiralling
around the magnetic fields of the galaxy. Only in the early 1960's
did it become possible to directly observe these particles as they
arrive at the earth. These observations were made almost simultaneously by Earl of the University of Minnesota and Vogt and
myself of the University of Chicago. Since that time a major
effort has been undertaken to accurately measure the spectrum and
the flux of these particles and indeed during the past decade the
electron spectrum has become known with an accuracy comparable to
that of the nuclear cosmic ray components. Fig. 1 summarises our
present knowledge of the electron spectrum over a wide range of
energies from about 100 KeV up to almost 1000 GeV. The solid line
in this figure represents the spectrum of protons for comparison.

   I shall divide these lectures into five parts. In the first
part I will discuss the instrumentation and methods of detection
that lead to the determination of the electron spectrum and to
the measurements of the fraction of positrons. The second, third
and fourth parts will be devoted to the potential sources of
cosmic ray electrons, to the mechanisms by which they lose energy,
and to their propagation within the galaxy. In the last paragraph
I shall give a summary of the results and their interpretation.

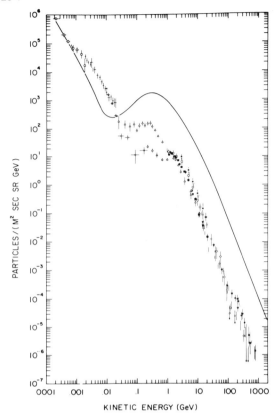

Fig. 1: Summary of electron spectrum measurements. The spread of data in the 0.1 to 1 GeV interval is due to changes in solar modulation. The line represents the proton spectrum for comparison.

II INSTRUMENTATION AND EXPERIMENTAL TECHNIQUES

Reliable measurements of the spectrum of cosmic ray electrons at the low energy end have now been extended to a few hundred KeV. While initial attempts to obtain spectral data in this region had lead to a surprisingly high flux, very recent work by the Californian Institute of Technology Group (Hurford et al. 1973) has shown that the spectrum, without much change in slope, extends below the 5 to 20 MeV region that was observed already several years ago (Simnett and McDonald 1969). Fig. 2 summarises results from this work.

The basic experimental method which led to the spectrum in

this low energy region is also used, in several variants, for observations at higher energies which we will discuss shortly. It consists of the simultaneous measurements of the energy loss dE/dX of the particles, their total energy, E, and/or their range. In this manner it is possible to not only determine the particle's energy but to uniquely identify its type.

At low energies electrons and protons are clearly distinguishable by this method since the energy loss of electrons is much smaller than that of protons. At higher energies, however, other ways of discriminating between the electron and the nuclear components have to be found. Cerenkov detectors and in particular gas Cerenkov detectors, with their low index of refraction play an important role in achieving this discrimination. These counters have a velocity threshold, and discrimination against particles with a Lorentz factor $\gamma = \frac{E}{mc^2}$ below 10 and 20, can readily be

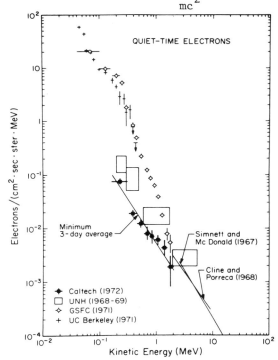

Fig. 2: Measurements of the quiet time electron spectrum below 10 MeV (Hurford et al. 1973).

---

achieved. Added to the energy loss vs. total energy measurement this type of detector has been used for several experiments. For example, it forms the basis of a satellite instrument designed by the University of Chicago for the OGO5 space craft, in which electrons in the energy range from 5-200 MeV were measured (L'Heureux et al. 1972). Similar instruments have been built and flown by other groups notably by the Milano/Saclay co-operation (Bland et al. 1969), the group at Leiden University (Burger and Swanenburg 1973) and the University of Leeds (Marsden et al. 1971).

The method is also used in observations with balloon borne instruments and as an example I show in figure 3 a cross-section of a detector system that has been designed at the University of Chicago. It permits the measurement of electron energies ranging

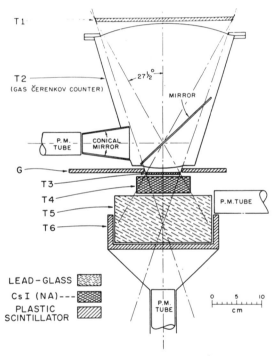

Fig. 3: Detector system for cosmic ray electrons at energies between 10 MeV and 20 GeV (Hovestadt et al. 1970).

from 5 MeV to about 20 GeV. This instrument which is described by Hovestadt et al. uses an active target, a Cesium Iodide crystal in which an electron-photon shower is initiated by the incoming electron. The total energy is measured by the sum of the outputs of this detector (T4), the lead glass Cerenkov counter (T5) and the penetration counter (T6). Through careful calibrations with electrons from accelerators linearity of response and energy resolution are determined. A typical measurement of the electron spectrum at balloon altitude (under 2.5 g/cm² of atmosphere) obtained with this instrument is shown in figure 4 (G. Fulks, 1974).

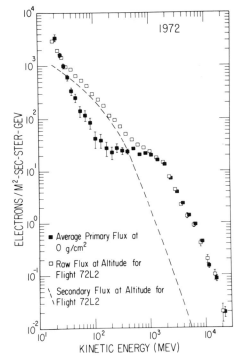

Fig. 4: A typical measurement of the electron spectrum at altitude under 2.5 g/cm² of residual atmosphere (Fulks, 1974).

This figure displays the difficulty of measuring the electron spectrum under even a small amount of residual atmosphere since, particularly at energies between about 50 and 300 MeV, the flux of atmospheric secondary electrons is high and becomes comparable to the flux of the primary particles. Careful measurements of the altitude dependence of the electron spectrum make it possible to empirically determine the energy spectrum of secondary atmospheric electrons, as shown by the dashed curve in figure 4. To obtain the primary electron spectrum the secondary electron spectrum is subtracted from the measured spectrum, and an extrapolation to the top of the atmosphere is carried out which includes corrections for various energy losses.

Similar instruments have been built and flown by other groups, and particularly by those of the Universities of New Hampshire and Leiden. I must apologise that I mainly describe instruments developed at the University of Chicago, using them as examples of particular techniques since I have greatest familiarity with these instruments.

It may at this point be worth while to insert a few remarks on the properties of gas Cerenkov counters and in particular the choice of suitable gases. In focussing Cerenkov counters the detection of Cerenkov light is usually much favoured over any possible contribution by scintillation in the gas since the Cerenkov light moves closely in the direction of the incident particle. One is therefore rarely troubled by the residual gas scintillation, although this effect did deteriorate our data on the OGO5 satellite slightly. Non-focussing gas Cerenkov counters which have no directional response may, however, suffer badly from scintillation of the gas. We have therefore studied the properties of some gases and also used the results of other groups. In Table

PROPERTIES OF CERENKOV COUNTER GASES

Gas	$(n-1) \times 10^5$ NTP	Threshold γ (NTP)	Flammable?	Critical T (°C)	Vapor Pressure at 21°C (atm)	Scintillatim (Freon-12 = 100)
He (10%) Ne (90%)	3.5	89	No	-229	-	>1100
O_2	27.2	42	No*	-119	-	4
N_2	29.2	40	No	-147	-	1100
CO_2	45.0	32	No	31	57	100
Ethylene	69.6	26	Yes	9	-	0
SF_6	79.4	25	No	46	22	200
Freon-12	117	21	No	112	5.8	100
Isobutane	147	19	Yes	135	3.1	1

Table I   * High pressure oxygen is however a fire hazard

I list properties of several gases at normal temperature and pressure in ascending order of their index of refraction. This table shows that there are two highly suitable gases, ethylene and isobutane, both unfortunately flammable. Ethylene, in particular, lends itself to work at relatively high pressures and low temperatures. Both these gases scintillate much less than the commonly used Freon 12 and Sulphur Hexaflouride.

Moving on toward higher energies, and measurements of the electron spectrum above about 50 GeV, I should first mention the nuclear emulsion technique. This technique was used by Daniel and Stephens (1966) at the Tata Institute for their pioneering work in the high energy regime. The major difficulties encountered with the emulsion technique are the time consuming effort required to search for and measure the proper events, and the small exposed area. Since the flux of electrons is only around 1% that of protons and since the energy spectrum is approximately inversely proportional to the third power of the energy, the number of events decreases roughly by a factor of 1000 for each decade in energy.

The obvious advantage of the emulsion technique is the possibility to uniquely identify individual incident particles as electrons. The problem of obtaining sufficient statistical accuracy with nuclear emulsions has lead several experimenters to attempt measurements of the high energy electron spectrum with counter devices. Freier et al. (1974) have tried to use the desirable features of both techniques, and employed an emulsion target to identify the incident particle and its first interaction. Spark chambers are used in their instrument to locate the origin of the resulting particle shower. Counter experiments have most recently been carried out by Silverberg et al. (1973), Meegan and Earl (1973) and by Muller and Meyer (1973). As an example for a counter experiment, in which one attempts to extend the measurement of the electron spectrum up to almost 1000 GeV, I will describe the detector system that has been used by the University of Chicago group (Muller and Meyer 1973). At the high energies it is no longer advisable to attempt containment of the energetic primary particle and its secondaries within the detector. Hence one must analyse the shower development in some detail. The cross section of the Chicago instrument is shown in figure 5. Counters that determine

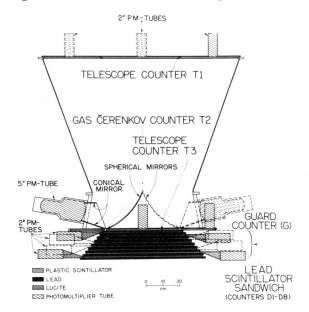

Fig. 5: Counter telescope for measuring the electron spectrum at high energies (Muller and Meyer 1973).

the energy loss of the incident electrons (T1, T3) and a gas Cerenkov counter (T2) are used as in the previously described experiment. The particle then enters a twenty radiation length deep stack of alternating lead plates and scintillation counters to permit a detailed measurement of the shower development. At the very high energies the gas Cerenkov counter is no longer as helpful in the discrimination against protons as it was in the low energy instruments. One therefore measures in addition the time of flight of the particles between counter T1 and counter T3. Through this time of flight measurement about 50% of the triggering events can readily be eliminated as not being due to particles entering the instrument from the top. Careful calibrations of the instrument are made using particle accelerators up to the highest available energies. Figure 6 shows the observed shower

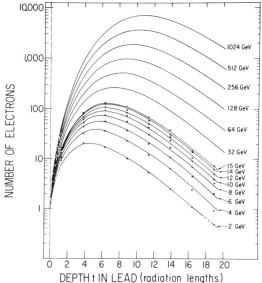

Fig. 6: Average shower profiles for electrons entering the detector along its axis. The data points have been measured at SLAC at energies between 2 and 15 GeV. The solid lines represent fits to the measured data with the primary energy as the only free parameter, therefore providing a method of extrapolation up to energies around 1000 GeV. (Muller and Meyer 1973).

---

profiles between 2 GeV and 15 GeV measured with electrons from the Stanford Linear Accelerator. Semi-empirical fits have been made to these observations and these have then been extrapolated up to energies of about 1000 GeV. Excellent linearity between the sum of the outputs of all shower counters and the energy of the incident electron is obtained over the range of calibration. This is to be expected since in that range most of the shower energy is contained within the detector. The energy resolution turns out to be well below 10% for energies beyond 10 GeV.

The major problem in this work is the discrimination against the very large flux of protons. The shower form and the starting point of the shower are used for this discrimination. Protons interacting in the detector stack may easily masquerade as electrons by producing π^0 mesons which in turn produce an electron-photon shower which is indistinguishable from an electron induced shower. However, an electron shower will almost always start to develop within the first radiation lengths of the detector material, in contrast to cascades that are induced by protons. The interaction length of protons is larger than the depth of the stack and hence protons have almost the same probability of generating a shower anywhere in the stack. One therefore uses a starting point analysis to discriminate, on a statistical basis, between incident electrons and protons. The results from this analysis can best be seen in figure 7 which shows the distribution

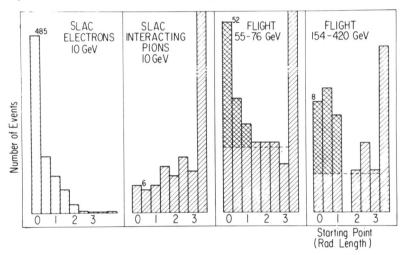

Fig. 7: Starting point distributions as measured for mono-energetic electrons and for interacting pions at SLAC, and for samples of flight data. (Muller and Meyer 1973).

of starting points for 10 GeV electrons and for 10 GeV pions from the SLAC accelerator and for particles observed at balloon altitudes in two energy intervals. One can clearly see the distinct difference in the depth of the interactions between the electrons and the strongly interacting pions as well as the contribution of electron and proton components in the flight data. For each electron event a depth of the starting point was determined which leads to an optimum fit to the shower curves. It is possible in this manner to reliably distinguish between proton induced and electron induced shower events, although only on a statistical basis.

Clearly, it would be highly desirable to devise a method which is free of the statistical analysis and which could, even at the high energies, uniquely distinguish between electrons and protons for each event. This has been the stimulus for attempts toward developing new and different methods for future measurements of the high energy electron spectrum. My colleague, Dr. Muller and his team are therefore exploring the use of transition radiation detectors and their suitability for high energy cosmic ray studies.

There is the hope that by using such detectors the gap between the energy range now covered with balloon and satellite borne instruments and the range now covered with air shower or other ground based instruments will be narrowed or even be closed.

The options for instruments covering energies up to and beyond 10^{12} eV are limited: low pressure gas Cerenkov counters yielding indices of refraction close to 1 have to be very large to provide a tolerable photon yield. Calorimeters and shower detectors require enormous weight because of the needed depth of heavy materials and exposure area.

The transition radiation detector promises the possibility to develop a low mass instrument sensitive to particles with a Lorentz factor, γ, of 10^3 or greater. We plan to use this type of detector, firstly for a study of very energetic electrons, discriminating effectively against all nuclear particles, and secondly to investigate very energetic nuclei.

Here I shall briefly summarise the properties of a TRR detector using the recent work of Cherry et al. (1974a, 1974b). The effect of transition radiation was first predicted by Ginzburg and Franck in 1946. It consists of the generation of photons when an energetic particle traverses the interface between two media of different index of refraction. In physical terms, this production of real photons may be understood as a consequence of matching boundary conditions across the interface for the electromagnetic field which accompanies the charged particle. This radiation is mainly in the X-ray region and is strongly peaked in the direction of motion of the particle.

If ω_1 and ω_2 are the plasma frequencies of the two media and if $\omega_2 < \omega_1$, one finds that the differential frequency spectrum from a single interface has the form:-

$$\frac{dS_o}{d\omega} \begin{cases} \sim \text{const for } \omega < \gamma \omega_2 \\ \sim \log 1/\omega \text{ for } \gamma\omega_2 < \omega < \gamma\omega_1 \\ \sim \omega^{-4} \text{ for } \gamma\omega_1 < \omega \end{cases}$$

$\omega \approx \gamma \omega_1$ is essentially the highest frequency contained in the TRR spectrum i.e. it is a cut-off frequency. This behaviour is shown in figure 8 by the upper dashed curve.

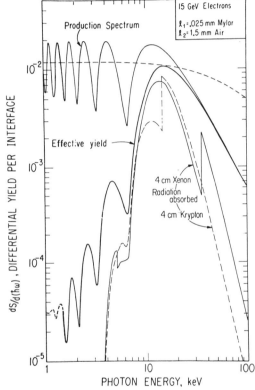

Fig. 8: The X-ray spectrum of transition radiation. Upper dashed curve: from a single radiator. Upper solid curve: from a periodic radiator. Lower solid curve: X-ray yield after absorption in radiator. Curves labelled Xenon and Krypton: radiation absorbed in Xenon and Krypton multiwire chambers respectively. (Cherry et al. 1974b).

The radiation yield is extremely low, ($S_o \approx 10^{-2} \gamma$ eV for a single interface), and several interfaces are required for detection of the X-rays. As soon as several interfaces are used the spectrum becomes complicated due to interference effects and saturation of the radiation yield for high γ sets in. A practical arrangement for a radiator may consist of many foils of a low Z material. To obtain the total yield for this periodic medium one must calculate the superposition in amplitude and phase of the radiation from the individual interfaces. If the gap between foils, ℓ_2, is >Z_2, the formation zone in say air, then the multifoil yield is very closely the sum of the yields from the individual slabs. (The formation zone is physically understood as the distance along the particle trajectory in the particular medium after which the separation between particle and generated photon is of the order of the photon wavelength). In order to get a feel for the numbers in a practical situation, the plasma frequencies of some materials are as follows:-

For Mylar $\quad\hbar\omega_1 = 24.4$ eV
For Air $\quad\hbar\omega_2 = 0.71$ eV
For Polypropylene $\quad\hbar\omega_1 = 20.9$ eV

The most essential part of the spectrum is the last maximum, since the softer radiation is absorbed more readily by the radiator foils themselves (see figure 8, effective yield).

The X-rays may be detected using multiwire chambers filled with either Krypton or Xenon for efficient X-ray absorption in the region of the X-ray spectrum of interest. The effective yield of each radiator, including absorption by the radiation material and the efficiency of the detector is also shown in figure 8 for both gases, using a 4 cm sensitive thickness of Xenon or Krypton at atmospheric pressure. It is clear from this figure that for practical purposes the important part of the X-ray spectrum is in the 10 to 100 KeV range.

The desirable property of the X-ray yield from a single interface, namely, the proportionality between the total radiation yield and γ, is not maintained as one uses multiple foil radiators. The interference leads to saturation, and unfortunately measurements of γ can only be made over a limited energy region. Cherry et al. (1974a, 1974b) have investigated the details of the photon yield as a function of foil thickness, foil spacing and foil number and have described ways on how to optimize a TRR detector for practical purposes.

We may then go back to the initial problem: the measurement of high energy electrons and their spectrum. Since a TRR detector is sensitive to γ (like Cerenkov radiation but at much higher γ) it can clearly be used for discrimination between electrons and protons in some energy intervals. How well can this discrimination be achieved?

Runs on the S.L.A.C. accelerator with electrons and pions, which traverse radiators as well as multiwire chambers, were made to illuminate this situation. The discrimination between these particles becomes very efficient if one utilises the information contained in the distribution of n pulse heights produced by a particle traversing n sets of radiators and chambers. Figure 9 shows pion versus electron interpretation probability in a pictorial way. More quantitatively, the results show that, by using 7 chambers less than 0.1% of the π-meson flux is confused with electrons at a 90% efficiency for electron detection. This is precisely the type of discrimination that is needed in a cosmic ray experiment to clearly distinguish between electrons and protons for each individual particle.

With this information it is not too difficult to design a cosmic ray instrument for measuring the electron spectrum. Due to the small flux of electrons an instrument of large size is required. We are primarily interested in energies from 10^2 GeV to 10^4 GeV i.e. γ from 10^5-10^7. At these energies electrons are in TRR saturation while protons of comparable energy have γ from 10^2 to 10^4 and either produce none or very little TRR. A great simplification can be used, since it has been found (Prince et al. to be published) that some commercial foams are almost as efficient TRR radiators as stretched foils and hence the practical construction of detectors can be very much simplified.

In the cosmic ray work it is necessary to analyse individual wires or wire groups to determine the particle trajectory and reject contributions from background. This type of equipment is presently under construction in Chicago.

As the final paragraph in this chapter I wish to discuss the methods which yield separate spectra for electrons and positrons.

At very low energies, that is in the range from about 200 KeV to 3 MeV where electrons and positrons can be stopped in a detector without the production of many secondaries, the identification of positrons has been attempted through the observation of the gamma rays, which follow the annihilation of the positron. The contributors to this work have been Cline and Hones (1970a) and Cline and Porecca (1970b) and very recent work has been carried out by the Caltech group (Hurford et al. 1973) and has lead to upper limits for the positron flux. It is unfortunate that this detection method is quite limited and it is my opinion that at the present time we have no definitive measurements of finite fluxes of positrons at very low energies (<3 MeV).

At higher electron energies magnetic separation of the electrons and positrons becomes possible and one is able to cleanly measure the two species. The first work in this direction was done by the group at the University of Chicago using a balloon borne magnetic spectrometer (Fanselow et al. 1969) and this work was soon followed by studies of Beuermann et al. (1970)

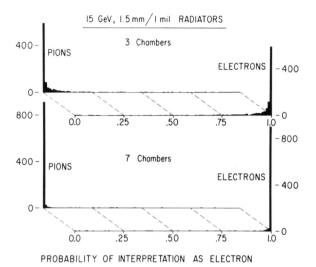

Fig. 9: Discrimination between pions and electrons of 15 GeV by transition radiation detectors. (Cherry et al. 1974a)

and more recently by Daugherty (1974) and by Buffington et al. (1974).

In order to discuss the principles of the method I will again choose the University of Chicago instrument as the prototype. Its cross section is seen in figure 10. The apparatus consists of a counter telescope including a gas Cerenkov counter, which ensures that only particles are accepted which pass through the gap of a 3 x 12 x 12cm permanent magnet. The magnetic field has a strength of about 5.5 kilo gauss, deflecting electrons and positrons in opposite directions. The particle trajectories are determined by a set of 4 spark chambers, 2 above and 2 below the magnet. At the bottom of the instrument a fifth spark chamber with plates of a high Z material is located. This chamber is used to diagnose the traversing particle as being an electron by observing the development of a characteristic electron/photon shower in the plates. This instrument covered an energy range between a few hundred MeV and about 8 GeV. The Caltech experiment observed a lower range of energies, using a very similar method, employing for the first time digitised spark chambers.

An attempt to extend this work to the interesting regime of higher energies has recently been made by the Berkeley group (Buffington et al. 1974). These investigators constructed a detector of much larger geometric factor and produced a considerably stronger field through the use of a superconducting magnet. An

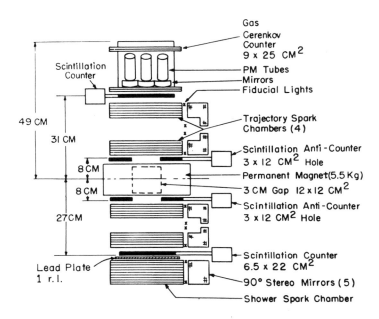

Fig. 10: Detector system for separate measurement of electrons and positrons in the 300 MeV to 8 GeV range. (Fanselow et al. 1969).

interesting new feature was used to improve the difficult discrimination between positrons and protons as one moves to higher energies. They introduced a Bremsstrahlung radiator, as shown in Figure 11. The traversing electron may produce a bremsstrahlung photon so that one simultaneously observes showers produced by the deflected electron and by the photons which, of course, travel along a straight line. This method uniquely identifies an electron or a positron, although one pays for this welcome feature by a reduction in efficiency.

I should conclude by remarking that several years ago attempts were made to use the east-west asymmetry of the particle trajectories in the geomagnetic field to discriminate between electron and positron fluxes (Bland et al. 1966). This method does no longer appear to be competitive with the more recently developed direct methods.

III ELECTRON SOURCES

The sources of electrons must be galactic as the energy loss processes, which will be discussed in the next paragraph, prohibit

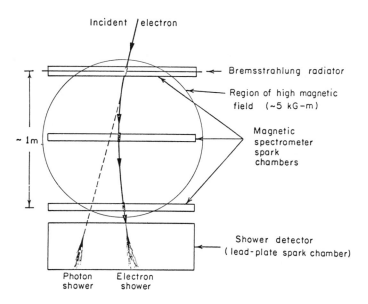

Fig. 11: Schematic diagram illustrating the method used by Buffington et al. (1974) to discriminate between protons and positrons at high energy.

an extragalactic origin.

Just as in the case of the nuclear components, the electron component is made up of particles directly accelerated in sources and of particles which are secondaries resulting from interactions of cosmic ray nuclei. We shall now discuss these separately.

1. Secondary Electron Sources

(a) <u>Knock-on electrons</u>. The spectrum of knock-on electrons was first calculated by Abraham, Brunstein and Cline (1966) who showed it to follow a power law of the form $\sim E^{-2}$. Hurford, Mewaldt, Stone and Vogt (1974) have extended these calculations to energies below a few MeV where ionisation loss plays a major role. These calculations are displayed in fig. 12 together with the spectrum measured by the Caltech group. One sees that the flux can be well accommodated up to 15 MeV if solar modulation does not suppress the electron flux by more than a factor of ~ 3.

(b) <u>Positrons from Radioactive Decay</u>. This possibility was investigated by Ramaty, Stecker and Misra (1970) and by Verma (1969). Table II shows some potential positron emitters

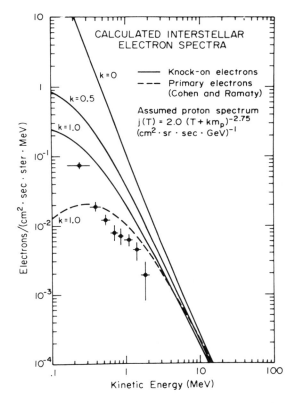

Fig. 12: A Comparison of calculated interstellar electron spectra with the observed quiet time electron spectrum (Hurford et al. 1974).

and their production processes. Calculations of the positron production rate in the galaxy were compared with presently available data and upper limits by the Caltech group (Hurford et al. 1973) (Figure 13). The evidence for a finite flux of positrons at low energies is not on solid experimental ground at the present time. In any case, there appears to be no need to invoke any positron sources beyond the β^+ decay source.

(c) π-μ-e Decay Source. This source becomes increasingly important as one moves towards energies above 10 MeV. Here electrons are produced by collisions of high energy protons with the interstellar gas, leading to the production of π-mesons and subsequent π-μ-e decay of the charged pions. The producing protons have roughly one order of magnitude higher energy than the resulting electrons, and hence calculations of the source spectrum are not handicapped by a lack of knowledge of the proton spectrum due to a poorly known amount of solar modulation. The

POSITRON PRODUCTION MECHANISMS

Emitter and Decay Mode	Maximum Positron Energy (MeV)	Half-life (min)	Production Modes	Production Threshold Energies (MeV/n)
$^{11}C \rightarrow {}^{11}B + \beta^+ + \nu$	0.97	20.5	$^{12}C(p,pn){}^{11}C$	20.2
			$^{14}N(p,2p2n){}^{11}C$	13.1
			$^{14}N(p,\alpha){}^{11}C$	2.9
			$^{16}O(p,3p3n){}^{11}C$	28.6
$^{13}N \rightarrow {}^{13}C + \beta^+ + \nu$	1.19	9.96	$^{14}N(p,pn){}^{13}N$	11.3
			$^{16}O(p,2p2n){}^{13}N$	5.54
$^{14}O \rightarrow {}^{14}N + \beta^+ + \nu$	1.86	1.18	$^{14}N(p,n){}^{14}O$	6.4
$^{15}O \rightarrow {}^{15}N + \beta^+ + \nu$	1.73	2.07	$^{16}O(p,pn){}^{15}O$	16.54

Table II

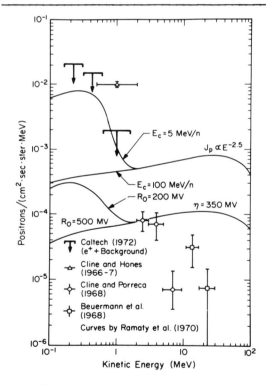

Fig. 13: Low energy positron measurements compared to the theoretical calculations of Ramaty et al. (1970) (Hurford et al. 1973).

requirements for calculating the electron source spectrum are a knowledge of (a) The proton flux and energy spectrum, (b) The production cross-section for pions, (c) the density of the interstellar gas.

Calculations of the π-μ-e electron spectrum have progressively improved with time (partly because of a better knowledge of the high energy cross-sections). Early work was carried out by Ramaty and Lingenfelter (1966) and by Perola, Scarsi and Sironi (1967), and a computer analysis was recently made by Ramaty (1974). I shall not go into any details of these calculations, but rather present some results of the various steps. The detailed work involves kinematics, transforming the particle spectra into the frame of the galaxy, etc. First the π-meson spectrum is calculated taking into account angular distributions, and the electron spectrum that finally results after π-μ-e decay is seen in figure 14. In this figure positron and electron

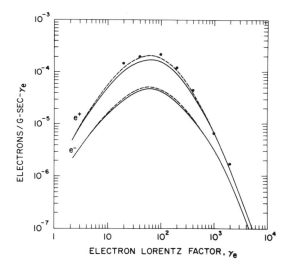

Fig. 14: Electron and positron production spectra in interstellar space, originating in the π-μ-e source (Ramaty 1974).

spectra are shown separately, and it should be noted that at all energies the fraction of positrons

$$\frac{e^+}{e^+ + e^-}$$ is $\geqslant 0.5$. Hence the observation of

approximately 90% of negatrons at E > 500 MeV constituted the proof for a major primary source contribution to the electron flux.

2. Primary Electron Sources

Here one enters a much more speculative region. It is known that many galactic as well as extra galactic objects contain highly relativistic electrons but little is understood about the details of the acceleration process. Dr. Pacini and Dr. Colgate will further discuss this matter and I shall not attempt to say much on this subject.

I may just add two things. It has been proposed that low energy positrons may be copiously produced in some cosmic ray sources. Colgate pointed out a few years ago that if Silicon burning shells in supernovae produced enough of the chain $Ni^{56} \to Co^{56} \to Fe^{52}$ one may have a substantial positron yield. Sturrock considers the possibility that high energy γ's originating in a pulsar's magnetosphere produce electron-positron pairs in the high magnetic field, eventually yielding e^+ and e^- in the energy region of about 1 MeV.

The second remark concerns the electron proton ratio of about 10^{-2} that is observed above a few hundred MeV and that has not found an obvious explanation. Consider two very simpleminded alternatives. If both protons and electrons were accelerated by a shock wave from a supernova explosion, they should end up with about the same velocity. Therefore, at the same energy, if they have a power law spectrum, their intensity should be in the ratio $\{m_e/m_p\}^{\gamma-1}$, where γ is the power law index. For $\gamma = 2.8$ this yields $e/p \approx 10^{-6}$ which is much lower than that observed.

In pulsar models, usually the same energy is imparted to proton and electron, hence $e/p \approx 1$. This is much larger than the observed ratio. The causes for the observed electron to proton ratio are obviously much more complex.

IV ENERGY LOSS PROCESSES

The energy loss processes bring out the most important distinction in the behaviour of electrons and of nuclei: the former interact with photons and magnetic fields in the galaxy, the latter with interstellar matter.

1. Synchrotron Radiation

The interaction of relativistic electrons with interstellar magnetic fields leads to the emission of synchrotron radiation, sometimes called magnetic bremsstrahlung. As the electrons spiral around the lines of force, their gyrofrequency ω_G is given by $\frac{eB}{mc}\frac{1}{\gamma}$. The synchrotron radiation is emitted in a narrow beam around the instantaneous velocity vector and has an opening angle given by $<\theta> \approx 1/\gamma$. For example a 1 GeV electron has

$$<\theta> = \frac{5 \times 10^5}{10^9} \approx 5 \times 10^{-4} \approx 1 \text{ minute of arc.}$$

The frequency for a monoenergetic electron has the form shown in the figure with $\omega_m \approx \frac{e\beta}{mc} \gamma^2$

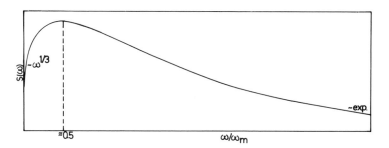

Fig. 15: Synchrotron radiation: the frequency spectrum for a monoenergetic electron.

In interstellar space $B \approx 5 \times 10^{-6}$ Gauss and hence for a 1 GeV electron $\omega_m \approx 400$ MHz i.e. ω_m lies in the radio range. The total power radiated by an electron is given by

$$S(E) = \frac{-dE}{dt} = \frac{2c}{3}\left(\frac{e^2}{mc^2}\right)^2 B_\perp^2 \left(\frac{E}{mc^2}\right)^2 \approx 4 \times 10^{-6} B_\perp^2 \text{ GeV sec}^{-1}$$

where $B_\perp$ is measured in gauss.

The connection with the radio spectrum may be made in the following way. If $S(\nu)$ is the power radiated in **frequency** per electron of energy E then, if there is an isotropic distribution of electrons, the observed intensity is given by

$$I(\nu) = \frac{1}{4\pi} \int_0^L \int_0^\infty S(\nu) N(E,r) \, dE \, dr$$

where $N(E,r)$ is the density of electrons as a function of E and This leads to a power law in frequency if the electron spectrum has the form of a power law $N(E)dE = N_0 E^{-\gamma} dE$

i.e. $I(\nu) \sim \nu^{-\frac{\gamma-1}{2}} = \nu^{-\alpha}$ cf. a power law spectrum in frequency.

2. Inverse Compton Scattering

This is a bit of a misnomer as the interesting interaction is Thomson scattering i.e. the interaction of light on free electrons. In the Thomson approximation the electron loses little energy in each collision. There is an upper energy limit for the electrons (depending on the photon energy) in the frame of the galaxy for which the following holds:-

$$\frac{dE}{dt} = \frac{4}{3} c \, \sigma_{Th} \, \rho_{ph} \left(\frac{E}{mc^2}\right)^2 \approx 10^{-16} \, \rho_{ph} \, E^2 \, (\text{GeV sec}^{-1}),$$

where $\sigma_{Th} \approx 6.6 \times 10^{-25}$ cm^2 is the Thomson cross-section and ρ_{ph} the energy density of photons (in eV/cm^3). If $\bar{\varepsilon}$ is the average energy of the photons in the frame of the galaxy then this energy loss expression is valid up to

$$E \leqslant E_{tr} = \frac{(mc^2)^2}{\bar{\varepsilon}}$$

For higher energies one must use the Klein-Nishima cross section.

For the case of cosmic ray electrons we deal with two photon fields: the photons from starlight, $\rho_{pn} \approx 0.5$ eV/cm^3, $\bar{\varepsilon} \approx 3$eV and those from the universal blackbody radiation $\rho_{ph} \approx 0.25$eV/cm^3, $\bar{\varepsilon} \approx 10^{-3}$eV. These give respectively values of 80 GeV and 2×10^5 GeV for the transition energy E_{tr} corresponding to γ's of 10^5 and 10^8.

3. Ionisation Loss

For relativistic electrons the ionisation loss is almost constant, showing only the relativistic increase of the Bethe-Bloch expression:-

$$\frac{dE}{dt_i} \approx 7.6 \times 10^{-18} \, n_H \, (3 \log E/mc^2 + 18.8) \text{ GeV/sec}.$$

This is the expression for neutral hydrogen, the energy loss becomes somewhat larger if He is present and larger by a factor of 3 or 4 in ionised Hydrogen. n_H is the number density of hydrogen per cm^3.

4. Bremsstrahlung

This may, in a good approximation be considered a continuous energy loss, proportional to the electron energy. In atomic hydrogen

$$\frac{dE}{dt}\bigg|_B \approx 8 \times 10^{-16} \, n_H \, E \text{ GeV sec}^{-1}$$

This is slightly modified in the case of an ionised medium.

These various energy loss rates of electrons in the galaxy are displayed as a function of energy in figure 16.

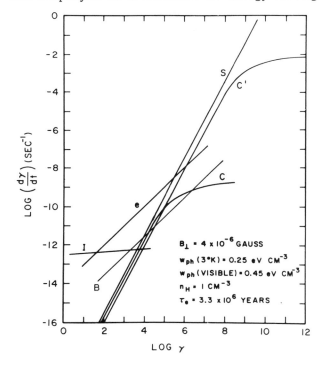

Fig. 16: Energy loss processes for electrons in the galaxy. S - synchrotron loss, C - Compton loss, B - bremsstrahlung, I - ionisation, e - escape.

V PROPAGATION IN THE INTERSTELLAR MEDIUM

With these facts in mind, I now wish briefly to discuss the

question of particle propagation and transport in the galaxy. This will then readily lead us to a summary of the present status of the field, to the solved and unsolved problems and to the ways in which one may tackle the remaining questions.

Particle propagation in the galaxy is certainly a very comlex problem which is not understood in detail and which has in the past been attacked by rather simple-minded models. The usual starting point is already a tremendous simplification: although one knows that the particles are tied to the interstellar magnetic lines of force one usually sets up a transport equation in terms of a random spatial motion - an isotropic diffusion-energy loss equation of the form:-

$$\frac{\partial n}{\partial t} - \vec{\nabla} \cdot (D \vec{\nabla} n) + \frac{\partial}{\partial E} (\dot{E} n) + \frac{n}{T} = Q (\vec{r}, t, E)$$

where $n(\vec{r}, t, E)$ is the particle density per unit energy interval as a function of time and position, D the diffusion coefficient, $\dot{E}$ the rate of energy loss or gain, T the lifetime against catastrophic loss and $Q(\vec{r}, t, E)$ the source per unit volume and unit energy interval at $\vec{r}$ and time t. In general, solutions can only be obtained after making drastic simplifying assumptions.

For the subject of our discussion, the propagation of electrons in the galaxy, Ramaty (1974) has numerically carried through this formulism with some further assumptions and I shall later present some of his results. In order to demonstrate the main points, I shall use a much simplified approach to obtain qualitative results which one may compare with observations and the more detailed calculations.

For electrons no catastrophic losses occur and hence one may assume $T \to \infty$. I shall further replace the diffusion term by $\frac{n}{T_e}$ postulating an energy independent exponential escape probability and I shall assume that the source is time and space independent. This leads to the simple expression

$$q(E) - \frac{n}{T_e} - \frac{\partial}{\partial E} (\dot{E} n) = 0$$

Since for a steady state situation

$$\frac{\partial n}{\partial t} = 0$$

Assuming a power law spectrum for the source, $q(E) = q_o E^{-\Gamma}$, and knowing the energy loss rate to be of the form

$$\dot{E} = \frac{dE}{dt} = (4 \times 10^{-6} \, B_\perp^2 + 10^{-16} \, \rho_{ph}) \, E^2 = bE^2$$
$$\text{Gauss} \quad \text{eV/cm}^3$$

on obtains solutions for the equilibrium electron spectrum of the form

$$n(E) = T_e \, q_o \, E^{-\Gamma} \quad \text{for } E \ll E_{cr} = \frac{1}{bT_e(\Gamma - 1)}$$

$$\simeq \frac{q_o \, E^{-(\Gamma-1)}}{b \, (\Gamma - 1)} \quad \text{for } E \gg E_{cr}$$

and since $b \simeq 10^{-16}$ (GeV sec)$^{-1}$ one obtains numerically $E_{cr} \approx 200$ GeV. On the basis of this simple model, one expects the electron spectrum to steepen in the vicinity of 200 GeV.

VI SUMMARY OF THE RESULTS AND THEIR IMPLICATIONS

We now turn to the experimental results, compare them with expectation, and discuss their implications.

At very low energy, 0.1 to 20 MeV we seem to observe an electron energy spectrum and flux that may be entirely due to the knock-on source, under the proviso, however, that solar modulation is relatively small. The evidence for the presence of positrons below 3 MeV is not well established at the present time, and further experiments are needed. The presently known upper limits of the positron flux do not require any peculiar sources. One should keep in mind that the very recent observations of energetic electrons that appear to be emitted by the planet Jupiter (Teegarden et al. 1974) may contribute to the low energy flux of electrons at the earth and that the spectrum may in part be of planetary origin. The planetary particles are expected to consist exclusively of negatrons.

Between about 50 MeV and several hundred MeV, the π-μ-e source substantially contributes to the electron flux, but at higher energies additional sources of negatrons become the major contributors. This is demonstrated in figure 17 which summarises the present evidence for the fraction of positrons in the electron flux as a function of energy, and covers the work from 10 MeV to about 30 GeV. One notes that the positron fraction is significantly below expectations for a pure π-μ-e source at all energies. This constitutes unambiguous evidence that the electrons are not entirely of secondary origin, but that sources of electrons contribute significantly to this component. The low values of the positron fraction show that sources must be responsible for about 80% of the electron component above about 1 GeV.

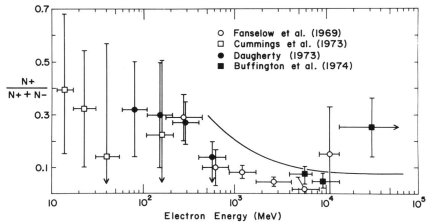

Fig. 17: The fraction of positrons in the cosmic radiation as a function of energy as measured by several observers.

It is interesting to ask, how much one can learn about the electron spectrum from observations of the non thermal radio emission. In spite of considerable uncertainties, one may draw some conclusions, keeping in mind, however, that the cosmic ray observations sample the local interstellar electron spectrum, while the radio data provide an average over major portions of the galaxy.

The range for which the galactic radiospectrum is accessible is limited – on the low frequency end by interstellar absorption, on the high frequency end by the rise of the Rayleigh-Jeans portion of the universal black-body radiation. The observable range extends from about 1 MHz to 1000 MHz corresponding to 100 MeV and 7 GeV electrons respectively. There is considerable ambiguity in translating the radio spectrum into an electron spectrum as was pointed out most clearly by the Cal. Tech. group in a recent paper (Cummings et al. 1973). Much of the uncertainty is due to a lack of knowledge of the relevant astrophysical parameters. If one fits the radio data with a high, nominal and low spectrum model which are all compatible with observations, one obtains possible interstellar electron spectra that are displayed in figure 18 and compared to the measured spectrum near earth. One may, however, confirm one of the predictions of the radio observations: all measurements of the radio spectrum insist on a flattening of the electron spectrum at energies below about 2 GeV.

There seems to be a model independent confirmation of this fact,

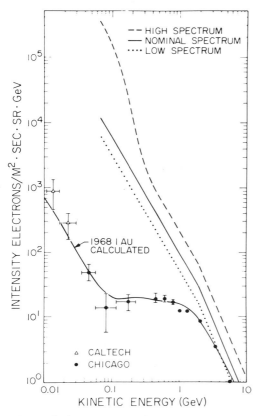

Fig. 18: Limits on the interstellar electron spectrum from radio observations and the measured and calculated spectrum at 1.A.U. (Cummings et al. 1973).

from the electron-positron observations. The increase of the positron-fraction with decreasing energy (see Fig. 17) implies that the total interstellar electron spectrum in the 0.1 to 1 GeV range is flatter than the interstellar positron spectrum. The latter can be calculated from π^+-decay, and such calculations show that the spectral index of the positron spectrum from 0.2 to 1 GeV is $\simeq$ 2.0, and certainly less than the index of the proton spectrum of 2.6. Therefore below ~2 GeV the electron spectrum must have an index appreciably smaller than 2.8, the value measured at higher energies. The change in slope of the electron spectrum at a few GeV is therefore not entirely due to solar modulation. At least part of it is intrinsic to the interstellar electron spectrum.

Let me finally turn to the problem of the high energy spectrum of the electrons and the question of the cosmic ray confinement time. We use for this discussion the most recent measurements of the high energy portion of the spectrum that are shown in figure 19. Our own results can be readily fitted by a

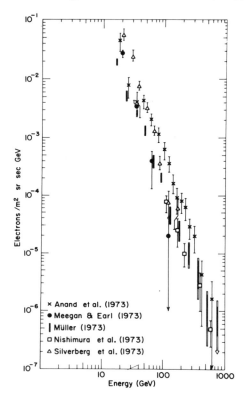

Fig. 19: Recent measurements of the high energy portion of the electron spectrum.

---

simple power law spectrum up to almost 1000 GeV, but the uncertainty is large and we cannot claim that a steepening of the spectrum around

$$E_{cr} = \frac{1}{b\, T_e (\Gamma - 1)}$$

~200 GeV could not be compatible with the data. We would also

claim that none of the other observations that today are available
can be used to make a more decisive statement.

None of the measurements indicate a definite steepening of
the spectrum or exclude it. Should one be surprised if no steepening were observed? This would imply that one is dealing with
particles that are younger than the customary 10^6 years deduced
from low energy nuclear composition measurements. Two obvious
reasons could cause this behaviour: either one observes
particles from nearby sources, or the energetic particles have a
smaller containment time than those of low energy.

Table III shows a list of supernova remnants compiled by
Milne (1970) at distances of less than 1 Kpc from the solar system

Distances and Ages of Supernova Remnants

Galactic Source Number	Distance (kpc)	Age (10^4 yr)	Name
G 41.9 - 4.1	0.7	3.2	CTB 72
G 74.0 - 8.6	0.6	3.5	Cygnus Loop
G 89.1 + 4.7	0.8	2.3	HB 21
G117.3 + 0.1	0.9	4.7	CTB 1
G156.4 - 1.2	0.6	3.2	CTB 13
G160.5 + 2.8	0.8	2.7	HB 9
G180.0 - 1.7	0.7	4.3	S149
G205.5 + 0.2	0.6	4.6	Monoceros
G263.4 - 3.0	0.4	1.1	Vela X
G330.0 + 15.0	0.4	3.8	Lupus Loop

Table III

which might possibly have injected energetic electrons into the
interstellar medium. The other alternative, a decreasing containment time, with increasing energy, was introduced through
studies of the nuclear composition at high energies. Measurements
of the ratio of galactic secondary nuclei to source nuclei show a
marked decline with increasing energy (see figure 20). This
indicates that the total amount of matter traversed by the source
nuclei is smaller at high than at low energies. Among the
possible explanations for this effect is a decreased containment
time in the galaxy for the cosmic ray particles at high energy.

It can be seen from figure 20 that a plot of the ratio of
secondary nuclei over primary nuclei as a function of energy can
be fitted with a power law in total energy. If this is interpreted in terms of an energy dependent leakage path length
$\lambda(E) \sim E^{-\delta}$, the containment time, which is related to this path
length via $\lambda = \rho \beta c \tau_e$, is then also of the form of a power law

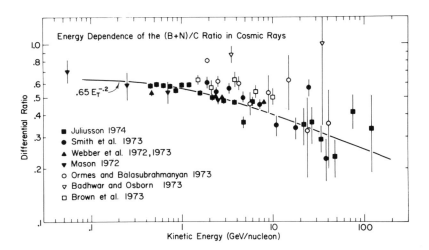

Fig. 20: The ratio of the secondary nuclei, B and N, to the source nucleus, C, as a function of energy. (Juliusson 1974)

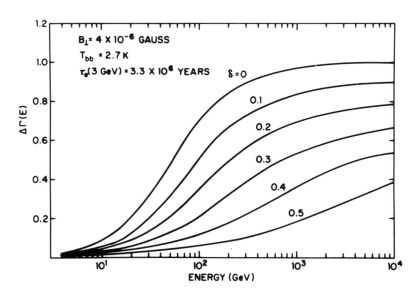

Fig. 21: The change of power law exponent of the electron spectrum as a function of energy under the assumption of an energy dependent containment time.

$$\tau_e = \tau_o \left(\frac{E}{E_o}\right)^{-\delta}$$

The experiments show that δ is approximately 0.3. Juliusson Meyer and Muller (1974) put this energy dependence of τ_e into the simple transport equation and obtain solutions of the form:-

$$n(E) = \tau_o q_o E^{-(\Gamma - \delta)} \quad E \ll E_{cr}$$

$$= \frac{q_o E^{-(\Gamma + 1)}}{b(\Gamma - 1)} \quad E \gg E_{cr}$$

where

$$E_{cr} = \left(\frac{1}{b_o(\Gamma - 1)}\right)^{1/1-\delta}$$

hence giving a value of E_{cr} larger than 1000 GeV. More quantitative calculations were made by Silverberg and Ramaty (1973b) who calculate the energy dependent spectral index

$$\Gamma(E) = -\frac{d \log n}{d \log E}$$

and give quantitatively the increment in spectral index $\Delta\Gamma(E)$ as a function of energy, for various values of δ. This is shown in figure 21. One may conclude that if the interpretation of the nuclear data as an energy dependent containment time is correct, then one need not be surprised to find no change in the slope of the electron spectrum within the energy range that has so far been explored. In this case the source spectrum of electrons is flatter than the measured spectrum at low energy and the total change in spectral index will be smaller than 1. It is obvious that a more detailed measurement of the energy spectrum at very high energies is most desirable to gain answers to those important astrophysical questions.

I wish to express my thanks to Messrs. D.K. French and D. Dodds of the University of Durham who have written most of this manuscript on the basis of my lecture notes and transcripts.

REFERENCES

Abraham, P.B., K.A. Brunstein and Cline, T.L. Phys. Rev. 150, 1088, 1966.

Beuermann, K.P., Rice, G.J., Stone, E.C. and Vogt, R.E., Acta. Physica Academiae Scientiarum Hungaricae, suppl. 1, 173, 1970.

Bland, C.J., Boella, G., Degli Antoni, G., Dilworth, C., Scarsi, L., Sironi, G., Agrinier, B., Koechlin, Y., Parlier, B. and Vasseur, J., Phys. Rev. Letts., 17, 813, 1966.

Bland, C.J., Degli Antoni, G., Dilworth, C., Maccagni, D., Tanzi, E.G., Koechlin, Y., Raviart, A. and Treguer, L., Acta. Physica Academiae Scientiarium Hungaricae, suppl. 1, 195, 1969.

Buffington, A., Orth, C.D. and Smoot, G.F., Phys. Rev. Letts., 33, 34, 1974.

Burger, J.J. and Swanenburg, B.N., J. Geophys. Res. 78, 292, 1973.

Cherry, M.L., Muller, D. and Prince, T.A., Nuc. Inst. and Meth., 115, 141, 1974a.

Cherry, M.L., Hartmann, G., Muller, D. and Prince, T.A. To be published Phys. Rev., 1974b.

Cline, T.L. and Hones, E.W., Jr. Acta Physica Academiae Scientiarum Hungaricae, suppl. 1, 159, 1970a.

Cline, T.L. and Porreca, G., Acta. Physica Academiae Scientiarum Hungaricae, suppl. 1, 145, 1970b.

Cummings, A.C., Stone, E.C. and Vogt, R.E., Proc. 13th Int. Conf. on Cosmic Rays, Denver, 1, 335, 1973.

Daniel, R.R. and Stephens, S.A., Phys. Rev. Letts., 17, 935, 1966.

Daugherty, J.K., Goddard Space Flight Centre preprint, X-660-74-16, 1974.

Fanselow, S.L., Hartmann, R.C., Hildebrand, R.H. and Meyer, P., Ap. J., 158, 771, 1969.

Freier, P.S., Gilman, G. and Waddington, C.J., Bull. Am. Phys. Soc., 19, 581, 1974.

Fulks, G. To be published J. Geophys. Res., 1974.

Hovestadt, D., Meyer, P. and Schmidt, P.J., Nuc. Inst. and Meth., 85, 93, 1970.

Hurford, G.J., Mewaldt, R.A., Stone, E.C. and Vogt, R.E., Proc. 13th Int. Conf. on Cosmic Rays, Denver, 1, 324, 1973. Also 1, 330, 1973.

Hurford, G.J., Mewaldt, R.A., Stone, E.C. and Vogt, R.E. To be published Ap. J. 1974.

Juliusson, E., Meyer, P. and Muller, D., Proc. 13th Int. Conf. on Cosmic Rays, Denver, 1, 373, 1973.

Juliusson, E., Ap. J., 191, 331, 1974.

L'Heureux, J., Fan, C.Y. and Meyer, P., Ap. J., 171, 363, 1972.

Marsden, P.L., Jakeways, R. and Calder, I.R., Proc. 12th Int. Conf. on Cosmic Rays, Hobart, 1, 110, 1971.

Meegan, C.A. and Earl, J.A., Proc. 13th Int. Conf. on Cosmic Rays, Denver, 5, 3067, 1973.

Milne, D.K., Austr. J. Phys. 23, 425, 1970.

Muller, D. and Meyer, P., Ap. J., 186, 841, 1973.

Perola, G.C., Scarsi, L. and Sironi, G., preprint parts I and II, 1967.

Ramaty, R. and Lingenfelter, R.E., J. Geophys. Res., 71, 3687, 1966.

Ramaty, R., Stecker, F.W. and Misra, D., Acta. Physica Academiae Scientiarum Hungaricae, suppl. 1, 165, 1970.

Ramaty, R. and Lingenfelter, R.E., Goddard Space Flight Centre preprint X-660-73-14, 1973.

Ramaty, R., "High energy particles and quanta in astrophysics". F.B. McDonald and C.E. Fichtel, Ed. M.I.T. Press, Ch. III, 1974.

Schmidt, P.J., J. Geophys. Res. 77, 3295, 1972.

Silverberg, R.F., Ormes, J.F. and Balasubrahmanyan, V.K., Proc. 13th Int. Conf. on Cosmic Rays, Denver, 1, 347, 1973a.

Silverberg, R.F. and Ramaty, R., Nature Phys. Sci., 243, 134, 1973b.

Simnett, G.M. and McDonald, F.B., Ap. J. 157, 1435, 1969.

Teegarden, B.J., McDonald, F.B., Trainor, J.H., Webber, W.R. and Roelof, E.C., Goddard Space Flight Centre, preprint X-660-74-197, 1974.

Verma, S.D., Ap. J. 156, 479, 1969.

GAMMA RAY ASTROPHYSICS

F.W. Stecker

Theoretical Studies Group
NASA Goddard Space Flight Center
Greenbelt, Maryland 20771

I   INTRODUCTION

In the context of an advanced study institute on the origin of cosmic rays, one can look at the subject of theoretical gamma-ray astrophysics from two points of view. The first seeks to answer the question, "What is the origin of the observed cosmic gamma radiation?", considering this radiation as a component of cosmic radiation. The second seeks to answer the question, "What does the observed cosmic gamma radiation tell us about the origin of cosmic rays?" with the term "cosmic rays" meant to be the primary nuclear component of cosmic radiation (or perhaps the electronic component) and the gamma radiation considered to be a secondary product of various interactions between primary radiation and fields, photons and nuclei in the cosmos. This latter question is the historically older question, but it is inherently a much more difficult one to answer and is one which cannot be unambiguously answered at this point in time, given our relatively primitive observations and the context of our present theoretical understanding. Thus, of necessity, we will spend most of our time here discussing the origin of the gamma radiation itself, bearing in mind the relation of this problem with that of the origin of the nuclear component of cosmic radiation.

The possible existence of a secondary component of cosmic gamma radiation first appeared in the literature almost incidental to problems bearing on the question of the origin of the primary cosmic radiation. After it had been determined that the overwhelming component of primary cosmic radiation impinging on the upper atmosphere consists of high-energy protons, Feenberg and Primakoff (1947) addressed themselves to an explanation of the conspicuous absence of electrons in significant quantities in the primary radiation. To do this, they examined the various interactions which cosmic ray electrons and protons could be expected to undergo with low-energy starlight photons in interstellar space and found that electrons could be effectively depleted of their energy by Compton interactions in a fraction of the age of the universe. In this process, the energy of the electrons is transferred to the photons which can be boosted to X-ray or gamma ray energy. Similarly, Hayakawa (1952), in examining the propagation of cosmic radiation through interstellar space, pointed out the effect of meson-producing nuclear interactions between cosmic rays and interstellar gas. Hayakawa noted that the neutral pions produced would decay to produce cosmic gamma radiation. The production of bremsstrahlung radiation by cosmic rays was discussed by Hutchinson (1952). However, the idea of establishing a science of gamma ray astronomy itself as a tool for answering questions in high-energy astrophysics and cosmology appears to have been stimulated in an important article by Morrison (1958). Morrison (1958) and Felten and Morrison (1963) pointed out that the processes of most significance for producing cosmic gamma rays were (1) electron bremsstrahlung, (2) electron-photon interactions (Compton effect), (3) cosmic-ray produced pion decay, (4) synchrotron radiation, (5) annihilation produced pion decay, and (6) line emission from electron-positron annihilation and nuclear deexcitation.

We will first review here these various processes for producing cosmic gamma radiation, as well as the significant processes for absorption of cosmic gamma radiation, discussing the basic physics of these processes. We will then attempt to place these processes in their astrophysical context in the galaxy and the universe as a whole. We will then turn to the interpretation of the present data on cosmic gamma radiation and its implications for cosmology and cosmic ray origin.

II GAMMA RAY PRODUCTION PROCESSES

1. Compton Scattering Between Cosmic-Ray Electrons and Starlight Microwave Background Photons

Compton scattering is the relativistic limit of the electron-photon scattering process which is referred to as Thomson scattering in the non-relativistic case. The traditional Compton effect is one in which a γ-ray scatters off an electron at rest in the laboratory. In the case of astrophysical interest, the electron is a cosmic-ray of energy $E = \gamma mc^2$ and velocity $v = \beta c$. A photon of initial energy ϵ then scatters off the electron, coming out of the interaction with an energy ε'. Denoting quantities in the electron rest system by asterisks and denoting the scattering angle of the photon by $\theta*$, the initial and final energy of the photon in the electron rest system are related by

$$\varepsilon'* = \frac{\varepsilon*}{1 + \frac{\varepsilon*}{mc^2}(1 - \cos\theta*)} \qquad (2.1)$$

Denoting the angle between the electron and the photon by α, it then follows that

$$\varepsilon* = \gamma\varepsilon(1 + \beta\cos\alpha)$$

$$\varepsilon' = \gamma\varepsilon*'(1 - \beta\cos\alpha*')$$

and

$$\tan\alpha* = \frac{\sin\alpha}{\gamma(\cos\alpha+\beta)} \qquad (2.2)$$

so that

$$\varepsilon' = \frac{\gamma^2\varepsilon(1 + \beta\cos\alpha)(1 - \beta\cos\alpha*')}{1 + \frac{\gamma\varepsilon}{mc^2}(1 + \beta\cos\alpha)(1 - \cos\theta*)} \qquad (2.3)$$

and, on the average, the final photon energy in the observers' frame is

$$\varepsilon' \sim \gamma^2\varepsilon \qquad (\gamma\varepsilon \ll mc^2) \qquad (2.4)$$

For the astrophysical conditions we will primarily be concerned with here, the condition $\gamma\varepsilon \ll mc^2$ holds and the cross section for the scattering is

$$\sigma_c \rightarrow \sigma_T = \frac{8}{3}\pi\left(\frac{e^2}{mc^2}\right)^2 = 6.65 \times 10^{-25} \text{cm}^2 \qquad (2.5)$$

(Heitler 1960).

In the other extreme, the Klein-Nishina form of the cross section holds, which has the asymptotic form.

$$\sigma_c \rightarrow \pi\left(\frac{e^2}{mc^2}\right)^2 \left[\frac{mc^2}{\varepsilon}\right]\left[\frac{1}{2} + \ln\left(\frac{2\varepsilon}{mc^2}\right)\right] \qquad (\gamma\varepsilon \gg mc^2) \qquad (2.6)$$

Also in this extreme

$$\varepsilon' \sim \gamma mc^2 \sim E \quad (2.7)$$

However in most cases, i.e. whenever $\varepsilon E_\gamma \ll (mc^2)^2$, the energy of the γ-ray produced is on the average

$$\langle E_\gamma \rangle = \frac{4}{3} \gamma^2 \langle \varepsilon \rangle \quad (2.8)$$

In the important case of astrophysical interest where the cosmic ray electrons are assumed to have an energy distribution of power-law form

$$I_e(E) = KE^{-\Gamma} \quad (2.9)$$

if we make the delta-function approximation for the differential production function

$$\sigma(E_\gamma | \varepsilon, E) \simeq \sigma_T \, \delta \left[E_\gamma - \frac{4}{3} \langle \varepsilon \rangle \gamma^2 \right] \quad (2.10)$$

the γ-ray production rate as a function of energy is given by

$$q(E_\gamma) = 4\pi n_{ph} \sigma_T \int dE \, KE^{-\Gamma} \, \delta \, (E_\gamma - \frac{4}{3} \langle \varepsilon \rangle \gamma^2)$$

$$= 4\pi n_{ph} K \left[\left(\frac{3}{4} \right)^{\frac{1}{2}} E_\gamma^{-\frac{1}{2}} \langle \varepsilon \rangle^{-\frac{1}{2}} mc^2 \right]^{-\Gamma} \frac{dE}{dE_\gamma}$$

$$= 2\pi n_{ph} \, \sigma_T K (mc^2)^{1-\Gamma} \left(\frac{4}{3} \langle \varepsilon \rangle \right)^{\frac{\Gamma-1}{2}} E_\gamma^{-\frac{\Gamma+1}{2}} \quad (2.11)$$

$$= \frac{8\pi}{3} \sigma_T \rho_{ph} (mc^2)^{1-\Gamma} \left(\frac{4}{3} \langle \varepsilon \rangle \right)^{\frac{\Gamma-3}{2}} KE_\gamma^{-\frac{\Gamma+1}{2}}$$

where n_{ph} is the number density of target photons of low energy in the interstellar or intergalactic medium (usually in the form of starlight photons of ~ 1 eV energy or universal microwave blackbody photons of energy $\sim 0.6 \times 10^{-3}$ eV) and $\rho_{ph} = n_{ph} \langle \varepsilon \rangle$ is the energy density of the radiation.

In particular, for blackbody radiation of temperature T

$$<\tfrac{4}{3}\varepsilon> \simeq 3.6\ kT \simeq 3.1 \times 10^{-10}\ T\ \text{MeV}$$

and
$$\rho_{ph} = 4.75 \times 10^{-9} T^4\ \text{MeV/cm}^3 \qquad (2.12)$$

2. Synchrotron Radiation

Synchrotron radiation, or magnetic bremsstrahlung, is the radiation emitted by a relativistic particle spiraling in a magnetic field. Its mathematical description has been given by Schwinger (1949). An electron suffers energy losses by synchrotron radiation at a rate

$$\left.\frac{dE}{dt}\right)_{\text{synch}} = -\frac{4}{3}\sigma_T c \gamma^2 \rho_H \qquad (2.13)$$

where the magnetic energy density ρ_H is equal to $H^2/8\pi$. This rate can be compared with the energy loss suffered by an electron through Compton interactions with a photon field of energy density ρ_{ph}. That rate is

$$\left.\frac{dE}{dt}\right)_c = -\frac{4}{3}\sigma_T c \rho_{ph} \qquad (2.14)$$

The equivalence of equations (2.13) and (2.14) can be shown to be a direct consequence of electromagnetic theory (Jones 1965). The photons emitted as synchrotron radiation have a characteristic frequency given by

$$\omega_c = \frac{3}{2}\gamma^2 \left(\frac{eH_\perp}{mc}\right) \qquad (2.15)$$

The resultant γ-rays have a characteristic energy from equation (2.15)

$$E_{\gamma,c} = \frac{3}{2}\gamma^2 \left(\hbar/mc\right) eH_\perp \qquad (2.16)$$

Because this is the same type of energy dependence as that for Compton radiation, a power-law cosmic-ray electron spectrum of the form $KE^{-\Gamma}$ will again generate a synchrotron radiation spectrum of the form

$$q(E_\gamma) \propto E_\gamma^{-\frac{\Gamma+1}{2}} \tag{2.17}$$

The above equations can be used to find some useful numerical relations involving γ-ray production by synchrotron radiating electrons, with the synchrotron radiation being determined from radio observations (e.g., in the 10-1000 MHz range) and the γ-radiation being produced from Compton interactions of the same electrons responsible for the synchrotron radio emission. Typical galactic magnetic fields are of the order of a few μG. From equation (2.15) it follows that the frequency of a synchrotron radiating electron ν_s as a function of electron energy is given by

$$\nu_s \simeq 4.2 \, \gamma^2 \, H_\perp \text{ MHz} \tag{2.18}$$

with the magnetic field strength given in gauss. It then follows from equations (2.12) and (2.18) that the energy of a Compton γ-ray produced by an electron which synchrotron radiates at frequency ν_s is given by

$$E_\gamma = \frac{4}{3} <\epsilon> \gamma^2 = 0.74 \times 10^{-10} \left[T/H_\perp \right] \nu_s \text{ MeV} \tag{2.19}$$

To determine the relative importance of synchrotron radiation and Compton scattering in producing γ-radiation under astrophysical conditions, we note that the energy of a synchrotron photon is

$$E_{\gamma,s} = h\nu_s = 2.8 \times 10^{-15} \, \gamma^2 \, H_\perp \text{ MeV} \tag{2.20}$$

We then specify that an electron of energy $\gamma_c mc^2$ will radiate a Compton photon at the same mean energy as an electron of energy $\gamma_s mc^2$ will radiate a synchrotron photon. It then follows from equations (2.8), (2.12) and (2.20) that the ratio

$$\frac{\gamma_s}{\gamma_c} \simeq 10^2 \left(\frac{T}{H_\perp}\right)^{\frac{1}{2}} \tag{2.21}$$

The relative production rates from synchrotron and Compton radiation are then related by

$$\frac{Q_s}{Q_c} = \frac{\rho_H}{\rho_{ph}} \frac{I_e(\gamma_s)}{I_e(\gamma_c)} \tag{2.22}$$

In the galaxy, for example, $\rho_{ph} \simeq \rho_H$, but $\gamma_s \gg \gamma_c$ so that $I_e(\gamma_s) \ll I_e(\gamma_c)$ and synchrotron radiation is negligible compared to Compton scattering as a γ-ray production mechanism.

3. Bremsstrahlung Interactions

Bremsstrahlung, which is a German word meaning "braking radiation" is the radiation emitted by a charged particle accompanying deceleration. The cross sections for γ-ray production from bremsstrahlung are derived in Heitler's book (Heitler 1954). In the case of bremsstrahlung from non-relativistic electrons radiating in the field of a target nucleus, the cross sections for production of a γ-ray of energy E_γ by an electron of energy E in the field of a nucleus of charge Z is given by

$$\sigma_b(E_\gamma|E) = \sigma_b(E, E_\gamma) f(E_\gamma|E) \qquad (2.23)$$

where $f_b(E_\gamma|E)$ is the normalized distribution function for γ-ray production, and where, in terms of the kinetic energy $T = E - mc^2$

$$\sigma_b(E, E_\gamma) = \frac{\alpha Z^2}{\pi} \sigma_T \left(\frac{mc^2}{T}\right) \ln\left(\frac{\sqrt{T} + \sqrt{T-E_\gamma}}{E_\gamma}\right) \left(\frac{T}{E_\gamma}\right) \qquad (2.24)$$

for $2\pi Z \alpha \ll \beta \ll 1$

The cross section for bremsstrahlung by relativistic electrons is given by

$$\sigma_b(E, E_\gamma) = \frac{4\alpha Z^2}{\pi} \sigma_T \ln\left(\frac{2E}{mc^2} - \frac{1}{3}\right)\left(\frac{E}{E_\gamma}\right)$$

for $mc^2 \ll E \ll \alpha^{-1} Z^{-1/3} mc^2$ \qquad (2.25)

In the ultrarelativistic case, where the cross section is calculated by taking into account the screening of the charge of the atomic nucleus by atomic electrons, the resultant cross section is given by

$$\sigma_b(E, E_\gamma) = \frac{4\alpha Z^2}{\pi} \sigma_T \ln\left(183 Z^{-1/3} + \frac{1}{18}\right)\left(\frac{E}{E_\gamma}\right) \qquad (2.26)$$

for $E \gg \alpha^{-1} Z^{-1/3} mc^2$

Of course, in the case of an ionized gas, equation (2.25), which holds for the case when screening effects are unimportant, is applicable at all relativistic energies. The cross sections given in equations (2.24) and (2.25) may be corrected for additional contributions from interactions between cosmic ray electrons and atomic electrons by the replacement $Z^2 \to Z(Z+1)$.

It is immediately obvious from equations (2.25) and (2.26) that for relativistic particles the bremsstrahlung cross sections have little or no dependence, except for a linear one, on E. Indeed, equation (2.26) may be written in the form

$$\sigma_b(E, E_\gamma) \simeq \frac{\langle M \rangle}{\langle X \rangle} \left(\frac{E}{E_\gamma} \right) \tag{2.27}$$

where $\langle M \rangle$ is the average mass of the target atoms in grams and $\langle X \rangle$ is the average radiation length for the gas in grams per cm^2. The average radiation length for interstellar matter is

$$X = 65 \text{ g/cm}^2 \tag{2.28}$$

based on the values given for pure hydrogen and pure helium by Dovshenko and Pomanskii (1964) of

$$X_H = 62.8 \text{ g/cm}^2 \tag{2.29}$$

and

$$X_{He} = 93.1 \text{ g/cm}^2 \tag{2.30}$$

To a good approximation, especially in the case of relativistic bremsstrahlung (see Heitler 1954) the normalized distribution of gamma-rays produced may be taken to be a square distribution given by

$$f(E_\gamma | E) = \begin{cases} E^{-1} & \text{for } 0 \leq E_\gamma \leq E \\ 0 & \text{otherwise} \end{cases} \tag{2.31}$$

so that the gamma-ray production spectrum is given by

$$I_b(E_\gamma) = \frac{\langle M \rangle}{\langle X \rangle} \left[\int_{E_\gamma}^{\infty} dE \, I_e(E) \right] E_\gamma^{-1}$$

$$= \frac{1}{\langle X \rangle} \int d\vec{r} \, \rho(\vec{r}) \, \frac{I_e(>E_\gamma)}{E_\gamma} \tag{2.32}$$

where $\rho(\vec{r})$ is the matter density of the gas in grams per cm^3. For bremsstrahlung between cosmic-ray electrons and interstellar gas we may use equation (2.31) to write equation (2.32) in the form

$$q_b(E_\gamma) = 4.3 \times 10^{-25} n \frac{I_e(>E_\gamma)}{E_\gamma} \text{ cm}^{-3}\text{s}^{-1}\text{MeV}^{-1} \quad (2.33)$$

where n is the number density of nuclei in the production region.

4. Cosmic-Ray Produced π^0 Meson Decay

Cosmic-rays of high enough energy can produce various secondary particles of short lifetime upon collision with interstellar or intergalactic gas nuclei. Of these secondaries, the most important for production of γ-radiation is the π^0 meson which decays almost 100% of the time into 2 gamma rays. An extensive treatment of these various secondary production processes leading to cosmic γ-ray production has been given elsewhere (Stecker 1970, 1971a,1973a) and the reader is referred to these references for more detail of the material outlined here.

The collision processes of highest frequency occurring in interstellar and intergalactic space are those between cosmic-ray protons and hydrogen gas nuclei and are therefore p-p interactions. The threshold kinetic energy which a cosmic-ray proton must have to produce a secondary particle of mass m in such an interaction is

$$T_{th} = \left(2 + \frac{m}{2M_p}\right) m \quad (2.34)$$

Thus, to produce a single π^0 meson of rest mass 135 MeV/c^2 requires a threshold kinetic energy of 280 MeV/c^2. Within the past two decades, many measurements have been made of the cross sections for inclusive production of various secondary particles in high energy interactions using proton accelerators. These data are summarized in Figure 1, taken from Stecker (1973a) wherein the references may be found. Figure 1 shows the product of the production cross section σ_{π^0} and multiplicity ζ_{π^0} given as a function of kinetic energy T. Utilizing the data shown in Figure 1, and a demodulated cosmic ray proton spectrum given by Comstock, et al. (1972), Stecker (1973a) calculated a total γ-ray production rate from this process in the solar region of the galaxy to be

$$q_\gamma = (1.3 \pm 0.2) \times 10^{-25} n_H \text{ cm}^{-3}\text{s}^{-1} \quad (2.35)$$

taking account of the effect of p-He and α-H interactions as well as than of p-p interactions.

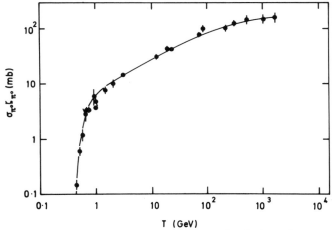

Fig. 1: Cross section times multiplicity for neutral pion production in p-p interactions as a function of incident kinetic energy (Stecker 1973a).

Figure 2, which shows the product of the cosmic-ray energy distribution I(T) and the function $\sigma_{\pi^0}\zeta_{\pi^0}$ of figure 1, is a measure of the kinetic energy of the cosmic ray protons which are most effective in producing π^0 mesons in the galaxy. It is found from this figure that the average proton which produces a π^0 meson in the galaxy has an energy of about 3.3 GeV and that 85% of the π^0 mesons produced in the galaxy are produced by cosmic ray protons in the energy range between 1 and 10 GeV.

When a π^0 meson decays into 2 γ-rays, because of conservation of momentum they both are produced with an energy of $m_\pi c^2/2$ = 67.5 MeV in the rest system of the pion. However, in the system of a terrestrial observer, these γ-rays have unequal energies which are determined by the relativistic Lorentz transformation. To determine the observed γ-ray energy, let γ_π and β_π refer to the energy and velocity of the parent π^0 meson in the observer's system. Let us also define the energy $\nu = m_\pi c^2/2$. Then the energies of the γ-rays in the observer's system are given by the Doppler relation

$$E_{\gamma 1,2} = \nu \gamma_\pi (1 \pm \beta_\pi \cos \theta) \qquad (2.36)$$

where θ is the angle between the γ-ray and the axis of the transformation in the pion rest system. The ± sign applies because the γ-rays come off at opposite directions in the pion rest system.

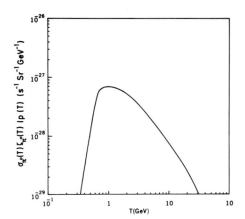

Fig. 2: Differential neutral pion production function from p-p interactions (Stecker 1973a).

Since there is no preferred direction to the decay in the pion rest system, the distribution in cos θ is isotropic, i.e. the distribution function

$$f(\cos\theta) = \text{const.} \tag{2.37}$$

and therefore, for a given value of $\gamma_\pi \beta_\pi$

$$f(E_\gamma | E_\pi) = \text{const.} \tag{2.38}$$

The extreme upper and lower limits on the γ-ray energy in the observers system, E_u and E_ℓ respectively, are defined by letting cos θ = 1 in equation (4.36). We then find

$$E_u = \nu\gamma_\pi(1 + \beta_\pi)$$

and

$$E_\ell = \nu\gamma_\pi(1 - \beta_\pi) \tag{2.39}$$

and since $\gamma_\pi = (1 - \beta_\pi^2)^{-1/2}$, it follows that $E_u E_\ell = \nu^2$ or

$$\ln \nu = \frac{\ln E_u + \ln E_\ell}{2} \tag{2.40}$$

From equations (2.38) and (2.39), it follows that the normalised γ-ray energy distribution function from the decay of pions of energy E_π is given by

$$f(E_\gamma \mid E_\pi) = \begin{cases} (E_\pi^2 - m_\pi^2 c^4)^{\frac{1}{2}} & \text{for } E_\ell < E_\gamma < E_u \\ 0 & \text{otherwise} \end{cases} \quad (2.41)$$

and thus the total production rate from cosmic ray protons with an energy spectrum $I(E_p)$ is

$$q(E_\gamma) = 4\pi n \int_{E_{th}}^{\infty} dE_p \, I(E_p) \cdot 2\zeta(E_p)$$

$$\times \int_{E_{\pi,min}}^{E_{\pi,max}} dE_\pi \, \frac{\sigma(E_\pi \mid E_p)}{(E_\pi^2 - m_\pi^2 c^4)^{\frac{1}{2}}} \quad (2.42)$$

To find the limits on the E integration in equation (2.42), we note that the relation between the maximum γ-ray energy and E_π can be written as

$$E_\pi = E_u + \frac{\nu^2}{E_u} \quad (2.43)$$

and we may therefore reverse the criteria to note that γ-rays of energy E_γ may be produced by pions of energy as low as

$$E_{\pi,min} = E_\gamma + \frac{\nu^2}{E_\gamma} \quad (2.44)$$

In the upper limit, we note as $E_\pi \to \infty$, $\beta_\pi \to 1$ and from equation (2.39), $E_\ell \to 0$. This implies that as we increase the pion energy without limit, the allowed range of γ-rays produced expands in the lower limit to allow for all energies and pions of arbitrarily high energy can produce γ-rays of energy E_γ, i.e.,

$$E_{\pi,max} \to \infty \quad (2.45)$$

and therefore equation (2.45) becomes

$$q(E_\gamma) = 8\pi n \int_{E_{th}}^{\infty} dE_p I(E_p) \zeta(E_p) \int_{E_\gamma + \frac{\nu^2}{E_\gamma}}^{\infty} dE_\pi \frac{\sigma(E_\pi | E_p)}{(E_\pi^2 - m_\pi^2 c^4)^{\frac{1}{2}}} \quad (2.46)$$

It follows from equations (2.40) and (2.41) that the γ-ray spectrum from an arbitrary distribution of energies in a decaying pion beam can be looked on as the superposition of a number of rectangular distributions centered about $\lg \nu$ on a **logarithmic plot** Also, because the energy ν is included within the allowed range of γ-ray energies, there is a maximum in the energy spectral distribution at this value. Intuitively, one can also see from equation (2.40) that the full spectrum from any arbitrary pion energy distribution should be symmetric about the maximum at $\lg \nu$ on a logarithmic plot. These characteristics can be rigorously proven (Stecker 1971a). The differential γ-ray energy spectrum calculated for galactic cosmic ray interactions per unit path length and target nucleon density is shown in figure 3 calculated from a two-component kinematical model for pion production in high-energy cosmic ray interactions (Stecker 1970). The integral γ-ray energy spectrum is shown in figure 4 along with calculations by Cavallo and Gould (1971) based on a different kinematical model. These spectra have been normalized to compare the shapes obtained. The wiggles in both spectral calculations represent artifacts of the kinematical models assumed and should not be taken too seriously. The shapes of the two spectra are in good agreement and probably represent an accurate approximation to reality within the uncertainty indicated by the wiggles.

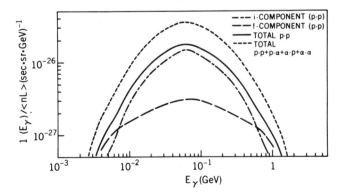

Fig. 3: The calculated differential production spectrum of γ-rays produced in cosmic ray interactions in the galaxy based on the "isobar (i)-plus-fireball (f)" model of Stecker (1970).

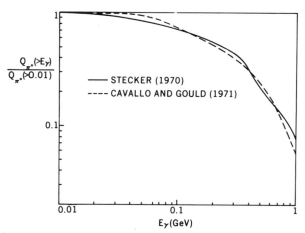

Fig. 4: A comparison of the shapes of the integral galactic pion-decay energy spectra calculated by Stecker (1970) and Cavallo and Gould (1971). The total production rate is normalized to unity.

Various other secondary products of high energy cosmic ray interactions such as other mesons, hyperons and excited nucleon states contribute to the overall cosmic γ-ray spectrum, particularly at high energies. However, their total contribution to the production rate is relatively minor. The kinematics of these reactions is, in general, much more complicated than in the case of pion decay and has been thoroughly discussed elsewhere (Stecker 1971a).

5. Nucleon-Anti Nucleon Annihilations

If antimatter exists in significant quantities in the universe, secondary particles can result from the annihilation of nucleons and anti-nucleons. Of these secondaries, the most significant for γ-ray production again are the π^o mesons. Thus, the resultant γ-ray production spectrum is also symmetric about 67.5 MeV on a logarithmic plot, but in this case it is also bounded by the limited rest-mass energy which can be released in the annihilation. Frye and Smith (1966) using accelerator data, and independently Stecker (1967, 1971a) using a theoretical pion production model for nucleon-antinucleon annihilation have calculated the resultant γ-ray spectrum from annihilations at rest. There is excellent agreement between the two calculations and the resultant spectrum is shown in figure 5.

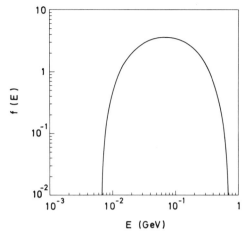

Fig. 5: Normalized local differential γ-ray spectrum from p-p̄ annihilation at rest (Stecker 1971a).

6. Form of the Spectrum from Pion Production at High Energies

It is interesting to examine the asymptotic form of the γ-ray spectrum to be expected from the decay of π^o mesons produced at energies above a few GeV by cosmic rays having a power-law spectrum $\propto E_p^{-\Gamma}$. In this case, we assume that the pion production cross section is constant and that the pion multiplicity rises as a power of the primary energy $\propto E^a$. We also assume that the average pion energy rises as E_p^b. We can then write the production function in the form

$$\sigma(E_\pi | E_p) = \sigma_o E_p^a \, \delta(E_\pi - \chi_o E_p^b) \qquad (2.47)$$

where the coefficients σ_o and χ_o and the exponents a and b are taken to be constants. Equation (2.43) then reduces to

$$q(E_\gamma) = 8\pi n \, K_p \sigma_o \int_{E_{th}}^{\infty} dE_p \, E_p^{a-\Gamma} \int_{E_\gamma}^{\infty} dE_\pi \, \frac{\delta(E_\pi - \chi_o E_p^b)}{E_\pi} \qquad (2.48)$$

$$= 8\pi n\, K_p \sigma_o \int_{(E_\gamma/\chi_o)^{1/b}}^{\infty} dE_p\, \frac{E_p^{-[(\Gamma+b)-a]}}{\chi_o}$$

$$= \frac{8\pi n}{g}\, K_p \sigma_o \chi_o^{[(g/b)-1]}\, E_\gamma^{-g/b}$$

where $\quad g \equiv [(\Gamma + b)-(a + 1)]$

As can be seen from equation (2.48), the high energy γ-ray spectrum produced by cosmic rays having an energy spectrum with index Γ is also a power-law, but one which is different and, in general, steeper than the primary spectrum. For example, in the case where $\Gamma = 2.5$ and $a = 1/4$, $b = 3/4$ (the Fermi model for meson production), the γ-ray index $g/b = 2.67$. The primary γ-ray spectra have the same index in the important case where a constant and large fraction of the primary energy of the cosmic ray is carried off by a single secondary which decays to produce a π^o meson. This corresponds to the case $a = 0$, $b = 1$ and therefore $g/b = \Gamma$.

7. γ-rays from Photomeson Production at Ultrahigh Energies

A process which may be important in producing γ-rays of energies in the 10^{19}eV range is that of photomeson production, i.e. reactions of the type

$$\gamma + p \to \pi^o + p \qquad (2.49)$$

It is now generally believed that the universe is filled with thermal microwave radiation of temperature 2.7K which is a remnant of the primordial big-bang. This temperature corresponds to an average photon energy of $2.7kT = 6.4 \times 10^{-4}$eV and an average photon number density of about 400 photons per cm^3. To an ultrahigh energy cosmic ray of energy in the 10^{20}eV range, these microwave photons look like γ-rays of energy equal to hundreds of MeV. These energies are above the threshold for reaction (2.49) so that various ultrahigh energy mesons can be produced. One very important effect of this is the attenuation of ultrahigh energy cosmic rays in intergalactic space in 10^8-10^9 years (Greisen 1966, Zatsepin and Kuz'min 1966, Stecker 1968). A typical γ-ray produced by reaction (2.49) carries off about 10% of the primary energy of the cosmic ray (Stecker 1973b) The γ-rays produced may themselves be attenuated by pair production processes of the type.

$$\gamma_{2.7} + \gamma \to e^+ + e^- \qquad (2.50)$$

and this process, together with Compton scattering of the electrons and positrons by the 2.7 K radiation may lead to a cascade process which has been treated in detail by Wdowczyk, et al. (1972) and Stecker (1973b).

III γ-RAY ABSORPTION MECHANISMS

Various processes are of astrophysical importance in depleting γ-rays in the galaxy and the universe. By "absorption" we will mean not only those processes in which the γ-ray completely disappears, such as process (2.50), but also those processes in which the γ-ray is scattered out of the energy range of interest as can occur in the case of Compton scattering.

We will consider two basic categories of absorption processes: 1. absorption in matter and 2. absorption through interactions with radiation. The later process is of importance because of the existence of the 2.7 K thermal universal radiation field.

1. Absorption Through Interactions with Radiation

Let us first consider the effects of the universal radiation field on the intensity of cosmic γ-rays. The attenuation process of importance here is the pair-production process (2.50). This process can only take place if the total energy of the photons in the C.M.S. of the interaction is greater than or equal to $2mc^2$. The cross section for reaction (2.50) can be calculated using quantum electrodynamics and a derivation may be found in Jauch and Rohrlich (1955). The importance of reaction (2.50) was first pointed out by Nikishov (1961) in considering interactions of γ-rays with ambient starlight photons and with the discovery of the universal radiation. Gould and Schreder (1966, 1967a, b) and Jelley (1966) were quick to point out the opacity of the universe to γ-rays of energy above 10^{14} eV. Stecker (1969b) and Fazio and Stecker (1970) generalized these calculations by including cosmological effects.

In discussing reaction (2.50), we will generally follow the discussion of Gould and Schreder (1967a, b) with one important difference. At the time Gould and Schreder published their papers, it was generally thought that the universal radiation field had a somewhat higher temperature than the presently accepted 2.7 K. Therefore, the Gould and Schreder results have been corrected here to correspond to a 2.7 K radiation field.

Denoting C.M.S. quantities by primes and noting that for reaction (2.50).

$$E'_{e^+} = E'_{e^-} = E'_e \tag{3.1}$$

we can determine the threshold energy for the reaction by noting that the relativistic four-momentum invariance condition in this case reduces to the form

$$(2E'_e)^2 = (E_{2.7} + E_\gamma)^2 - |(\vec{p}_{2.7}c + \vec{p}_\gamma c)|^2 \tag{3.2}$$

$$= (E_{2.7}^2 + 2E_{2.7}E_\gamma + E_\gamma^2) - (E_{2.7}^2 + E_\gamma^2 - 2E_{2.7}E_\gamma \cos\theta)$$

$$= 2E_{2.7}E_\gamma(1-\cos\theta)$$

At threshold, both the electron and positron are produced at rest in the C.M.S. of the interaction. The minimum energy required corresponds to a head-on collision ($\cos\theta = -1$). Equation (3.2) then reduces to the relation

$$E_{\gamma,th} = \frac{(m_e c^2)^2}{E_{2.7}} \tag{3.3}$$

If we consider a typical blackbody photon to have an energy of approximately 10^{-9} MeV, then from (3.3) we find a threshold energy of approximately 2.5×10^8 MeV for reaction (2.50). However, this threshold is somewhat blurred due to the fact that the blackbody photons are not all at the same energy but have a Bose-Einstein distribution given by the well-known relation

$$n(E_{2.7}) = \frac{1}{\pi^2 \hbar^3 c^3} \frac{E_{2.7}^2}{1-e^{-E_{2.7}/kT}} \tag{3.4}$$

and also various possible values of $\cos\theta$ must be allowed for.

The cross section for reaction (2.50) is given by (Jauch and Rohrlich, 1955)

$$(E_{2.7}, E_\gamma) = \frac{\pi}{2}\left(\frac{e^2}{m_e c^2}\right)^2 (1-\beta_e'^2)\left[(3-\beta_e'^4)\ln\frac{1+\beta_e'}{1-\beta_e'} - 2\beta_e'(2-\beta_e'^2)\right] \tag{3.5}$$

where the C.M.S. velocity of the electron (positron) is given by

$$\beta'_e = \left(1 - \frac{m_e^2 c^4}{E_{2.7} E_\gamma}\right)^{1/2} \qquad (3.6)$$

The absorption coefficient is then

$$\kappa_{\gamma\gamma}(E_\gamma) = \iint dE_{2.7} d\theta \frac{\sin\theta}{2} n(E_{2.7}) \sigma(E_{2.7}, E_\gamma) (1-\cos\theta) \qquad (3.7)$$

Gould and Schréder (1967a,b) have reduced equation (3.7) to the form

$$\kappa_{\gamma\gamma}(E_\gamma) = \alpha^3/\pi \left(\frac{m_e c^2}{e^2}\right)^2 \left(\frac{kT}{m_e c^2}\right)^3 f(\nu) \qquad (3.8)$$

where

$$\nu \equiv \frac{(m_e c^2)^2}{E_\gamma kT} \qquad (3.9)$$

They find that the function $f(\nu)$ has a maximum value $\simeq 1$ at $\nu \simeq 1$ and that $f(\nu)$ has the asymptotic forms given by

$$f(\nu) \rightarrow \frac{\pi^3}{3} \nu \ln\left(\frac{0.117}{\nu}\right) \quad \text{for } \nu \ll 1$$

and $\quad f(\nu) \rightarrow \left(\frac{\pi\nu}{4}\right)^{1/2} e^{-\nu} \left(1 + \frac{75}{8\nu} + \ldots\right) \quad \text{for } \nu \gg 1$

$\qquad (3.10)$

They have also calculated $\kappa_{\gamma\gamma}$ for γ-ray interactions with various photon fields in interstellar space. The results of their numerical calculation are shown in figure 6.

2. Absorption by Interactions with Matter

There are two types of interactions of importance to consider here. The first is the Compton scattering interaction

$$e^- + \gamma \rightarrow e^- + \gamma \qquad (3.11)$$

which we have discussed in connection with γ-ray production.

Electrons play the dominant role in the Compton scattering of γ-rays. For γ-rays of energy $E_\gamma \gg mc^2$, almost all of the energy of the γ-ray is absorbed and then we can consider that the γ-ray has disappeared.

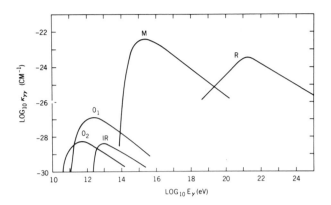

Fig. 6: Absorption probability per unit distance by
$\gamma + \gamma \to e^+ + e^-$ as a function of photon energy for γ-rays passing
through various universal radiation fields. The contributions
from the optical (O_1 and O_2) infrared (IR) 2.7°k blackbody,
microwave (M) and radio (R) fields are shown. Absorption at
lower energies from X-rays is negligible. O_1 represents the
contribution from the light of population II stars; O_2 that from
population I stars (from Gould and Schreder 1967a,b); M is
corrected to 2.7°K (Stecker 1971a).

In some cases it is useful to define an "absorption cross section", σ_a, such that

$$\sigma_a = \left(\frac{\Delta E_\gamma}{E_\gamma}\right) \sigma_c \qquad (3.12)$$

where ΔE_γ is the average amount of energy transferred from the gamma-ray to the electron. It is then found that, (Heitler 1954), with $\varepsilon = E_\gamma/m_e c^2$

$$\sigma_a = \varepsilon \sigma_c \quad \text{for } \varepsilon \ll 1 \qquad (3.13)$$

and

$$\sigma_a = \sigma_c \left(\frac{\ln 2\varepsilon - \frac{5}{6}}{\ln 2\varepsilon + \frac{1}{2}}\right) \quad \text{for } \varepsilon \gg 1 \qquad (3.14)$$

The second type of gamma-ray absorption process in matter that we must consider involves the conversion of a gamma-ray into an electron-positron pair in the electrostatic field of a charged particle or nucleus. If we designate such a charge field by the symbol CF, such an interaction may be symbolically written as

$$\gamma + CF \rightarrow e^+ + e^- + CF \tag{3.15}$$

The conversion interaction, or pair-production as it is usually called, has a cross section that involves an extra factor of the fine structure constant, $\alpha = e^2/\hbar c$, since it involves an intermediate interaction with an electrostatic field. At non-relativistic energies, this cross section is a complicated function of energy which must be determined numerically (see Heitler (1954) for further details) however a closed analytic approximation for this cross section may be given for energies greater than 1 MeV, which happily corresponds to the energy region where the pair-production cross section becomes more important than the Compton scattering cross section in determining the gamma-ray mass absorption coefficient for hydrogen gas.

For pair-production in the field of a nucleus of atomic number Z, the cross section for reaction (3.15) is given by

$$\sigma_p = \alpha \left(\frac{e^2}{m_e c^2}\right)^2 \left(\frac{28}{9} \ln 2\varepsilon - \frac{218}{27}\right) Z^2 \tag{3.16}$$

$$\text{for } 1 \ll \varepsilon \ll \alpha^{-1} Z^{-1/3}$$

which is the energy region where electron screening of the nuclear charge field may be neglected. The no-screening case, of course, also holds for an ionized gas (plasma).

In the energy region where the complete-screening approximation is valid,

$$\sigma_p = \alpha \left(\frac{e^2}{m_e c^2}\right)^2 \left(\frac{28}{9} \ln (183 Z^{-1/3}) - \frac{2}{27}\right) \tag{3.17}$$

$$\text{for } \varepsilon \gg \alpha^{-1} Z^{-1/3}$$

The threshold energy for pair production in the field of an atomic nucleus is, of course $2m_e c^2$. In the case of pair production in the field of atomic electrons, the threshold energy is $4m_e c^2$. Above this energy, the pair production cross section must be modified to include the additional contribution of the electrons and this may be done approximately by making the replacement

$$Z^2 \rightarrow Z^2 (1 + \xi^{-1} Z^{-1}) \tag{3.18}$$

in equations (3.13) and (3.14), where the quantity ξ varies

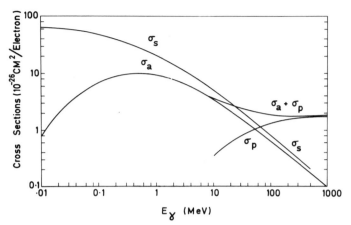

Fig. 7: Compton scattering (σ_s), Compton absorption (σ_a), pair-production (σ_p) and total ($\sigma_a + \sigma_p$) cross sections as a function of γ-ray energy for absorption of γ-rays in hydrogen gas (based in part on work of Nelmes, 1953) from Stecker (1971a).

from $\simeq 2.6$ at $E_\gamma \simeq 6.5$ MeV to $\simeq 1.2$ at $E_\gamma \simeq 100$ MeV. For gamma-ray energies above 200 MeV, $\xi \simeq 1$ and the pair production cross section has the approximately constant value of 1.8×10^{-26} cm^2, according to the results of Trower (1966). The values of the various cross sections for γ-ray absorption in matter, as discussed in this section, are shown in figure 7.

IV. γ-RAYS OBSERVED FROM THE GALAXY

In this section, we will now turn our attention to the interpretation of presently existing observations of cosmic γ-radiation. Because of lack of time, we will restrict ourselves to two important topics, galactic diffuse γ-radiation and the diffuse γ-ray background. This means we are omitting a discussion of a number of important topics which will be touched upon by others in these proceedings. I would also like to recommend the discussion of some of these topics in the reviews published in the Proceedings of the International Symposium and Workshop on Gamma Ray Astrophysics held at NASA Goddard Space Flight Center last May (Stecker and Trombka 1973).

Radio astronomers have for years been mapping accumulations and features involving relatively high densities of atomic hydrogen in the galaxy. They do this by studying the 21cm emission line of atomic hydrogen caused by the hyperfine splitting of the

1^2S ground state of the hydrogen atom as a result of the interaction between the magnetic moments of the proton and the electron in the atom. The separation between these levels is very small because the magnetic moment of the proton is almost 2000 times smaller than that of the electron owing to its proportionally larger mass. The frequency of the emission corresponding to this energy-level separation is in the radio range at 1420 MHz.

Radio source intensities are usually given in terms of brightness temperature T_b (Kelvin) corresponding to the emission temperature of a blackbody on the Rayleigh-Jeans (low frequency) side of the Planck distribution. Thus

$$T_b(\nu) = \frac{I(\nu)c^2}{2\nu^2 k} \qquad (4.1)$$

Sometimes these intensities are given in terms of antenna temperature which is equal to the brightness temperature multiplied by antenna efficiency. For a gas radiating at constant temperature T but with opacity κ, the brightness temperature measured along the line-of-sight in a given direction is

$$T_b = \int ds\, T\, e^{-\int_0^1 \kappa ds'} = \int ds\, T\, e^{-\tau(s)} = T(1-e^{-\tau}) \qquad (4.2)$$

Thus, at small optical depths $\tau \ll 1$,

$$T_b \simeq T\tau \qquad (4.3)$$

and at large optical depths $\tau \gg 1$,

$$T_b \simeq T \qquad (4.4)$$

The opacity κ can be expressed as the atomic opacity α times the atomic density n of absorbing atoms.

Consider an atom with two state levels, the upper level u and a lower level ℓ. Then according to Boltzmann's law the proportion in each level as a function of T is

$$\frac{n_u}{n_\ell} = \frac{g_u}{g_\ell} e^{-h\nu_0/kT} \qquad (4.5)$$

where ν_0 is the transition frequency and g_u and g_ℓ are the statistical weights of the two levels. Thus, for the 21 cm transition where $g_u = 3 g_\ell$ (triplet to singlet)

$$\frac{n_u}{n_\ell} = 3\, e^{-h\nu_0/kT} \tag{4.6}$$

The total absorption coefficient

$$\kappa = \alpha\, n_\ell \left(1 - \frac{n_u}{n_\ell}\frac{g_\ell}{g_u}\right) \tag{4.7}$$

where the second term takes account of stimulated emission. From equation (4.1) this reduces to

$$\kappa = \alpha\, n_\ell\, (1 - e^{-h\nu_0/kT}) \simeq \alpha\, n_\ell\, \frac{h\nu_0}{kT} \text{ for } \frac{h\nu_0}{kT} \ll 1 \tag{4.8}$$

Thus, in the optically thin case, from equation (3.18)

$$T_b = T\tau = \alpha\, \frac{h\nu_0}{k} \int n_\ell ds$$

which is proportional to the total number of atoms along the line-of-sight. In the case of 21 cm emission

$$T_b(21\text{cm}) = 5.49 \times 10^{-19} \int n_\ell ds \tag{4.9}$$

In general, however, it should be kept in mind that equation (4.9), which is valid for an optically thin gas, in general only provides a <u>lower limit</u> to the amount of gas along the line-of-sight.

In general, line emission is broadened in frequency by the Doppler shift effect of gas moving with various radial velocities $v_r = \beta_r c$ along the line of sight. Then

$$\nu = \nu_0(1+\beta_r) \tag{4.10}$$

What is usually measured is a velocity profile determined by the frequency distribution of the line emission $f(\nu')$ so that a distribution in brightness temperature versus velocity or frequency is constructed. Then from equation (4.9)

$$N_H = \int n_\ell ds = 1.82 \times 10^{18} \int T_b(\nu)\, d\nu \tag{4.11}$$

From the two dimensional distribution $n(\ell,\nu)$ where ℓ is galactic longitude, one can use a rotation model of the galaxy to construct a positional map of dense regions of atomic hydrogen. Such a map is shown in figure 8 with some of the 21 cm hydrogen "arms" of the galaxy named.

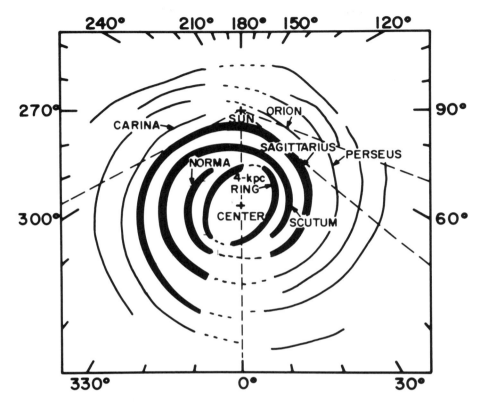

Fig. 8: A smoothed spatial diagram of the locations of the maxima of the matter density deduced from the 21-cm neutral H line measurements and the density-wave theory by Simonson (preprint).

One can also plot a graph of the average hydrogen density seen in 21 cm emission as a function of galacto-centric distance in kpc. Such a plot is shown in figure 9.

The galactic distribution of γ-ray emission was first mapped by Kraushaar, et al. (1972) on OSO-3 with rather limited resolution. More recently, the galactic distribution was remeasured with better angular resolution by Kniffen et al. (1973) on SAS-2. The results of both measurements are shown in figure 10. The map of Kniffen et al. has recently been updated by the addition of new data having been analyzed and the updated distribution is shown in figure 11 (Fichtel 1974).

The data from SAS-2 of γ-rays above 100 MeV shows radiation

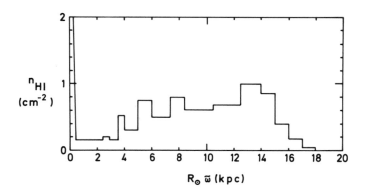

Fig. 9: The neutral hydrogen surface density based on the discussion of Kerr (1969), Shane (1972), and Sanders and Wrixon (1973).

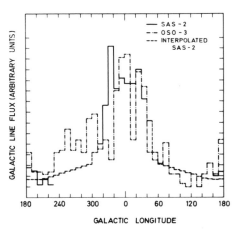

Fig. 10: Distribution in galactic longitude of the galactic γ-ray line flux as observed by OSO-3 (Kraushaar et al. 1972) and SAS-2 (Kniffen et al. 1973), normalized arbitrarily for purposes of comparison. Statistical uncertainties quoted for these results (not plotted) are significantly greater for the OSO-3 results as compared with the SAS-2 results. The interpolation of the SAS-2 data was obtained on the assumption of a smooth average variation in the γ-ray production rate (Puget and Stecker 1974).

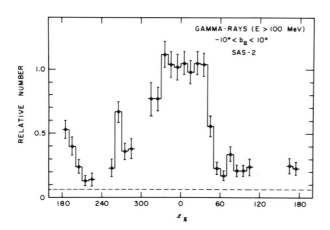

Fig. 11: Distribution of high-energy (> 100 MeV) gamma-rays along the galactic plane. The diffuse background level is shown by a dashed line. The SAS-2 data are summed over $|b^{II}| < 10°$. The ordinate scale is approximately in units of 10^{-4} photons cm^{-2} rad^{-1} sec^{-1} (Thompson et al. 1974).

---

coming from the galactic disc. The galactic latitude distribution of the emission is shown in figure 12. Figure 12 indicates that the radiation in the anticenter direction comes from within 6° of the galactic plane whereas that in the direction of the galactic center also is limited to within 6° with perhaps a particularly intense component restricted to within 3° of the plane. Since 3° was the angular resolution of the instrument, this narrow source could have a true width of less than 3° corresponding to a source distance of at least 2 to 4 kpc from the sun toward the inner galaxy. A two dimensional SAS-2 map of the galactic disk is shown in figure 13.

Data obtained by Samimi et al. (1974) on the latitude distribution of the galactic γ-radiation also indicate a width of 3° for most of the radiation.

The longitude data in figure 13 shows a broad flat region of intense emission within 30° to 40° of the galactic centre on either side. This indicates that there is a large emission rate within 5 to 6 kpc of the galactic center.

Puget and Stecker (1974) have geometrically unfolded the SAS-2 distribution shown in figure 10 and the resultant emissivity

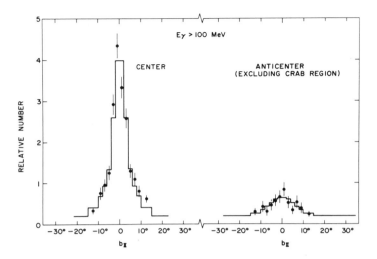

Fig. 12: (A) Distribution of high-energy ($E_\gamma > 100$ MeV) gamma-rays summed over $335° < \ell^{II} < 25°$ as a function of b^{II}. The solid line represents the sum of two Gaussians, one with $\sigma = 3°$ (the detector resolution), the other a Gaussian with a width $\sigma = 6°$ (Thompson et al. 1974).
(B) Distribution of >100 MeV gamma-rays summed from $90° < \ell^{II} < 170°$ and $200° < \ell^{II} < 270°$ where data exists, (Thompson et al. 1974).

---

as a function of galacto-centric distance is shown in figure 14. Strong (1974) has performed a corresponding unfolding of combined data from SAS-2 and OSO-3. He obtained a similar distribution to that shown in figure 14. The results show a maximum in the γ-ray emissivity at ∼ 5 kpc and a reduced emissivity within 3 kpc of the galactic center. An unfolding of figure 11 would probably show a less pronounced maximum at 5 kpc, but this unfolding has not been attempted here because the SAS-2 group will be publishing their final and complete longitude data in the near future.

Data have also been obtained on the energy spectrum of the galactic γ-radiation in the direction of maximum intensity by Kniffen et al. (1973). These spectral measurements on the integral γ-ray spectrum are consistent with a two-component origin with ∼ 70 percent of the radiation above 100 MeV due to π^0-decay as shown in figure 4 and ∼ 30 percent due to Compton radiation from cosmic ray electrons interacting with the higher radiation field in the inner galaxy (Stecker et al. 1974). Under this interpretation, one would expect that almost all of the γ-radiation above 100 MeV in the outer regions of the galaxy

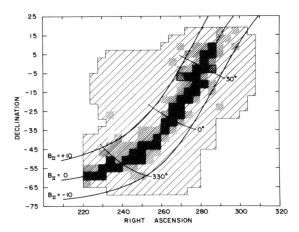

Fig. 13: Schematic diagram of relative intensity of the gamma radiation above 100 MeV as deduced from preliminary data from SAS-2. Increasing shading indicates higher intensity, but no single ($4°$ x $4°$) box has sufficient numbers of gamma rays to justify statistically significant conclusions to be deduced for it alone. It is clear however, that the galactic plane stands out sharply (Fichtel, Hartman, Kniffen and Thompson 1973).

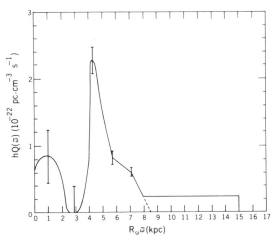

Fig. 14: The value for hQ ($\tilde{\omega}$) (γ-ray emissivity times disk width) given by Puget and Stecker (1974). A constant value for hQ is assumed for $R_\odot \tilde{\omega} >$ kpc. Uncertainties in the determination of the value of hQ($\tilde{\omega}$) as evaluated from an integral equation of Puget and Stecker grow quite large in the central region $\tilde{\omega} <$ 4kpc.

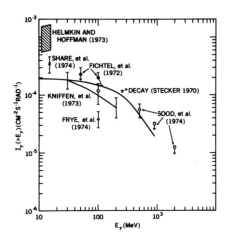

Fig. 15: Summary of the integral flux measurements for the galactic center region given by various experimental groups.

would be due to π^0-decay. This interpretation would also be consistent with the γ-ray production rate deduced by Kraushaar et al. of $(1.6 \pm 0.5) \times 10^{-25}$ n_H cm^{-3} s^{-1} above 100 MeV which is only slightly larger than the theoretical calculated value of $(1.3 \pm 0.2) \times 10^{-25}$ n_H cm^{-3} s^{-1} for the solar vicinity (Stecker 1973) with $n_H \sim 0.7$ cm^{-3} for atomic hydrogen alone.

While this interpretation presents a coherent and plausible model for discussion, one should bear in mind that the empirical situation is still in a state of flux and one measure of uncertainty is indicated by the various different measurements of the inner-galactic flux as shown in figure 15 and the differential data of Samini et al. (1974) as shown in figure 16. In particular, the data summarized in figures 15 and 16 appear to support the presence of a soft component of radiation at low energies possibly due to Compton interactions or bremsstrahlung.

It is instructive to estimate Compton and bremsstrahlung intensities in the galaxy relative to π^0-decay γ-radiation. First a crude estimate of the Compton radiation in the galactic disk. Starlight photons in the galaxy have an average energy of ~ 1 eV and a number density of ~ 0.4 cm^{-3} (Allen 1973). Let us consider the total gas density of atomic and molecular hydrogen to be ~ 1 cm^{-3}. As pointed out previously, the median energy of cosmic ray protons which produce π^0-mesons leading to γ-rays in the observed energy range $\gtrsim 100$ MeV is ~ 3 GeV. By a curious

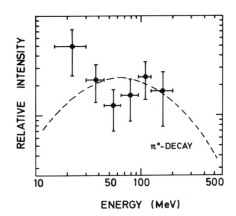

Fig. 16: The differential photon spectrum of γ-rays coming from $-3° < b^{II} < 0°$ as determined from the emulsion measurements of Samimi et al. (1974) with the detector response unfolded.

---

coincidence, a 3 GeV electron with $\gamma \simeq 6 \times 10^3$ interacting with a 1 eV photon will produce a γ-ray of energy

$$E_\gamma = \frac{4}{3} \gamma^2 \text{ eV} \simeq 50 \text{ MeV} \tag{4.12}$$

so that electrons of comparable (or slightly higher) energies will produce γ-rays in the same energy range as protons which produce π^0 decay γ-rays. The electron to proton ratio at these energies is $\sim 10^{-2}$. The ratio of total γ-ray intensities is then

$$\frac{I_c}{I_\pi} \simeq \left(\frac{I_e}{I_p}\right) \left(\frac{\sigma_c}{2\sigma_{\pi^0} \zeta_{\pi^0}}\right) \left(\frac{n_{ph}}{n_H + 2n_{H_2}}\right) \tag{4.13}$$

$$\simeq (10^{-2}) \left[\frac{4.4 \times 10^{-25}}{3 \times 10^{-26}}\right] \left(\frac{0.4}{1}\right) \simeq 0.06$$

More involved calculations using equations (2.9) and (2.10) give similar results. Equations (2.9) and (2.10) can also be used to calculate the γ-ray spectrum produced by interactions of electrons with the 2.7K universal background radiation. The ratio of 2.7K blackbody Compton γ-rays to starlight Compton γ-rays is

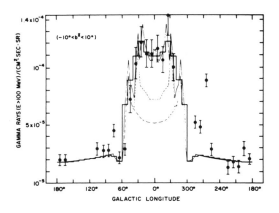

Fig. 17: Longitudinal distributions of galactic gamma-flux integrated over $\pm 10°$ in b^{II} from Bignami and Fichtel (1974). SAS-II points are given together with their error bars (Kniffen et al. 1973). The thick line represents the model of Bignami and Fichtel smoothed in $10°$ ℓ^{II} intervals. The thin line represents the model in $2°$ intervals. The dashed line (---) gives the contribution of the Sagittarius and Norma-Scutum arms and dash-dot (-.-.-.) the contribution of the Sagittarius arm alone.

$$\frac{I_{2.7,c}}{I_{*,c}} = \frac{\rho_{2.7}}{\rho_*} \left(\frac{E_{2.7}}{E_*}\right)^{\frac{\Gamma-3}{2}} \approx \left(\frac{0.25}{0.4}\right) (6.4 \times 10^{-4}) \approx 2.7$$

(4.14)

As can be seen from figure 11, the γ-ray intensity in the inner galaxy is about 5 times that in the anticenter direction when one subtracts out the peaks due to the crab nebula at $180°$ and the Vela source at $270°$. A similar large ratio was found by Kraushaar et al. (1972) on OSO-3. The OSO-3 ratio was speculated by Stecher and Stecker (1970) to be due to two effects: 1) $\pi°$ production off H_2 not seen in 21 cm emission, and 2) a Compton source from a large infrared radiation field at the galactic center (Hoffman and Frederick 1969). However, the Compton source would produce a clear peak at $\ell = 0°$ whereas the new SAS-2 data do not indicate such a peak but rather a broad flat region of relatively intense radiation (see figure 11).

Bignami and Fichtel (1974) have attempted to explain the SAS-2 longitude distribution as due to an n^2 enhancement of the product of gas density n and cosmic-ray intensity I with large density contrasts in galactic spiral arms as defined by 21 cm measurements of atomic hydrogen. Their results are shown in figure 17.

Stecker et al. (1974) and Puget and Stecker (1974) pointed out that the apparent peak in γ-ray emissivity at ~ 5 kpc implied by the data of Kniffen et al. corresponded to a maximal dissipation of the kinetic energy of outward gas motion in the galaxy and that possible cosmic-ray compression in that region and first order Fermi acceleration could enhance the cosmic-ray density in that region to a large enough extent to explain the enhancement in γ-ray emissivity.

An enhancement of cosmic-rays in a localized region of the galaxy can be due to three factors: (1) an increase in the density of cosmic-ray sources (or alternatively the production rate) in the region, (2) an increase in the trapping time (escape time) of cosmic-rays in the region, and (3) acceleration and compression of cosmic-rays in the region. We may write this as follows:

$$\ln (I_{cr}/I_\odot) = \ln (Q_{cr}/Q_\odot) + \ln (T_{cr}/T_\odot) + \delta \qquad (4.15)$$

where the first term on the right hand side of equation (4.15) represents the enhancement in the production rate, the second term represents the trapping factor and the third term represents the effect of acceleration and compression. The first term can only be guessed at, given our present lack of knowledge of the ultimate origin of cosmic rays. However, as an indication of the possible effect of source density, one can note the distribution of supernova remnants in the galaxy which may be proportional to the density of cosmic-ray sources if we assume that cosmic-rays are produced by supernova explosions (Ginzburg and Syrovatsky 1964) or in remnant pulsars (Gunn and Ostriker 1969). According to Ilovaisky and Lequeux (1972) the density of supernova remnants increases roughly by a factor of 2 over the local galactic value for $\tilde{\omega} < 0.8$ and drops sharply for $\tilde{\omega} > 1.2$. A more recent study by Clark et al. (1973) is probably less susceptible to selection effects because their survey included remnants down to lower luminosity levels. The results obtained by Clark et al. confirm the earlier results of Ilovaisky and Lequeux regarding the general distribution of supernova remnants in the galaxy. We will, therefore estimate here that $Q_{cr}/Q_\odot \approx 2$ for the enhanced region $0.4 < \tilde{\omega} < 0.6$. This factor, by itself, cannot account for the order-of-magnitude enhancement deduced for $I_{cr}/I_\odot$.

The second factor, $T_{cr}/T_\odot$ is so difficult to estimate that we will treat it as a free parameter > 1 to be solved for. It is not unreasonable to expect more effective trapping in the inner galaxy (i.e. $T_{cr}/T_\odot > 1$) due to compression and a resultant stronger magnetic field strength.

The factor δ can be broken up into two parts, an acceleration factor δ_{ACC} which accounts for acceleration of the more numerous lower energy cosmic-rays (given an assumed power-law differential energy spectrum of the form $\propto E_{cr}^{-\Gamma}$) to an energy above the threshold for π^0 production, and a simple density enhancement δ_D due to the lower volume of the compressed region in which the trapped particles find themselves. If we designate the specific volume compression rate by

$$\alpha = (1/V)(dV/dt) \qquad (4.16)$$

then obviously

$$\delta_D = \alpha T_{cr} \qquad (4.17)$$

The acceleration factor can be estimated thermodynamically. Regardless of the exact details of this process, which may be coherent, first order Fermi acceleration which can transfer the momentum of moving "clouds" with trapped magnetic irregularities to cosmic-rays trapped in the compressed region (Fermi 1954, Stecker et al. 1974). The increase in the energy of the individual cosmic-rays can be treated as an adiabatic compression heating of a "cosmic-ray gas molecule". The energy enhancement factor is

$$\ln(E/E_0) = (\gamma - 1)\alpha T_{cr} \qquad (4.18)$$

where $\gamma = \frac{4}{3}$ for relativistic cosmic-rays, $\gamma = \frac{5}{3}$ for sub-relativistic cosmic rays. For a cosmic ray differential energy spectrum of the form $I_{cr} \propto E^{-\Gamma}$, we then find

$$\delta_{ACC} = (\Gamma - 1)(\gamma - 1)\alpha T_{cr}$$

and therefore $\qquad (4.19)$

$$\delta = [(\Gamma - 1)(\gamma - 1) + 1]\alpha T_{cr}$$

(Stecker et al. 1974). The value of the specific volume compression rate $\alpha\,(\overset{\cdot}{\omega})$ can be given in terms of the radial expansion velocity v_r of the gas deduced from the 21 cm observations of the "expanding arm" feature. Under this interpretation

$$R_\odot \alpha = \frac{dv_r}{d\tilde{\omega}} - \frac{v_r}{\tilde{\omega}} \qquad (4.20)$$

where the galactocentric distance $R = R_\odot \tilde{\omega}$.

The function $\alpha(\tilde{\omega})$ is deduced from the observations of $v_r(\tilde{\omega})$ Shane 1972, Sanders and Wrixon 1972, 1973). It is positive and maximal in the region of observed maximum γ-ray emission (see Figure 14) and is negative in the inner region $\tilde{\omega} < 3$ kpc where there may be a significant drop in γ-ray emission, although the uncertainties in that region are very large.

Because of the large uncertainties in T_{cr} and $T_\odot$ Puget and Stecker (1974) considered a range of values for T_{cr} such that $3 \times 10^6 < T_{cr} < 3 \times 10^7$ yr where the upper limit on T_{cr} is taken to be the deduced age of the expanding feature (Oort 1970 Van der Kruit 1971). The cosmic-ray lifetime in the local region, $T_\odot$, is given by O'Dell et al. (1973) and Brown et al. (1973), to be 10^6 yr $\lesssim T \lesssim 10^7$ yr. Taking $\alpha = 2 \times 10^{-15}$ s^{-1} in the region of maximum compression and assuming $I_{cr}/I_\odot \simeq 15$ and $Q_{cr}/Q_\odot \simeq 2$, $\Gamma = 2.5$, $\delta = 1.75$, $\alpha T_{cr} = 1.5 \times 10^{-15} T_{cr}$, the values given for the parameters $T_{cr}/T_\odot$, e^δ and T_{cr} and $T_\odot$ consistent with equation (4.15) are given in Table I. The factor e^δ represents the enhancement effect due to acceleration and compression.

Aside from the possibility that cosmic-ray enhancement accounts for the large γ-ray emissivity in the 5 kpc region, recent carbon monoxide emission measurements in a 2.6 mm microwave survey of the galaxy by Scoville and Solomon have given new life and strong emphasis to the idea that much if not all of the increase in the γ-ray emissivity could be due to interaction between cosmic rays and large amounts of H_2 not seen in 21 cm emission (Stecker 1969a, 1971a; Stecher and Stecker 1970). However, the recent measurements of Scoville and Solomon point up this possibility in a much more specific and meaningful way, since their measurements indicate an increased enhancement in CO and by

$T_{cr}/T_\odot$	e^δ	$T_{cr}(10^6 Y)$	$T_\odot(10^6 Y)$
2	3.7	12.6	6.3
3	2.5	8.7	2.9
4	1.9	6.0	1.5

Table I: Lifetime of Cosmic-Rays in the 4 to 5 kpc Region Calculated from the Acceleration-Compression Model.

implication an increased H_2 density in the very region where the γ-ray emissivity appears to be enhanced. This has led Solomon and Stecker (1974) to suggest that both enhancements provide evidence for the existence of a galactic H_2 "ring" or "arm" feature at ∿ 5 kpc.

Molecular hydrogen is expected to be the predominant form of hydrogen in cool clouds of sufficient density (Solomon and Wickramasinghe (1969), Hollenbach and Salpeter (1971), Hollenbach, Werner and Salpeter (1971). However, it is difficult to measure its galactic distribution directly. Strong H_2 absorption lines have been seen in the UV in almost all nearby clouds by the Copernicus satellite (Spitzer et al. 1973), but UV observations of H_2 at distances greater than 1 kpc from us are not feasible because of the large extinction of UV radiation by interstellar dust.

Because H_2 has no permanent dipole moment, it cannot emit electric dipole radiation and, even though quadrupole vibration-rotation feature can in principle be observed in the infrared, such features are inherently very weak.

CO emission can be used as a tracer of H_2 in molecular clouds because the most importance source of CO excitation in these clouds is by collisions with H_2.

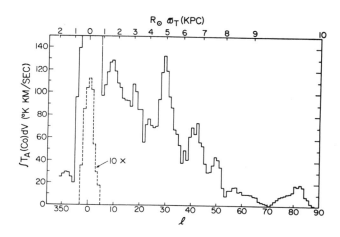

Fig. 18: The intensity distribution of 2.6mm line emission in the galactic plane from the $J = 1 \to 0$ transition of carbon monoxide integrated over velocity as a function of galactic longitude (Scoville and Solomon, to be published; Scoville, Solomon and Jefferts 1974)

The distribution obtained by Scoville and Solomon (to be published) is shown in figure 18. The data represent the antenna temperature integrated over the velocity profile of the 2.54 mm CO emission line and would therefore be proportional to the number of CO molecules along the line-of-sight in the optically thin case as we discussed previously in connection with the 21 cm line. Measurements were also obtained of the velocity profile function (ℓ, v) for each $1°$ of longitude from $0°$ to $90°$.

The function $T_A (\ell, v)$ can be converted into a galactocentric distance distribution by using a model for galactic rotation (Schmidt 1965). Such an unfolding is shown in figure 19 which can be compared with the γ-ray emissivity function given in figure 14.

Solomon and Stecker (1974) have used the CO emission data and the γ-ray emission data to place limits on the mean density of H_2 in the 5 kpc feature. They have obtained

$$1 \lesssim n_{H_2} (5 \text{ kpc}) \lesssim 5 \text{ molecules/cm}^3 \qquad (4.21)$$

The lower limit closely corresponds to the optically thin case for CO emission and the upper limit corresponds to optically thick emission of the type observed in nearby dark clouds. In the optically thick case for $C^{12}O$ in nearby clouds, emission from $C^{13}O$ is optically thin and one can use the abundance ratio of

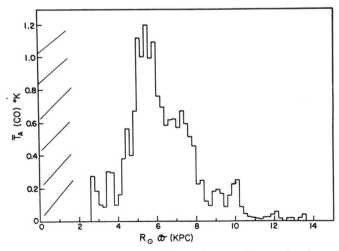

Fig. 19: The distribution of CO line emission as a function of galactocentric distance (labelled here $R_\odot \tilde{\omega}$) in kpc.

C^{13}/C^{12} to deduce that τ is in the range 10-20. It should be kept in mind that in order to obtain enough collisional excitation of CO to the $J = 1$ rotational state to produce a measurable amount of $J = 1 \to 0$ 2.64 mm emission, the H_2 density in the cloud must be in the range $10^3 \lesssim H_2 \lesssim 10^4$ cm^{-3}. Therefore, equation (4.21) represents only a smoothed out mean of a very uneven distribution of H_2. If one takes the upper limit in equation (4.21) for H_2 at 5 kpc, all of the γ-ray emissivity increase could be due to gas and the cosmic-ray intensity need not increase. This point has been stressed by Dodds et al. (1974) to argue the consistency of this data with the extragalactic cosmic-ray origin hypothesis.

It can thus be seen how γ-ray astronomy gives us another window on the study of cosmic rays and galactic structure and dynamics. The various implications of the SAS-2 data we have just discussed represent only a first step toward our understanding of γ-ray emission in the galaxy. Improved γ-ray telescopes will someday provide us with a clearer picture of the galaxy.

V. EXTRAGALACTIC γ-RAYS

Much data have recently been obtained on the diffuse γ-ray background radiation which appears to be isotropic in origin. These data have now defined a continuous background spectrum up to an energy of 200 MeV. They are summarized in figure 20*.

The cosmological nature of this radiation leads us to examine its origin in the context of the expanding universe model where such radiation is redshifted to lower energies as we see it because of the Doppler effect.

The simplest expanding universe model is one in which the universe is both homogeneous and isotropic. Such a model is quite adequate for our purposes here and support for it comes from the remarkable isotropy of the 2.7K microwave background radiation which is believed to have originated much earlier in the history of the universe than the observed γ-ray background. A homogeneous-isotropic universe can be described by the Robertson-Walker space-time metric of the form

$$ds^2 = c^2 dt^2 - R^2(t) du^2 \tag{5.1}$$

---

* Recent data reported by Tanaka (1974) in the energy range up to 7.5 MeV are not shown in the figure but are consistent with other data in this range.

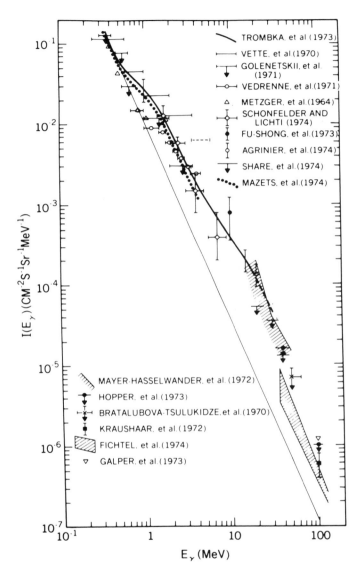

Fig. 20: Observational data on the gamma-ray background energy spectrum. The highest energy point of Vette et al. (1970), shown with a dashed line, and possibly the neighbouring point are now thought to be erroneously high due to an inefficiency in the anticoincidence circuit of their detector which should not significantly affect the points at lower energies (Vette, private communication).

where $R(t)$ is a scale factor which describes the expansion of the universe as a function of time according to the solution of the Einstein equations of general relativity.

Photons travel along the null geodetic which obeys the relation $ds^2 = 0$, i.e. from (5.1)

$$u = c \int_{t_e}^{t_r} \frac{dt}{R(t)} \tag{5.2}$$

with t_e being the time the photon was emitted and t_r being the time the photon is received.

The emitting and receiving points are embedded in the metric so that the distance between them is changing by the scale factor $R(t)$; the dimensionless metric distance, u, is a constant. If we therefore consider two successive wave crests of a light ray as being emitted at times $t_e + \Delta t_e$ respectively and being received at times t_r and $t_r + \Delta t_r$, then

$$\int_{t_e}^{t_r} \frac{dt}{R(t)} = \int_{t_e + \Delta t_e}^{t_r + \Delta t_r} \frac{dt}{R(t)} = u = \text{const.} \tag{5.3}$$

Thus

$$\int_{t_e + \Delta t_e}^{t_r + \Delta t_r} \frac{dt}{R(t)} - \int_{t_e}^{t_r} \frac{dt}{R(t)} = \int_{t_r}^{t_r + \Delta t_r} \frac{dt}{R(t)} - \int_{t_e}^{t_e + \Delta t_e} \frac{dt}{R(t)}$$

$$= \frac{\Delta t_r}{R(t_r)} - \frac{\Delta t_e}{R(t_e)} = 0 \tag{5.4}$$

or

$$\frac{\Delta t_e}{R(t_e)} = \frac{\Delta t_r}{R(t_r)}$$

Since the wavelength of the emitted wave is $c\Delta t_e$ and that of the wave when received is $c\Delta t_r$, the wavelength is shifted by the amount

$$z \doteq \frac{\lambda_r - \lambda_e}{\lambda_e} = \frac{\Delta \lambda}{\lambda} = \frac{R(t_r) - R(t_e)}{R(t_e)}$$

or

$$\frac{R(t_r)}{R(t_e)} = 1 + z \qquad (5.5)$$

We have observed this shift in the spectral lines of distant galaxies as always being toward longer wavelength so that $R(t_r) > R(t_e)$. From this evidence, it has therefore been deduced that our universe is expanding with time.

Gamma-Ray Fluxes

Let us now consider the effect of cosmological factors in calculating gamma-ray fluxes emitted at large redshifts, z. The number of photons received per second is reduced by a factor $R(t_e)/R(t_r)$ from the number produced per second at time, t_e. We consider here gamma rays produced in particle collisions between two components having densities $n_a(t_e)$ and $n_b(t_e)$ respectively. We specify the differential photon intensity produced per collision as

$$G(E_\gamma) \quad (cm^2 \text{ s Sr Mev } cm^6)^{-1}$$

Then the differential photon flux received at t_r is given by

$$dF_r = \frac{4\pi n_a(t_e) n_b(t_e) G(E_{\gamma,e}) dE_{\gamma,e} dV_e dt_e}{4\pi R^2(t_r) u^2} \qquad (5.6)$$

where the numerator represents the photon flux emitted at t_e, and the denominator indicates the fact that at t_r this flux is evenly distributed over a spherical wavefront of radius $R(t_r)$. We now define the three dimensional length,

$$d\ell = R(t) du \qquad (5.7)$$

so that the volume element

$$dV_e = d\ell [R^2(t_e) u^2 d\Omega] \qquad (5.8)$$

Since

$$dt_e = [R(t_e)/R(t_r)] dt_r$$

and $\qquad (5.9)$

$$dE_{\gamma,e} = [R(t_r)/R(t_e)] dE_{\gamma,r}$$

because the energy of a gamma-ray is inversely proportional to its wavelength, we may substitute (5.8) and (5.9) into (5.7) and obtain

$$dF_r = n_a(t_e)n_b(t_e)G\left[[R(t_r)/R(t_e)]\right]E_{\gamma,r}$$

$$\times \frac{4\pi R^2(t_e)u^2 d\Omega d\ell dE_{\gamma,r} dt_r}{4\pi R^2(t_r)u^2} \tag{5.10}$$

By making use of equation (5.5) and dropping the subscript, r, since we only measure gamma-rays when they are received, equation (5.10) reduces to

$$\frac{dF}{d\Omega dt dE_\gamma} = dI = \frac{n_a(z)n_b(z)G\left[(1+z)E_\gamma\right]d\ell}{(1+z)^2}$$

$$= \frac{n_a(z)n_b(z)G\left[(1+z)E_\gamma\right]}{(1+z)^2}\left(\frac{d\ell}{dz}\right)dz \tag{5.11}$$

Equation (5.11) is quite useful in evaluating the metagalactic gamma-ray spectra from various high-energy interactions. The results are obtained from numerical integration of the relation

$$I(E_\gamma) = \int_0^{z_{max}} dz\, n_a(z)n_b(z)\frac{\left[G\,(1+z)E_\gamma\right]}{(1+z)^2}\left(\frac{d\ell}{dz}\right) \tag{5.12}$$

where the factor $d\ell/dz$ is determined from the Einstein field equations to be

$$\frac{d\ell}{dz} = \frac{c}{H_o}(1+z)^2(1+\Omega z)^{-\frac{1}{2}} \tag{5.13}$$

where H_o is the Hubble constant and

$$\Omega = \frac{n_o}{n_c} \tag{5.14}$$

where n_o is the present mean density of all the matter in the universe and n_c is the critical density needed to gravitationally close the universe.

The value for cH_o^{-1} is in the range

$$10^{28} \lesssim cH_o^{-1} \lesssim 2 \times 10^{28} \text{ cm} \tag{5.15}$$

and

$$3 \times 10^{-6} \lesssim n_c \lesssim 10^{-5} \text{ cm}^{-3} \tag{5.16}$$

If absorption effects are included, equation (5.12) becomes

$$I(E_\gamma) = \frac{c}{H_o} \int_0^{z_{max}} dz \, n_a(z) n_b(z) \frac{G[(1+z)E_\gamma]}{(1+z)^4 \sqrt{1+\Omega z}} e^{-\tau(E_\gamma, z)} \tag{5.17}$$

Since the universal blackbody radiation plays an important role in various high-energy production processes, it is important to determine how this radiation behaves as a function of redshift. At the earliest stages in the evolution of the universe, when the matter in the universe was in thermal equilibrium with the universal blackbody radiation owing to Compton interactions between thermal photons and electrons, T_m was equal to T_r (the temperature of the radiation field). Zeldovich et al. (1969) have shown that even though the intergalactic gas has cooled to the 50% neutral point by the time corresponding to a redshift $z = 1200-1300$, the small fraction of ionized material left at lower redshifts is enough to sustain temperature equilibrium between matter and radiation field until a redshift of 150-200.

The photon density of a radiation field at temperature T_o is given from the Planck formula as

$$n_{r,o}(\epsilon, T_o) d\epsilon = \frac{8\pi}{h^3 c^3} \frac{\epsilon^2 d\epsilon}{e^{\epsilon/kT_{r,o}} - 1} \tag{5.18}$$

At a redshift $z \neq 0$, this distribution is then given by

$$n_r(z, \epsilon) d\epsilon = \frac{8\pi}{h^3 c^3} \frac{(1+z)^3 \epsilon^2 d\epsilon}{e^{(1+z)\epsilon/kT_{r,o}} - 1}$$

$$= (1+z)^3 n_{r,o}[\epsilon, T_r(z) d\epsilon] \tag{5.19}$$

where

$$T_r(z) = (1+z) T_{r,o} \tag{5.20}$$

so we find that the redshifted Planckian maintains its form with the parameter $T(z)$ in the exponent given by equation (5.20). The photon number density redshifts in the same manner as the density viz. $n \propto R(t)^{-3} \propto (1+z)^3$. The energy density in the radiation field then redshifts as

$$\rho_r = \rho_{r,o}(1+z)^4 \qquad (5.21)$$

in accord with the definition of $T_r(z)$ given by equations (2.11-2.12) and the relation $\rho \propto T^4$.

We can use these equations to take into account the cosmological effects in computing the extragalactic gamma-ray spectrum from Compton interactions between cosmic-ray electrons and photons of the universal blackbody radiation field in intergalactic space. The local ($z = 0$) spectrum is given from equation (2.11) as

$$I_{e,o}(E_\gamma) = \frac{2}{3}\sigma_T cH_o^{-1} f(\Gamma)(mc^2)^{(1-\Gamma)}(3.6k)^{\left(\frac{\Gamma-3}{2}\right)} K_{e,o}\rho_{r,o}$$

$$\times T_{e,o}^{\left(\frac{\Gamma-3}{2}\right)} E_\gamma^{-\left(\frac{\Gamma+1}{2}\right)} \qquad (5.22)$$

for Compton gamma-rays produced by cosmic-ray electrons having a power law energy spectrum of the form

$$I(E_e) = K_{o,e} E_e^{-\Gamma} \qquad (5.23)$$

Under the conditions where (5.23) is valid over all redshifts from 0 to z_{max} and where <u>electron energy losses other than those from the universal expansion are neglected</u>, the transformation from $K_{e,o}$ to $K_e(z)$ is given by

$$\begin{aligned} K(z) &= K_{e,o}(1+z)^3 (1+z)^{\Gamma-1} \\ &= K_{e,o}(1+z)^{\Gamma+2} \end{aligned} \qquad (5.24)$$

where the first factor of $(1+z)^3$ represents the density effect and the factor of $(1+z)^{\Gamma-1}$ represents the transformation of the power-law energy spectrum from a burst of relativistic cosmic-ray electrons occurring at $z_{max} > z$.

Using equations (5.21) and (5.24) and substituting them into the general formula (5.12), we find

$$I_c(E_\gamma) = I_{c,o}(E_\gamma) \int_0^{z_{max}} dz \frac{(1+z)^\Gamma}{\sqrt{1+\Omega z}} \qquad (5.25)$$

so that under these conditions, the power-law form for the Compton γ-ray spectrum is maintained but the cosmological flux is enhanced by a factor given by the integral over z in equation (5.25).

However, Brecher and Morrison (1967) showed that the true situation is not as simple as that given by equation (5.25) under the conditions when the high energy end of the electron spectrum is steepened owing to energy losses from the Compton interactions themselves. Since the blackbody radiation density is proportional to $(1+z)^4$, electrons lose energy primarily from Compton interactions with the energy loss rate being given by

$$\left(\frac{d\gamma}{dt}\right)_c = -\frac{4}{3}\frac{\sigma_T c \rho_\gamma}{m_e c^2}\gamma^2 = \frac{(1+z)^4}{\tau_o}\gamma^2 \quad (5.26)$$

with

$$\tau_o = \frac{3 m_e c^2}{4\sigma_T c \rho_{r,o}} = 7.7 \times 10^{19} \text{ s} \quad (5.27)$$

The energy loss from the universal expansion is given by

$$\left(\frac{d\gamma}{dz}\right)_{exp} = \frac{\gamma}{1+z} \quad (5.28)$$

From (5.28) and (5.13) we obtain

$$\left(\frac{d\gamma}{dz}\right)_c = \left(\frac{d\gamma}{dt}\right)_c \frac{dt}{dz} = \frac{(1+z)^2 \gamma^2}{H_o \tau_o \sqrt{1+\Omega z}}$$

The two energy loss rate terms are equal at the critical energy, $E_c = \gamma_c m_e c^2$, where

$$\gamma_c = \frac{H_o \tau_o \sqrt{1+\Omega z}}{(1+z)^3} \simeq \frac{256(1+\Omega z)^{\frac{1}{2}}}{(1+z)^3} \quad (5.30)$$

or

$$E_c \simeq 130 \frac{\sqrt{1+\Omega z}}{(1+z)^3} \text{ MeV} \quad (5.31)$$

For $E < E_c$, the electron spectrum maintains its power-law form since these electrons lose their energy through the universal expansion and equation (5.24) is therefore valid. However, for $E < E_c$, the equilibrium electron flux is given by the solution of a rate balance equation under the conditions when Compton losses dominate. Under these conditions,

$$\frac{\partial}{\partial \gamma}\left[K_e \gamma^{-\Gamma'} \frac{(1+z)^4 \gamma^2}{\tau_o}\right] = k_q \gamma^{-\Gamma} \quad (5.32)$$

(Brecher and Morrison, 1967)

where $k_q \gamma^{-\Gamma'}$ is the original (injection) electron spectrum and is the exponent of the resulting equilibrium electron spectrum. It follows from equation (5.32) that

$$\Gamma' = \Gamma + 1 \qquad (5.33)$$

and

$$K_e = \frac{k_q \tau_o}{\Gamma(1+z)^4} \qquad (5.34)$$

so that the equilibrium electron spectrum is depleted by a factor proportional to $(1+z)^4$ by Compton interactions with the universal radiation field and, in addition, the exponent of the electron spectrum is steepened by one power of E_e.

This steepening in the electron spectrum corresponds to a change of 1/2 in the exponent of the Compton gamma-ray spectrum at an energy

$$E_{\gamma,c} = <\tfrac{4}{3} E_o> \gamma_c^2 = 33 \left[\frac{1+\Omega z}{(1+z)^5}\right] \text{ eV} \qquad (5.35)$$

Thus, the Compton gamma-rays are expected to be produced by the steepened electron spectrum at all gamma-ray energies. From (5.34) Brecher and Morrison conclude that unless there is a large evolutionary factor in the electron production spectrum, i.e. $k_q \sim (1+z)^m$, where $m \gtrsim 4$, there will be no significant enhancement of Compton gamma-rays at large redshifts over those produced at the present epoch.

The local gamma-ray spectrum from relativistic bremsstrahlung interactions is given by (2.33) as

$$I_{b,o}(E_\gamma) = 3.4 \times 10^{-26} \, n_o \, cH_o^{-1} \, \frac{I_e(>E_\gamma)}{E_\gamma} \qquad (5.36)$$

Making use of equations (5.9) and (5.24), we obtain for the cosmological bremsstrahlung production spectrum

$$I_b(E_{\gamma o}, z_{max}) = \frac{3.4 \times 10^{-26} \, n_o \, cH_o^{-1} \, K_{e,o}}{\Gamma - 1}$$

$$\times \int_o^{z_{max}} dz \, \frac{(1+z)^3 \, (1+z)^{\Gamma+2} \, [(1+z)E_{\gamma o}]^{-\Gamma}}{(1+z)^4 \, \sqrt{1+\Omega z}} \qquad (5.37)$$

$$= \frac{3.4 \times 10^{-3}\Omega}{\Gamma-1} K_{e,o} E_{\gamma,o}^{-\Gamma} \int_0^{z_{max}} dz \frac{(1+z)}{\sqrt{1+\Omega z}}$$

$$= \frac{6.8 \times 10^{-3}}{\Gamma-1} K_{e,o} E_{\gamma,o}^{-\Gamma} \left[\left(1 + \frac{z}{3} - \frac{2}{3\Omega}\right)\sqrt{1+\Omega z} - \left(1 - \frac{2}{3\Omega}\right)\right]$$

In particular, for the Einstein-de Sitter universe where $\Omega = 1$,

$$I_b(E_{\gamma,o}, z_{max}) = \frac{2}{3} I_{b,o}(E_{\gamma,o}) \left[(1+z)_{max}^{3/2} - 1\right] \quad (5.38)$$

and for the low density model where $\Omega z \ll 1$,

$$I_b(E_{\gamma,o}, z_{max}) = \frac{1}{2} I_{b,o}(E_{\gamma,o}) \left[(1+z_{max})^2 - 1\right] \quad (5.39)$$

Where electron steepening by Compton interactions with the 2.7K radiation is important, the situation is more complicated and a more detailed treatment of this problem has been given by Stecker and Morgan (1972).

VI. GAMMA-RAY ABSORPTION PROCESSES AT HIGH REDSHIFTS

In Section III we discussed in detail the processes which result in the absorption of cosmic gamma-rays. In this section, we will show how the effectiveness of these absorption processes is enhanced at high redshifts due to increased photon and matter densities at these redshifts when the universe was in a more compact state. We will also derive the energy-dependences of these absorption processes, taking cosmological factors into account.

We begin our discussion by considering the absorption of cosmic gamma-rays by pair-production interactions with photons of the universal blackbody radiation field, viz., reaction (2.47) whose absorption coefficients are given by equations (3.8) - (3.10).

For cosmological applications, we must take into account the redshift dependences of T and E_γ in an expanding universe,

$$T = T_o(1+z) \quad (6.1)$$

and

$$E_\gamma = E_{\gamma,o}(1+z)$$

where the subscript zero refers to presently observed ($z = 0$) quantities, so that $T_o = 2.7K$.

Taking the z-dependence into account, we then find that the condition $\nu \gg 1$ in equation (3.7) is applicable in the energy range

$$E_\gamma \ll \frac{1.12 \times 10^6 \text{ GeV}}{(1+z)^2} \tag{6.2}$$

The optical depth of the universe to gamma-rays is then given by

$$\tau(E_\gamma, z_{max}) = \int_0^{\ell_{max}(z_{max})} d\ell \, \kappa_{\gamma\gamma}(E_\gamma, z) \tag{6.3}$$

$$= \int_0^{z_{max}} dz \, \kappa_{\gamma\gamma}(E_\gamma, z) \left(\frac{d\ell}{dz}\right)$$

where, from equation (5.10)

$$\frac{d\ell}{dz} \simeq \frac{10^{28} \text{ cm}}{(1+z)^2 (1+10^5 n_o z)^{\frac{1}{2}}} \tag{6.4}$$

n_o being the present mean atomic density of all the matter in the universe. We will consider here two types of model universes: 1) a "flat" or Einstein - de Sitter model with $n_o \sim 10^{-5}$ cm^3 and 2) an "open" model with $n_o \ll 10^{-5} z$.

For the flat model, equation (6.3) reduces to

$$\tau(E_{\gamma,o}, z_{max}) = 3.9 \times 10^8 \, E_{\gamma,o}^{-\frac{1}{2}} \int_0^{z_{max}} dz \frac{\exp-\left[1.12 \times 10^6/(1+z)^2 E_{\gamma,o}\right]}{(1+z)^{\frac{1}{2}}} \tag{6.5}$$

with $E_{\gamma,o}$ in GeV.

For $z_{max} \gg 1$, equation (6.5) can be further simplified to yield

$$\tau(E_{\gamma,o}, z_{max}) \simeq 1.7 \times 10^2 \, E_{\gamma,o}^{\frac{1}{2}} (1+z_{max})^{\frac{1}{2}} \exp-\left[\frac{1.12 \times 10^6}{(1+z_{max})^2 E_{\gamma,o}}\right] \tag{6.6}$$

A numerical solution found by setting equation (6.3) for $(E_{\gamma,o}, z_{crit}) = 1$, which defines the critical redshift where the universe becomes opaque to gamma-rays of energy $E_{\gamma,o}$ can be well approximated by the expression

$$1 + z_{crit} \simeq 2.60 \times 10^2 \, E_{\gamma,o}^{-0.484} \tag{6.7}$$

For the open model, we find

$$\tau(E_{\gamma,o}, z_{max}) = 3.9 \times 10^8 \, E_{\gamma,o}^{-\frac{1}{2}} \int_0^{z_{max}} dz \, \exp - \left[\frac{1.12 \times 10^6}{(1+z)^2 E_{\gamma,o}} \right] \tag{6.8}$$

For $z_{max} \gg 1$, equation (6.8) may be approximated by

$$\tau(E_{\gamma,o}, z_{max}) \simeq 1.7 \times 10^2 \, E_{\gamma,o}^{\frac{1}{2}} (1+z_{max}) \exp - \left[\frac{1.12 \times 10^6}{(1+z)^2 E_{\gamma,o}} \right] \tag{6.9}$$

Thus, there is no significant difference between the opacities of the open and flat model universes. This being the case, we may invert equation (6.7) to obtain an expression for the predicted cut-off energy E_c, above which gamma-rays originating at a redshift, z_{max}, cannot reach us. This relation is then given by

$$E_c \simeq \left(\frac{2.60 \times 10^2}{1 + z_{max}} \right)^{2.06} \tag{6.10}$$

and is graphed in figure 21.

In the other extreme, $\nu \ll 1$, we find that as we consider higher and higher energies, the universe will not become transparent to gamma-rays again until we reach an energy E_{tr}, where the optical depth, $\tau(E_{tr}, z_{max})$, again falls to unity. The expression for the optical depth when $\nu \ll 1$ in equation (3.10) is given for a flat universe by

$$\tau(E_{\gamma,o}, z_{max}) \simeq 4.4 \times 10^{13} \, E_{\gamma,o}^{-1} \int_0^{z_{max}} dz \, (1+z)^{-3/2} \tag{6.11}$$

$$\simeq 8.8 \times 10^{13} \, E_{\gamma,o}^{-1} (1+z_{max})^{-\frac{1}{2}} \quad \text{for } z_{max} \gg 1$$

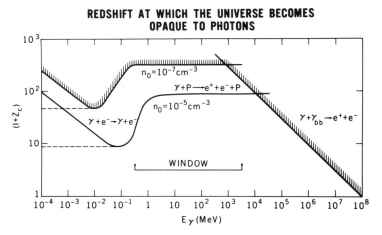

Fig. 21: The redshift at which the universe becomes opaque to photons given as a function of <u>observed</u> gamma-ray energy. Gamma-rays originating at all redshifts below the curve can reach us unattenuated with the energy indicated. The two curves on the left side of the figure are for attenuation by Compton scattering with intergalactic electrons having the densities indicated and for pair production and are based on the calculations of Arons and McCray (1969). The right-hand curve results from attenuation of gamma-rays by interactions with the microwave blackbody radiation and is based on the discussion of Fazio and Stecker (1970).

---

and for an open universe by

$$\tau(E_{\gamma,o}, z_{max}) \simeq 4.4 \times 10^{13} \, E_{\gamma,o}^{-1} \int_0^{z_{max}} dz \, (1+z)^{-1} \quad (6.12)$$

$$\simeq 4.4 \times 10^{13} \, E_{\gamma,o}^{-1} \, \ln(1+z_{max})$$

In both cases we find that $E_{tr} > 10^{13}$ GeV so that we may safely assume that the universe, due to the blackbody radiation field, is essentially opaque to gamma-rays of all energies greater than E_c.

Gamma-ray absorption by pair production and Compton interactions with intergalactic gas at high redshifts has been examined by Rees (1969) and by Arons and McCray (1969). Our discussion here essentially follows theirs. The absorption cross section for Compton interactions is given in equations (3.10) and (3.11); that for pair production in equation (3.16). The cross sections for these processes are graphed in figure 7. The total absorption cross section above 100 MeV energy is roughly constant and equals 1.8×10^{-26} cm^2. For the case of a constant absorption cross section, the optical depth as a function of redshift, z_{max}, is given by equations (6.3) and (6.4) as

$$\tau(z_{max}) = \int_0^{z_{max}} dz\, n(z)\, \sigma\left(\frac{d\ell}{dz}\right)$$

$$= \frac{n_c \sigma c \Omega}{H_o} \int_0^{z_{max}} dz\, \frac{(1+z)}{\sqrt{1+\Omega z}} \qquad (6.13)$$

$$= \frac{2}{3} \tau_c \left[\left(3+z - \frac{2}{\Omega}\right)\sqrt{1+\Omega z} - \left(3 - \frac{2}{\Omega}\right)\right]$$

where

$$\tau_c \equiv n_c \sigma c H_o^{-1} = 1.8 \times 10^{-3} \qquad (6.14)$$

a result which was first obtained by Gunn and Peterson (1965).

For an Einstein-de Sitter universe ($\Omega=1$),

$$\tau(z_{max}) = \frac{2}{3} \tau_c \left[(1+z_{max})^{3/2} - 1\right] \qquad (6.15)$$

and for a low density universe ($\Omega z \ll 1$)

$$\tau(z_{max}) = \frac{\Omega}{2} \tau_c \left[(1+z_{max})^2 - 1\right] \qquad (6.16)$$

At lower energies, when the cross section is not constant, the calculation is more complex.

Figure 21 shows the critical redshift for absorption of γ-radiation plotted as a function of observed energy. At lower energies absorption is due to Compton interactions with intergalactic matter, in the intermediate range absorption is due to pair-production interactions with intergalactic matter (Arons and McCray 1969, Rees 1969). At the higher energies absorption is due to pair production interactions with blackbody photons (Fazio and Stecker 1970). As one can see from the figure, there is a natural "window" between ∼ 1 MeV and ∼ 10 GeV which defines the optimal energy range for studying cosmological γ-rays.

In the case of pair production interactions, i.e. those which are important for $E_\gamma \gtrsim 1$ MeV as shown in figure 21, the photon completely disappears and the effect is a true absorption effect. In those cases, particularly above 10 GeV, z_c can be used as an upper limit on the integral given by equation (5.12) for evaluating the background energy spectrum. This immediately implies a steepening of the background spectrum above 10 GeV. However, in the case of Compton scattering, important for $E_\gamma \lesssim 1$ MeV, we do not have a pure absorption process but rather a process in which a photon of some initial energy E' is replaced by one at some lower energy E. Thus, in order to calculate the theoretical background spectrum properly one must solve an integrodifferential scattering equation (Arons 1971 a,b). We will refer to this equation as the cosmological photon transport (CPT) equation. We can write this equation in the form

$$\frac{\partial \mathcal{J}}{\partial t} + \frac{\partial}{\partial E}[-EH(z)\mathcal{J}] = \mathcal{Q}(E,z) - \kappa_{AB}(E,z) + \int_E^{\varepsilon(E)} dE' \, \kappa_{sc}(E,z) \, \mathcal{J}(E|E') dE' \quad (6.17)$$

where E is the photon energy, κ_{AB} and κ_{SC} are the photon absorption and scattering rates (which are a function of z because the intergalactic gas density is assumed to scale as $(1+z)^3$ because of the expansion of the universe). The script quantities for the γ-ray intensity and production rate

$$\mathcal{J}(E,z) \equiv (1+z)^{-3} I(E,z)$$

and (6.18)

$$\mathcal{Q}(E,z) \equiv (1+z)^{-3} Q(E,z)$$

are quantities with co-moving dependence cancelled out. $\varepsilon(E)$ is an upper limit on the scattering integral defined by the Compton process and H(z) is the Hubble parameter which, in terms of the Hubble constant H_0, is given by the relation

$$H(z) = H_0 (1+z) \sqrt{1+\Omega z}$$

where Ω is the ratio of the mean gas density in the universe to the density needed to close the universe gravitationally. The term

$$\frac{\partial \mathcal{J}}{\partial t} = -(1+z) H(z) \frac{\partial \mathcal{J}}{\partial z} \quad (6.20)$$

and the second term in equation (6.17) expresses the energy loss of the γ-rays because of the expansion redshift.

The third term in equation (6.17) is the source function or production function of γ-radiation produced with energy E at redshift z. The fourth term represents the true absorption due to pair-production processes as well as the total scattering cross section for Compton scattering as a function of redshift and energy. Because all of the scattered photons "disappear" at energy E' but reappear at some lower energy E, the integral term in the right-hand side of the equation has to be included. This term is equivalent to a source term of photons at lower energies. The upper limit on the integral is determined by the kinematics of Compton scattering and is given by

$$\varepsilon(E) = \begin{cases} E/[m_e c^2(1-2E/m_e c^2)], & E < \tfrac{1}{2} m_e c^2 \\ \infty, & E \geq \tfrac{1}{2} m_e c^2 \end{cases} \quad (6.21)$$

The CPT equation was solved numerically by Stecker, Morgan and Bredekamp (1971) in their study of the spectrum of background γ-radiation to be expected from cosmological matter-antimatter annihilation.

VII. INTERPRETATION OF THE DIFFUSE γ-RAY BACKGROUND OBSERVATIONS

It must have been evident very early, in an implicit way that the only source of matter large enough to give a significant background of <u>isotropic</u> radiation of a truly astronomical nature is the universe <u>itself</u>. Therefore, the connection with cosmology has been clearly the prime motivation for interest in an isotropic background radiation since Morrison's 1958 paper.

The mechanisms listed by Morrison (1958) to be of possible significance in producing continuum radiation were synchrotron radiation, cosmic-ray electron bremsstrahlung, π^0-meson decay (from cosmic ray-nucleon interactions) and antimatter annihilation. To these four, one more mechanism, viz. Compton interactions between cosmic-ray electrons and starlight (Felten and Morrison, 1963) was added. It later became apparent when the 3K blackbody radiation was discovered that these photons would be orders of magnitude more numerous than starlight photons in intergalactic space as targets for cosmic-ray electrons and should thus be the prime Compton radiation generators of cosmological interest. That the significance of this was readily grasped is obvious from the plethora of independent suggestions made immediately after the discovery of the microwave background radiation (Felten 1965, Gould 1965, Hoyle 1965, Fazio, Stecker and Wright 1966, Felten and Morrison 1966). The relative weakness of intergalactic magnetic fields, evidenced by data on the non-thermal radio background, eliminated synchrotron radiation as a prime contender in generating the diffuse gamma-ray background so that four mechanisms were left, (a) electron bremsstrahlung, (b) electron-photon interactions (Compton effect), (c) cosmic-ray produced π^o-decay, (d) decay of π^o-mesons from matter-antimatter annihilations.

Evidence for a diffuse background above 50 MeV was reported by Kraushaar and Clark (1962) from measurements from Explorer 11. The interpretation of Felten and Morrison (1963) that both the Ranger 3 and Explorer 11 results could be fitted reasonably well by a single power law of the type expected from Compton interactions seemed logical despite a possible flattening above 1 MeV reported by Arnold et al. (1962) and it was expected that data in the two energy decades between 1 and 100 MeV would exhibit nothing more exciting than a smooth power law spectrum as extrapolated from the sub-MeV ("X-ray") energies.

In the late 60's, the author after having made detailed thesis calculations of gamma-ray spectra from cosmic-ray produced secondary particles and from proton-antiproton annihilation (Stecker 1967), became interested in the effects cosmology might have on such spectra and on the implications of these effects for cosmology itself. Cosmic-ray π^o-decay was suspected to play a major role in generating galactic gamma-rays (Pollack and Fazio 1963) and it remained a viable possibility the extragalactic background flux above 100 MeV. But if such interactions are occurring in intergalactic space now, why not in the distant past when gas and cosmic-ray densities were higher (in an expanding universe)? If so, large fluxes of extragalactic cosmic-rays (comparable to galactic fluxes) need not exist now to explain the 100 MeV background (Stecker 1968b, 1969b). Also, the spectrum would be redshifted and would be softer than the galactic spectrum (Stecker 1969b). Similar ideas were being independently worked on by Ginzburg (1968) in the context of the Lemaitre cosmological model and by Rozental and Shukalov (1969) for the standard expanding universe model. In these models, various cosmological effects come into play to distort the spectrum from π^o-decay and redshift its characteristic peak from an energy of $m\pi c^2/2 \simeq 70$ MeV to lower energies (see extensive discussions in the monographs by Stecker (1971a) and Ozernoi, Prilutsky and Rozental (1973)). The result was the prediction of a possible enhancement in the gamma-ray background spectrum between 1 and 100 MeV deviating from the simple power law extrapolation of the X-ray background.

Figure 22 shows a two component model normalized for a best-fit to the observations involving the production of intergalactic gamma-rays from cosmic-ray interactions with intergalactic gas producing π^o-mesons out to a maximum redshift of 100 (Stecker

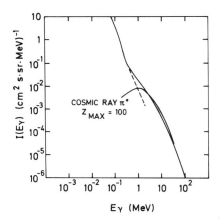

Fig. 22: Comparison of the observed background with a two-component model involving the production and decay of neutral pions produced in intergalactic cosmic-ray interactions at redshifts up to 100.

1969b,c, 1971b). Three problems arise with this explanation: 1) even with a relatively steep assumed cosmic-ray spectrum ($\sim E^{-2.7}$) the bulge in the theoretical spectrum may be too large to fit the observations, although this discrepancy may not be too serious considering observational uncertainties, 2) large amounts of energy are needed in cosmic-rays at high-redshifts, requiring the existence of strong primordial cosmic ray sources (protars) which are either pregalactic or protogalactic. Such objects may have been the "spinars" suggested by Morrison (1969), if spinars existed at such redshifts of about 70-100 (Stecker 1971b). If it is considered that each spinar produces approximately 10^{62} ergs over a time scale of 10^7-10^8 years, (Morrison 1969), a time comparable to the Hubble time at these redshifts, then at most 20 per cent of the presently observed galaxies are needed to have arisen from this early spinar state in order to provide the cosmic-ray energy needed to account for the diffuse γ-radiation above 1 MeV. At a redshift of about 70, the free-fall time for forming spinars from gas clouds is comparable to the Hubble time. This may provide a natural upper limit to the redshift, z_{max}, for primordial cosmic-ray production in the spinar model. (It should, however, be noted that such spinars may arise in other ways (see Stecker 1971b) and that they may now be a class of moribund objects unrelated to galaxies as we see them now). 3) The third problem with this hypothesis is that the maximum redshift for cosmic ray production z_{max} is a free parameter chosen to fit the observations.

A related hypothesis examined by Stecker et al. (1971) and Stecker and Puget (1972) is that the γ-ray background is from redshifted π^0-decay γ-rays but that π^0-mesons are the result of nucleon-antinucleon annihilation at an early epoch in the history of the universe. The annihilation hypothesis does not suffer from the above mentioned problems of the cosmic-ray protar hypothesis. The parameter z_{max} does not enter into the theory; annihilations occur at all redshifts and the 1 MeV-flattening is an absorption effect as discussed earlier. The transport equation (6.17) was solved to determine the exact form of the spectrum. Energy considerations do not present a problem. Another advantage of the theory is that it arises as a natural effect in a cosmology such as that suggested by Omnes (1972 and references therein). Figure 23 shows a detailed comparison of the annihilation hypothesis spectrum with present observations. The dashed line shows the effect of adding an additional component which is a power-law extrapolation of the γ-ray background. The two component model shown provides an excellent fit to the observational data.

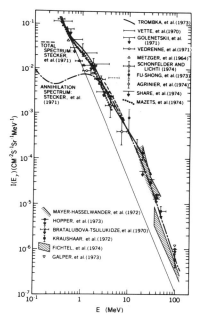

Fig. 23: A comparison of the data given in figure 20 with the annihilation model discussed by Stecker, Morgan and Bredekamp (1971) and Stecker and Puget (1972).

Early observations of gamma-radiation by Kraushaar and Clark (1962) had clearly indicated that if antimatter exists in the universe in large amounts, it must clearly be separated from matter so that the average annihilation rate is quite small. In 1969, Omnès suggested a baryon symmetric (equal amounts of matter and antimatter) cosmology based on a possible phase transition effect which could separate matter from antimatter at an early stage in the big-bang corresponding to nuclear density for the cosmic plasma. The phase transition effect was also studied by Aldrovandi and Caser (1972) and Cisneros (1973). Further work by Omnès (1972 and references therein) showed that the separate domains of matter and antimatter could grow to contain masses of the size of galaxies by the recombination epoch. This result has recently been refined by Aldrovandi et al. (1973). It was to be expected that boundary-region annihilations in this picture would also produce redshifted π^o-decay radiation and absorption effects would cut off the resultant flux below 1 MeV. Therefore, Stecker, Morgan and Bredekamp (1971) were motivated to make a detailed calculation of the resultant diffuse background spectrum to be expected, and the results agreed fairly well with the observations then available. The encouraging enhancement in the 1 to 100 MeV range is partially due to the existence of a "gamma-ray window" in this energy range as shown in figure 21. The results were encouraging enough to examine further the evolution of the Omnès Cosmology for redshifts less than 10^3 (Stecker and Puget 1972). This study had several exciting implications, (a) separate regions containing masses the size of galaxy clusters could be obtained, (b) turbulence produced by annihilation pressure could provide enough energy to trigger galaxy formation, (c) estimates obtained placed the galaxy formation stage at redshifts of the order of 60, (d) mean densities and angular momenta of galaxies could be estimated in this picture consistent with observation and related to the annihilation rates calculated by the model and implied by the observations.

The general scheme of the galaxy formation model is shown in figure 24. The observational implications of the model are outlined in figure 25.

Several other models of isotropic γ-ray production have been put forward. One suggestion is that the whole spectrum in the 10^{-3}-10^2 MeV range is due to Compton interactions of intergalactic electrons with the universal black-body radiation (Felten 1965, Gould, 1965, Hoyle 1965, Fazio, Stecker and Wright, 1966, Felten and Morrison 1966). In its most recent version Brecher and Morrison (1969) have attempted to explain the observed spectral features using the Compton hypothesis, however, Cowsik and Kobetich have done a more detailed calculation indicating that the Compton mechanism generates a smooth featureless power-law spectrum.

GENERAL SCHEME OF MODEL

- MATTER AND ANTIMATTER EXIST IN EQUAL AMOUNTS IN SEPARATE REGIONS.
- MIXING OCCURS ALONG BOUNDARY REGIONS OF THICKNESS $\sim \lambda_A$.
- RESULTING RAPID HEATING OF PLASMA WITHIN A DISTANCE $\sim \lambda_X$ OF BOUNDARY BY ANNIHILATION PRODUCTS PRODUCES EXPANSION AWAY FROM BOUNDARY.
- RESULTING EXPANSION OF PLASMA INDUCES HIGH-VELOCITY GAS MOTIONS.
- DURING THE NEUTRALIZATION ERA (POSSIBLY EVEN SOMEWHAT BEFORE) GAS MOTIONS BECOME TURBULENT ($\sim 500 \leq Z \leq \sim 3000$).
- WHEN GAS GOES FROM PLASMA TO ATOMIC STATE ($\sim 400 \leq Z_N \leq \sim 600$):
  - A) DUE TO DECOUPLING OF MATTER FROM RADIATION FIELD, VISCOSITY DROPS BY ALMOST 8 ORDERS OF MAGNITUDE AND SOUND VELOCITY DROPS BY FOUR ORDERS OF MAGNITUDE.
  - B) THIS CAUSES TURBULENCE TO BECOME SUPERSONIC AND TO EXTEND THE EDDY SPECTRUM DOWN TO A SCALE OF THE ORDER OF 10^{-3} pc.
  - C) THE SUPERSONIC TURBULENCE INDUCES DENSITY FLUCTUATIONS $\Delta\rho/\rho \sim 1$ OVER THE WHOLE RANGE OF EDDY SCALES.
- AT A REDSHIFT OF $\frac{2}{15} Z_N$ OR ABOUT 60, VIRIAL THEOREM BECOMES SATISFIED FOR BINDING OF GAS CLOUDS INTO PROTOGALAXIES DUE TO DENSITY FLUCTUATIONS INDUCED BY SUPERSONIC TURBULENCE.

Fig. 24: Outline of the galaxy formation theory of Stecker and Puget (1972).

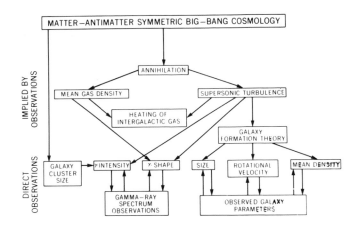

Fig. 25: Observational implications for baryon symmetric cosmology.

A critical test between the cosmic ray protar and annihilation hypotheses lies in a study of the energy spectrum. Figure 26 shows the present range of the data based on the review of Schwartz and Gursky (1973) and the data given in figure 20, indicated by the shaded region, along with the extrapolated power-law spectrum (X)

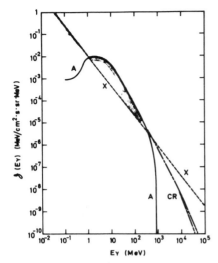

Fig. 26: Predicted energy flux spectra from the annihilation model (A) and cosmic ray (protar) model (CR) as discussed in the text. Also shown is the scatter area covered by the observational data (shaded) and the extrapolated x-ray background spectrum (X). The two curves shown for the CR spectrum above 7 GeV are for closed (Einstein-de Sitter) and open universes as discussed in the text (Stecker 1971a).

---

and the annihilation spectrum (A) and the high energy form of the spectrum predicted for redshifted cosmic-ray π^o-decay gamma-rays (CR). The annihilation spectrum should exhibit a sharp cutoff slightly below 1 GeV because the energy of the gamma-rays is limited by the rest-energy available to them from baryon-antibaryon annihilations. A detailed discussion of this may be found in Stecker (1971a). The cosmic-ray produced spectrum, on the contrary, can continue up to higher energies with a steepening induced around 10 GeV by pair-production losses through interactions with the microwave background (Fazio and Stecker 1970). This steepening should amount to an increase of 0.5 in the spectral index for a closed Einstein-de Sitter universe and an increase of 0.75 in the spectral index for a low-density open universe (Stecker 1971a). It should be kept in mind that the cutoff in the annihilation spectrum may be somewhat obscured by the presence of other background radiations having relatively lower intensities below 200 GeV.

Various other mechanisms which have been discussed in connection with the interpretation of the origin of the γ-ray background may be found in a review last year by the author (Stecker 1973c).

Another recent hypothesis for the origin of the background radiation which also involves primordial cosmic rays is that of Strong et al. (1973). This hypothesis is based on the model of Hillas (1967) that the observed cosmic rays, at least those above $\sim 10^6$ GeV, are universal and primordial. Hillas showed that of the observed cosmic rays originated at a redshift $z_{max} \simeq 15$, energy losses that these cosmic rays would undergo in pair-production interactions with the blackbody photons of the type

$$\gamma + p \to p + e^+ + e^- \tag{7.1}$$

would steepen the spectrum at the presently observed energy of $\sim 3 \times 10^6$ GeV. At a redshift of ~ 15, when these interactions took place, the blackbody photons had an energy of $\sim 10^{-2}$ eV and the cosmic rays had an energy of $\sim 5 \times 10^7$ GeV. The electron-positron pairs produced would then produce a shower of lower energy γ-rays through a repeated two-stage interaction process with the blackbody radiation field involving Compton scattering followed by pair production, i.e.

$$e^{\pm} + \gamma \text{ (blackbody)} \to e^{\pm} + \gamma \text{ (high energy)}$$

followed by (7.2)

$$\gamma \text{ (blackbody)} + \gamma \text{ (high energy)} \to e^+ + e^-$$

The resultant spectrum based on a complex model is shown in figure 27. Both the Hillas model and the hypothesis of Strong et al. require an exceedingly low intergalactic gas density, $n_0 \lesssim 10^{-9}$ cm^{-3}, in order that the γ-rays produced by the cosmic ray model of Stecker (1969b) do not exceed the observed background level.

If the intergalactic gas density should turn out to be of the order of 10^{-9} cm^{-3} which is only about 1% of the mean matter density in galaxies, we would have to radically alter our ideas about galaxy formation in order to account for 99% of the mass in the universe to be bound into galaxies.

Present observations give a background flux of γ radiation above 100 MeV of about

$$J = \frac{qL}{4\pi} = 3 \times 10^{-5} \text{ cm}^{-2} \text{ s}^{-1} \text{ sr}^{-1} \tag{7.3}$$

where the path length

$$L \simeq \frac{c}{H_o} \simeq 10^{28} \text{ cm} \tag{7.4}$$

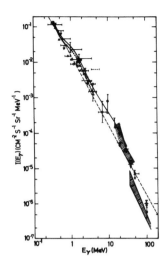

Fig. 27: The γ-ray background calculated from the Hillas model by Strong et al. (1973). (Dashed line). Data are same as in Fig. 20.

and the production rate

$$q \simeq 1.3 \times 10^{-25} n_H \xi$$
$$\simeq 10^{-30} \Omega \xi \qquad (7.5)$$

where ξ is the ratio of intergalactic to galactic cosmic ray intensity if part of the background is produced by noncosmological (contemporary $z = 0$) cosmic rays. If we call the fraction of the background produced by contemporary cosmic rays f, then from equations (7.3) to (7.5)

$$3 \times 10^{-5} f \simeq 10^{-3} \Omega \xi \qquad (7.6)$$

because the background spectrum is observed to be quite steep, conservatively $f \lesssim 1/3$ so that the limit on the product of intergalactic gas density and cosmic ray intensity becomes

$$\Omega \xi \lesssim 10^{-2} \qquad (7.7)$$

This excludes the possibility of both universal cosmic rays ($\xi = 1$) and a closed universe ($\Omega = 1$).

More stringent upper limits can be placed on primordial cosmic rays (Stecker 1971a) and the antimatter annihilation rate (Stecker et al. 1971), Stecker and Puget 1972). Thus it would seem that γ-ray astronomy at present can with certainty only put limits on our speculations about the origin of cosmic rays. But on the other hand we can say that speculations about the origin of cosmic γ-rays have broadened our conceptions about high-energy astrophysics and cosmology.

REFERENCES

Agrinier, B., Forichon, M., Laray, J.P., Parlier, B., Montmerle, T., Boella, G., Maraschi, L., Sacco, B., Scarsi, L., Da Costa, J.M., and Palmeria R., Proc. 13th Int. Conf. on Cosmic Rays, Denver, $\underline{1}$, 8, 1973.

Aldrovandi, R. and Caser, S., Nuc. Phys., $\underline{B38}$, 593, 1972.

Aldrovandi, R., Caser, S., Omnes, R. and Puget, J.L., Astron. and Astrophys., $\underline{28}$, 253, 1973.

Allen, C.W., Astrophysical Quantities, Athbone Press, London, 1973.

Arnold, J.R., Metzger, A.E., Anderson, E.C. and Van Dilla, M.A., J. Geophys. Res., $\underline{67}$, 4878, 1962.

Arons, J. and McCray, R., Astrophys. J., $\underline{158}$, L91, 1969.

Arons, J., Astrophys. J., $\underline{164}$, 437, 1971a.

Arons, J., Astrophys. J., $\underline{164}$, 457, 1971b.

Bignami, G.F. and Fichtel, C.E., Astrophys, J., 189, L65, 1974.

Boldt, E. and Serlemitsos, P., Astrophys. J., 157, 557, 1969.

Bratalubova-Tsulukidze, L.I., Grogorov, N.L., Kalinkin, L.F., Melioransky, A.S., Pryakhin, E.A., Savenko, I.A. and Yufarkin, V.Ya., Acta. Phys., $\underline{29}$, Suppl. 1, 1970.

Brecher, K. and Morrison, P., Astrophys. J., 150, L61, 1967.

Brecher, K. and Morrison, P., Phys. Rev. Lett., $\underline{23}$, 802, 1969.

Brown, J.W., Stone, E.C. and Vogt, R.E., Proc. 13th Int. Conf. on Cosmic Rays, Denver, 484, 1973.

Cavallo, G. and Gould, R.J., Nuovo. Cim. $\underline{2B}$, 77, 1971.

Cisneros, A., Phys. Rev., $\underline{D7}$, 362, 1973.

Clark, D.H., Caswell, J.L. and Green, A.J., Nature, 246, 28, 1973.

Comstock, G.M., Hsieh, K.C. and Simpson, J.A., Astrophys. J., 173, 691, 1972.

Cowsik, R. and Kobetieh E.J., Astrophys. J., 177, 585, 1972.

Dodds, D., Strong, A.W., Wolfendale, A.W. and Wdowczyk, J., Nature, 250, 716, 1974.

Fazio, G.G. and Stecker, F.W., Nature, 226, 135, 1970.

Feenberg, E. and Primakoff, H., Phys. Rev., 73, 449, 1948.

Felten, J.E. and Morrison, P., Phys. Rev. Lett., 10, 453, 1963.

Felten, J.E., Phys. Rev. Lett., 15, 1003, 1964.

Felten, J.E. Astrophys. J., 144, 241, 1966.

Fermi, E., Astrophys. J., 119, 1, 1954.

Fichtel, C.E., Hartman, R.C., Kniffen, D.A. and Sommer, M. Astrophy. J. 171, 31, 1972.

Fichtel, C.E., Hartman, R.C., Kniffen, D.A. and Thompson, D.J. "Gamma Ray Astronomy in the Time of SAS-2", invited talk 140 AAS Meeting, Columbio, Ohio, June 1973.

Fichtel, C.E., Kniffen, D.A. and Hartman, R.C., Astrophys. J., 186, L99, 1973.

Frye, G.M. and Smith, L.H., Phys. Rev. Lett., 17, 733, 1966.

Frye, G.M. Albats, P.A., Zych, A.D., Starb, J.A., Hopper, V.D., Rawlinson, W.R. and Thomas, J.A., Proc. ESLAB Symposium on the Context and Status of γ-Ray Astronomy, Frascati, Reidel, in press. 1974.

Golenetskii, S.V., Mazets, E.P., Il'insky, V.N., Aptekhar, R.L., Bredov, M.M., Gur'yan, Y.A. and Panov, V.N., Astrophys. Lett. 9, 69, 1971.

Garmire, G. and Kraushaar, W.L., Space Sci. Rev. 4, 123, 1965.

Ginzburg, V.L. and Syrovatsky, S.I., The Origin of Cosmic Rays, MacMillan, New York, 1964.

Ginzburg, V.L., Astrophys. and Space Sci. 1, 1, 1968.

Gould, R.J., Phys. Rev. Lett. 12, 511, 1965.

Gould, R.J. and Schréder, G.P., Phys. Rev. Lett. 16, 252, 1966.

Gould, R.J. and Schréder, G.P., Phys. Rev. 155, 1404, 1967a.

Gould, R.J. and Schréder, G.P., Phys. Rev. 155, 1408, 1967b.

Greisen, K., Phys. Rev. Lett. 16, 748, 1966.

Gunn, J.E. and Peterson, B.A., Astrophys. J. 142, 1633, 1965.

Hayakawa, S., Prog. Theor. Phys. 8, 571, 1952.

Heitler, W., The Quantum Theory of Radiation, Oxford Press, London, 1960.

Helmken, H. and Hoffman, J., Nature. Phys. Sci. 243, 6, 1973.

Hillas, A.M., Can. J. Phys. 46, 5623, 1968.

Hoffman, W.F. and Frederick, C.L., Astrophys. J. Lett. 155, L12, 1969.

Hollenback, D.J. and Salpeter, E.E., Astrophys. J., 163, 155, 1971.

Hollenback, D.J., Werner, M.W. and Salpeter, E.D., Astrophys. J., 163, 165, 1971.

Hoyle, F., Phys. Rev. Lett. 15, 131, 1965.

Hopper, V.D., Mace, O.B., Thomas, J.A., Albats, P., Frye, G.B., Thompson, G.B. and Staib, J.A., Astrophys. J. 186, L55, 1973.

Hutchinson, G.W., Phil. Mag. 43, 847, 1952.

Ilovaisky, S.A. and Lequeux, J., Astron. and Astrophys. 20, 347, 1972.

Jauch, J.M. and Rohrlich, F., The Theory of Photons and Electrons. (Addison-Wesley, Cambridge, Mass.), 1955.

Jelley, J.V., Phys. Rev. Lett., 16 479, 1966.

Jones, F.C., Phys. Rev. 137B, 1306, 1965.

Jones, F.C., Can. J. Phys. 46, S1003, 1967.

Kniffen, D.A., Hartman, R.C., Thompson, D.J. and Fichtel, C.E., Astrophys. J. 186, L105, 1973.

Kraushaar, W.L. and Clark, G.W., Phys. Rev. Lett. 8, 106, 1962.

Kraushaar, W.L., Clark, G.W., Garmire, G.P., Borken, R., Higbie, P., Leong, C. and Thorsos, T., Astrophys. J., 177, 341, 1972.

Kuo, F.S., Frye, G.M. and Zych, A.D., Astrophys. J. Lett, 186, L51, 1973.

Mayer-Hasselwander, H.A., Pfefferman, E., Pinkau, K., Rothermel, H., and Sommer, K., Astrophys. J. Lett., 175, L23, 1972.

Mazetz, E.P., Golenetskii, S.V., Il'inskii, V.N., Gur'uan, Yu. A. and Kharitonova, T.V., Pis 'ma v Zh.E.T.F., 20, 77, 1974.

Metzger, A.E., Anderson, E.C., Van Dilla, M.A. and Arnold, J.R., Nature, 204, 766, 1964.

Morrison, P., Il Nuovo Cimento 7, 858, 1958.

Morrison, P., Astrophys. 157, L75, 1969.

Nikishov, A.I., Sov. Phys. JETP, 14, 393, 1962.

O'Dell, F.W., Shapiro, M.M., Silberberg, R., and Tsao, C.H., Proc. 13th Int. Conf. on Cosmic Rays, Denver, 490, 1973.

Omnès, R., Physics Reports, 3C, 1, 1972.

Oort, J.H., Galactic Astronomy, (ed. H.Y. Chiu and A. Muriel), Gordon and Breach, New York, 129, 1970.

Ozernoi, L.M., Prilutsky, O.F. and Rozental, I.L., Astrofizika Visokikh Energy, Atomizdat, Moscow, USSR, 1973.

Pollack, J.B. and Fazio, G.G., Phys. Rev., 131, 2684, 1963.

Puget, J.L. and Stecker, F.W., Astrophys. J., 191, 323, 1974.

Rees, M.J., Astrophys. Lett., 4, 113, 1969.

Rozental. I.L. and Shukalov, I., Astron. Zh., 46, 779, 1969. (Trans.Sov. Astron, A.J., 13, 612, 1970).

Samimi, J., Share, G.H. and Kinzer, R.L., Proc. ESLAB
   Symposium on the Context and Status of γ-Ray Astronomy,
   Frascati, Reidel, in press, 1974.

Sanders, R.H. and Wrixon G.T., Astronomy and Astrophysics, 18,
   92, 1972.

Sanders, R.H. and Wrixon, G.T., Astron. and Astrophys., 26
   365, 1973.

Schmidt, M., Galactic Structure (ed. A. Blaauw and M. Schmidt),
   U. of Chicago Press, Chicago, Ill., 513, 1965.

Schwinger, J., Phys. Rev., 75, 1912, 1949.

Schonfelder, V. and Lichti, G., Astrophys. J., 191, L1, 1974.

Schwartz, D. and Gursky, H., Gamma Ray Astrophysics, (F. Stecker
   J.I. Trombka, ed.) NASA SP-339, U.S. Gov't. Printing Office,
   Washington, D.C., 15, 1973.

Scoville, N.Z., Solomon, P.M. and Jefferts, K.B., Astrophys. J.,
   187, L63, 1974.

Shane, W.W., Astron. and Astrophys., 16, 118, 1972.

Share, G.H., Kinzer, R.L. and Seeman, N., Astrophys. J.,
   187, 45, 1974.

Share, G.H., Kinzer, R.L. and Seeman, N., Astrophys. J., 187,
   511, 1974.

Solomon, P.M. and Wickramasinghe, N.C., Astrophys. J., 158,
   449, 1969.

Solomon, P.M. and Stecker, F.W., Proc. ESLAB Symposium on the
   Context and Status of γ-Ray Astronomy, Frascati,
   Reidel, in press, 1974.

Sood, R.K., Bennett, K. Clayton, P.G. and Rochester, G.K.,
   Proc. ESLAB Symposium on the Context and Status of
   γ-Ray Astronomy, Frascati, Reidel, in press, 1974.

Spitzer, L., Drake, J., Jenkins, E.B., Morton, D.C., Rogerson, J.B.
   and York, D.G., Astrophys. J., 181, L116, 1973.

Stecher, T.P. and Stecker, F.W., Nature, 226, 1234, 1970.

Stecker, F.W., Smithsonian Astrophys. Obs. Spec. Rpt. No. 261,
   1967.

Stecker, F.W., Phys. Rev. Lett., 21, 1016, 1968a.

Stecker, F.W., Nature, 220, 675, 1968b.

Stecker, F.W., Nature, 222, 865, 1969a.

Stecker, F.W., Astrophys. J., 157, 507, 1969b.

Stecker, F.W., Nature, 224, 870, 1969c.

Stecker, F.W., Astrophys. and Space Sci., 6, 377, 1970.

Stecker, F.W., Cosmic Gamma Rays, Mono Book Corp., Baltimore, Md., 1971a.

Stecker, F.W., Nature, 229, 105, 1971b.

Stecker, F.W., Astrophys. J., 185, 499, 1973a.

Stecker, F.W., Astrophys. and Space Sci. 20, 47, 1973b.

Stecker, F.W., Gamma Ray Astrophysics, (F.W. Stecker and J.I. Trombka, ed.) NASA SP-339, U.S. Gov't. Printing Office, Washington, D.C., 211), 1973c.

Stecker, F.W. and Morgan, D.L., Astrophys. J., 171, 201, 1972.

Stecker, F.W., Morgan, D.L. and Bredekamp, J., Phys. Rev. Lett., 27, 1469, 1971.

Stecker, F.W. and Puget, J.L., Astrophys. J., 178, 57, 1972.

Stecker, F.W., Puget, J.L., Strong, A.W. and Bredekamp, J.H., Astrophys. J., 188, L59, 1974.

Stecker, F.W. and Trombka, J.K., Gamma Ray Astrophysics, NASA SP-399, U.S. Gov't. Printing Office, Washington, D.C., 1973.

Strong, A.W., Wdowczyk, J. and Wolfendale, A.W., Nature, 241, 109, 1973.

Strong, A.W., J. Phys. A., in the press, 1974.

Tanaka, Y., Proc. ESLAB Symposium on the Context and Status of γ-Ray Astronomy, Frascati, Reidel, in press, 1974.

Thompson, D., Fichtel, C.E., Hartman, R.C., Kniffen, D.A. and Bignami, G.F., Proc. ESLAB Conf. on the Context and Status of γ-Ray Astronomy, Frascati, Reidel, in press, 1974.

Thompson, D.J., Fichtel, C.E., Kniffen, D.A. and Hartman, R.C., Proc. ESLAB Conf. on the Context and Status of γ-Ray Astronomy, Frascati, Reidel, in press, 1974a.

Trombka, J.I., Metzger, A.E., Arnold, J.R., Matteson, J.L., Reedy, R.C. and Peterson, L.E., Astrophys. J., 181, 737, 1973.

Trower, W.P., Univ. of Cal., Lawrence Rad. Lab. Rpt. UCRL-2426, Vol. 2., 1966.

Van der Kruit, P.C., Astron. and Astrophys. 13, 405, 1971.

Vedrenne, G.E., Albernhe, F., Martin, I. and Talon, R., Astron. and Astrophys. 15, 50, 1971.

Vette, J.I., Gruber, D., Matteson, J.L. and Peterson, L.E., Proc. IAU Symp. No. 37, Rome, (L. Gratton, Ed.), 335, Reidel Pub. Co., Dordrecht, Holland, 1970.

Vette, J.I., Gruber, D. Matteson, J.L. and Peterson, L.E., Astrophys. J. Lett., 160, L161, 1970.

Wdowczyk, J. Tkaczyk, W. and Wolfendale, A.W., J. Phys. A5, 1419, 1972.

Zatsepin, G. and Kuz'min, V., JETP Lett. 4, 78, 1966.

Zel'dovich, Y.B., Kurt, V.G. and Sunyaev, R.A., Sov. Phys. JETP, 28, 146, 1969.

OBSERVATION OF CELESTIAL GAMMA RAYS[*]

K. Pinkau

Max-Planck-Institut für Physik und Astrophysik,
Institut für Extraterrestrische Physik, 8046 Garching
bei München, Germany.

I. INTRODUCTION

For the last 15 years attempts have been made to observe celestial gamma rays but it has been possible only very recently to construct instruments that are sufficiently sensitive for the measurement of these very weak fluxes. The very large scientific interest for these observations comes from the fact that, although gamma rays originate also from interactions of electrons with particles and fields, they are both created in nuclear interactions via the decay of unstable particles and in nuclear de-excitation as line radiation. They may thus indicate places of nucleogenesis, and of intense nuclear interaction or annihilation now and in the past. Gamma rays have a very low absorption cross-section (Stecker, 1971), and may thus be detected over very large distances or at very large red-shifts.

This presentation will be sub-divided into three parts:

    A.   Instrumental Problems
          1. Energy resolution
          2. Angular resolution
          3. Efficiency
          4. Background

    B.   Observational Results on
          1. Lines
          2. The diffuse flux

---

[*] In this paper, results presented previously (Pinkau 1973, 1974) have partly been used.

3. The galactic emission
4. Localized sources

C. Discussion of the Vela Supernova remnant.

These topics are chosen because they promise to provide a minimum overlap with the presentations of F. Stecker at this Institute.

II. INSTRUMENTAL PROBLEMS

Gamma rays may be detected using their interactions with matter. The relative importance of, respectively, the photoelectric effect, the Compton effect, and pair production is shown in Fig. 1 (Evans, 1955). Thus, depending on photon energy, these three processes are used for instruments in gamma ray astronomy.

An incident photon cannot be absorbed by a free electron from considerations of energy and momentum conservation. However in the case of the photoeffect, momentum is conserved by the recoil of the entire atom, and practically the entire photon energy turns up in the kinetic energy of the electron, minus its binding energy

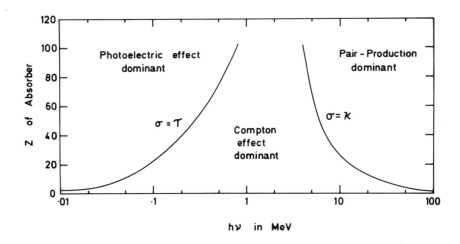

Fig. 1: Relative importance of the three major types of γ-ray interaction. The lines show the values of Z and hν for which the two neighbouring effects are just equal.

OBSERVATION OF CELESTIAL GAMMA RAYS

$$T = h\nu_o - B_e \tag{2.1}$$

In the case of the Compton effect, the photon $h\nu_o$ is scattered through an angle ϕ given by the equation

$$\cos \phi = \frac{m_e c^2}{h\nu'} + \frac{m_e c^2}{h\nu_o} \tag{2.2}$$

and the recoil electron obtains kinetic energy

$$T = h\nu_o - h\nu' \tag{2.3}$$

Finally, pair production again is possible only in the field of the atomic nucleus, but its recoil is small such that practically the entire photon energy turns up as kinetic energy of the two electrons, plus their rest energy.

$$h\nu_o = T_+ + T_- + 2 m_o c^2 \tag{2.4}$$

1. Energy Resolution

The energy resolution that can be obtained using the photoelectric effect is determined by the accuracy with which one can measure the recoil electron's kinetic energy. This is determined by the number N of photoelectrons if scintillator-photomultiplier combinations are used, or the number N of ion pairs or, respectively, electron-hole pairs in the case of proportional counters or solid state detectors. Typical energy loss values required for the formation of one quantum in these three types of detector are given in Table I.

For gamma-ray spectroscopy, Germanium is used rather than Silicon due to its higher value of Z. Fig. 2 (Neuert, 1966) shows instructively a comparison of the energy resolution of a Germanium and a NaI detector.

Type of Detector	Energy Required	Theor. Resolution for 500 keV
Scintillator	300-1000 eV/ photoelectron	4.4 %
Proportional Counter	20 - 30 eV/ion pair	2.4 °/oo
Semiconductor	3 eV/electron-hole pair	0.77 °/oo

TABLE I

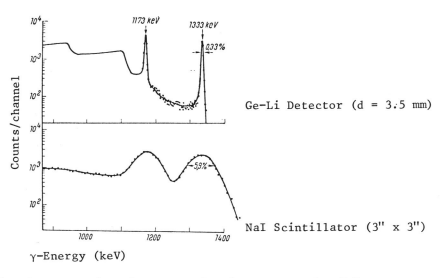

Fig. 2: Comparison between a Ge-Li counter and a NaI - scintillator recording a Co^{60} γ-spectrum.

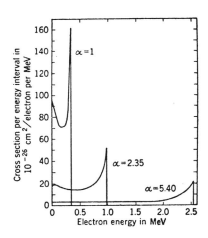

Fig. 3: Energy distribution of Compton electrons produced by primary photons whose energies are 0.51 MeV (α = 1), 1.2 MeV (α = 2.35), and 2.76 MeV (α = 5.40).

If a photon is absorbed by a Compton collision, the recoil electrons will show an energy distribution which is shown in Fig. 3. The maximum energy transfer to the electron is

$$T_{max} = \frac{(h\nu_o)^2}{h\nu_o + \frac{m_o c^2}{2}} \qquad (2.5)$$

Good energy resolution can thus only be obtained in the photopeak, or if the counter is so thick that the scattered photon can make more interactions.

An instrument designed specifically for the Compton interaction is the double Compton telescope (Schönfelder et al. 1973).

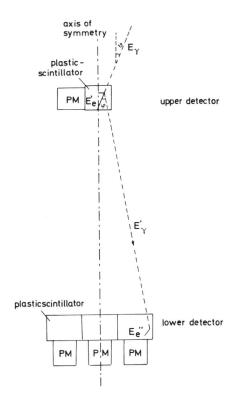

Fig. 4: Schematic drawing of a double-Compton telescope.

It is shown schematically in Fig. 4. The photon makes a Compton collision in the upper counter, where it is scattered through the angle ϕ and deposits the energy E' in the form of kinetic energy of the recoil electron. It then interacts a second time in the lower counter, depositing the energy E". A time of flight measurement makes sure that the time delay between both interactions corresponds to the light travel time between the two counters. One may use the sum of both energy deposits to approximate the photon energy

$$h\nu_0 \gtrsim E' + E''$$

Fig. 5 shows a calibration measurement of the energy resolution of such an instrument (Schönfelder et al., 1973) obtained using plastic scintillators for both upper and lower counters. Use of NaI in the lower counter could greatly improve the capability of the system.

At higher energies, in the region of pair production, energy measurement would again be possible in principle by measuring the kinetic energy of the two electrons.

If one uses spark chambers to record the paths of the two electrons, multiple scattering measurements are possible and

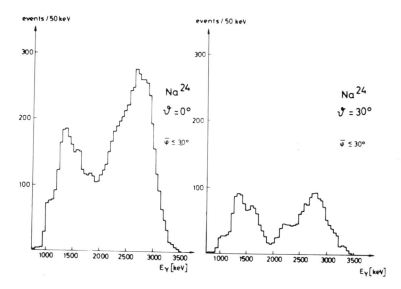

Fig. 5: Energy resolution obtained with the double Compton telescope with Na^{24} γ-rays measured under angles of incidence $\Theta = 0°$, and $\Theta = 30°$ respectively.

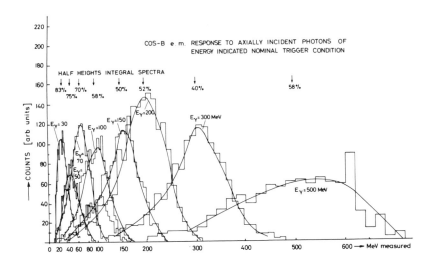

Fig. 6: Energy resolution obtained with the Cos-B γ-ray experiment.

---

have in fact been carried out in the SAS-II experiment (Fichtel et al., 1973). No experimental calibration results of multiple scattering measurements, vs. primary photon energy are as yet available.

If the two electrons of the pair are allowed to enter a thick detector, their energy can be measured by ionization loss. However, with increasing energy, electromagnetic cascades tend to develop. It is then increasingly difficult to absorb the entire cascade, and the energy resolution becomes worse. Since these instruments have to be large, their energy resolution at low energies is also not so good because of poor light collection. Fig. 6 shows the results of calibration measurements of the Cos-B collaboration (Caravane Coll., 1974) on a CsI crystal of thickness 4.7 rad. lengths.

At even higher energies, scintillator-lead sandwich arrangements may be used. Also, the radial distribution of electrons can be measured in electromagnetic cascades and the primary energy determined with fairly good precision (Pinkau, 1957, 1964).

2. Angular Resolution

Astronomy can be performed only with instruments of good

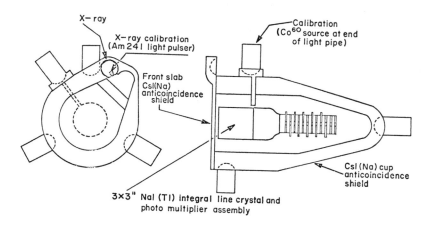

Fig. 7: The OSO-7 γ-ray instrument for the measurement of solar γ-rays.

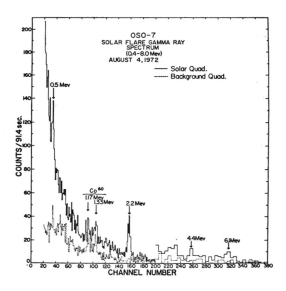

Fig. 8: Solar γ-ray spectrum measured with the instrument shown in Fig. 7.

angular resolution. In the gamma ray domain, angular resolution appears to be bad always.

In the high energy X-ray and low energy gamma ray region, modulation collimators are used to measure directions of point sources. With increasing energy, the collimation systems have to be more and more massive. This is due to the fact that the photon interaction cross-section goes through a minimum at $\sim$ 10 MeV (Fig. 12). Moreover the collimators have to be active in order to exclude secondary gamma ray production in the shield. Fig. 7 shows, as an example, the OSO-7 solar gamma ray instrument (Higbie et al., 1971). This detector actually observed solar gamma ray lines in the August 4, 1972, event, and this spectrum is shown in Fig. 8. For such measurements, only rough angular resolution is required.

One can improve angular resolution somewhat by placing the counter into a deeper well, but a value of approximately $7°$ full width half maximum (FWHM) seems to be a practical limit.

If more and more active material is used for collimation, anticollimation (occultation) of a source may be a more practical approach. Morfill et al. (1973) have made a comparison between both techniques, and their comparison of optimized instruments of both categories for the same weight is shown in Fig. 9. The anticollimation system thus seems to be more potent at energies exceeding 200 keV.

In the double Compton telescope, measurement of the angle of incidence would in principle be possible to an accuracy of the order of $4°$ if the direction of the recoiling electron could be measured and the only inaccuracy consisted in the energy measurement. However in practice one will measure only E' and E", and then

$$\cos \bar{\phi} = 1 - \frac{m_o c^2}{E''} + \frac{m_o c^2}{E'+E''} \tag{2.7}$$

and $\phi \stackrel{<}{\sim} \bar{\phi}$ always since $E'' + E' \stackrel{<}{\sim} h\nu_o$. One then can limit oneself to a cone of incidence $\phi \stackrel{<}{\sim} \bar{\phi}$, and practical values are cones of $\sim 30°$ FWHM.

The situation improves somewhat at higher energies. One attempts to measure the tracks of the two electrons in a spark chamber and to derive information about the gamma ray directions from such measurements. (See, for example, Pinkau, 1972; Caravane Coll., 1974). In practice, angular resolution is always worse than the theoretical value. Results of measurements on the Cos-B spark chamber are shown in Fig. 10, together with

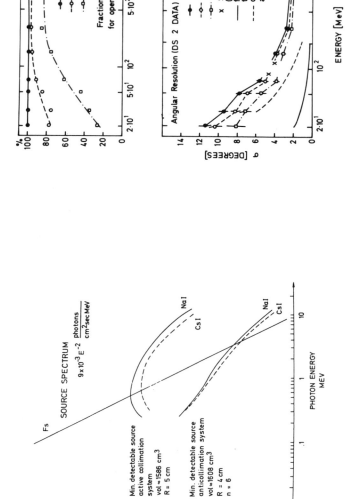

Fig. 9: Comparison of the minimum detectable source intensity that can be obtained with collimator and anticollimator systems, respectively, of the same weight.

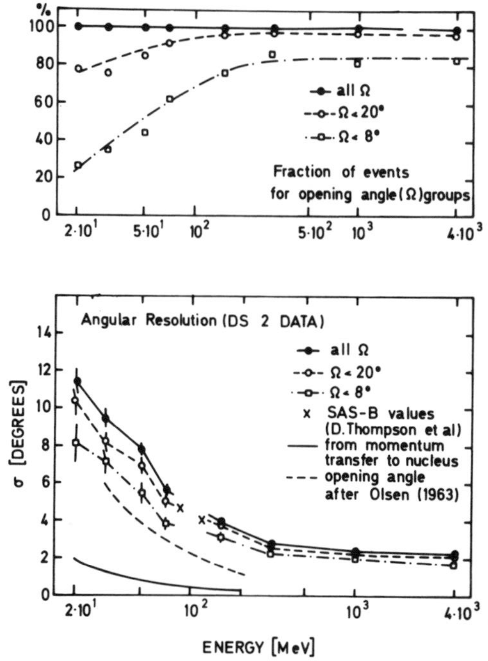

Fig. 10: Angular resolution obtained with the COS-B spark-chamber experiment. Also shown are results from SAS-2.

results obtained for SAS-II, and lower limits possible from the recoil to the nucleus.

Better angular resolution has been obtained using combinations of nuclear emulsions and spark chambers (Share, 1974). However, the scanning problem will always create severe problems with statistics.

Angular resolution at high energies is also statistics-limited rather than instrument resolution-limited. Assume that one wishes to see a 10% variation in bins of $1° \times 1°$ of an average flux of 10^{-5} quanta/cm^2 s ster. For an observation time of one week, one then needs an area of half a square meter of detection equipment with an efficiency of 10%.

3. Efficiency and Geometrical Factor

The efficiency of gamma astronomy instruments is determined

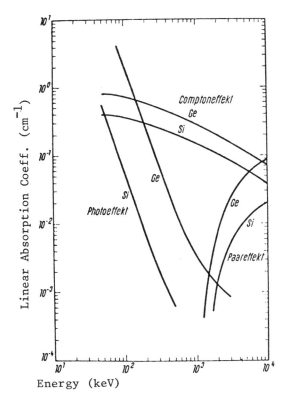

Fig. 11: Absorption coefficient for Ge and Si.

by the interaction cross-sections for gamma rays. These are shown for Si, Ge, and NaI in Figs. 11 and 12. It can be seen that typical efficiencies for solid state detectors, NaI-counters and spark chambers used at higher energies can be made to be several 10%.

The situation is different if one uses the double Compton technique. Here the efficiency of the instrument is typically a fraction of 1% due to the requirement that two interactions have to occur in two counters that are far ($\gtrsim$ 1m for time-of-flight measurements) apart. Although this can partly be offset by the large counter areas that can be employed, low statistics is one of the severest problems of this technique.

Gamma ray instruments show one typical difference from ordinary charged particle counter telescopes that is very important to appreciate: the gamma rays convert into charged particles in one of the telescope elements, or close to one

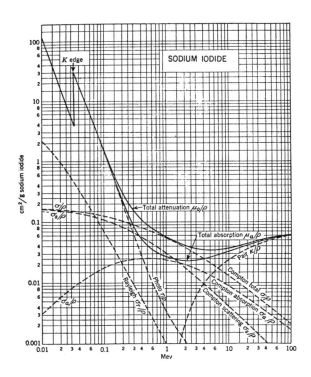

Fig. 12: Absorption coefficient for NaI.

of the elements. Since these secondary particles will show a spread around the incident gamma ray's direction, the solid angle is not well defined and also a function of energy.

Fig. 13 shows the normalized effective area of the Cos-B instrument (Caravane Coll., 1974) as a function of zenith angle and gamma ray energy. This is an eye-ball fit to the experimental data and shows that the angular sensitivity of the instrument widens with decreasing energy. Correspondingly, the effective solid angle of the instrument will increase with decreasing energy.

It is thus not correct to express the area-angle function

$$a(E, \theta) = A\, f(E)\, g(\theta) \qquad (2.8)$$

by the product of two functions $f(E)$ and $g(\theta)$ as was done by Kraushaar et al. (1973).

In addition, it must be realised that any pointing γ-ray instrument views certain areas of the sky only under certain fixed zenith angles. Thus, the exposure integral is known only to within the factor derived from a calibration like that shown in Fig. 13. If the energy spectrum of the radiation is not known,

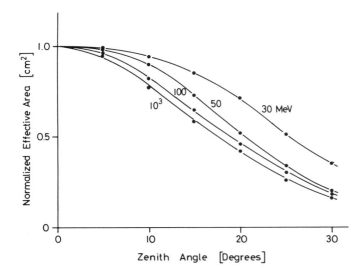

Fig. 13: The effective area of the COS-B instrument as a function of zenith angle. It is also dependent upon energy.

this is a factor of about 1.6 in the case of the COS-B instrument.

4. Background

Gamma rays are measured in a high radiation environment with flux ratios of the order of 10^5 or more. Background problems are thus always severe.

As an example of background problems, let us consider the measurement of the diffuse gamma ray spectrum on board of Apollo 15 (Trombka et al., 1973). These authors perform a background rejection while converting energy loss into photon spectra. First, a measured response library is established for the detector, and a matrix inversion technique applied (Trombka et al., 1970) in order to obtain the photon spectra. If one considers the convolution

$$Y(V) = \int_0^{E_{max}} T(E)\, S(E,V)\, dE \qquad (2.9)$$

of the pulse height spectrum $Y(V)$ resulting from a photon spectrum $T(E)$, one may represent this by a sum over a sufficiently large number of samples

$$Y_i = \sum_j T_j\, S_{ij} \qquad (2.10)$$

The problem then boils down to the determination of the S_{ij} and the inverse of this matrix. Once this is known, the T_j may be gotten directly from the Y_i. This method requires a very accurate knowledge of the S_{ij}.

Trombka et al. (1973) applied this technique to the energy. loss spectrum measured aboard Apollo 15 (Fig. 14). The spectrum shows lines that are believed to be due to weak radioactive sources and nuclear reactions in the spacecraft and local mass. If one transforms this spectrum to photon space as outlined above, the lines appear as discontinuities. They may then be subtracted by the requirement that the primary photon spectrum $T(E)$ be a smooth function. This results in the removal of 2.5 counts/sec over the 0.6-3 MeV range, or to about 16%, resulting in the photon spectrum shown in Fig. 15.

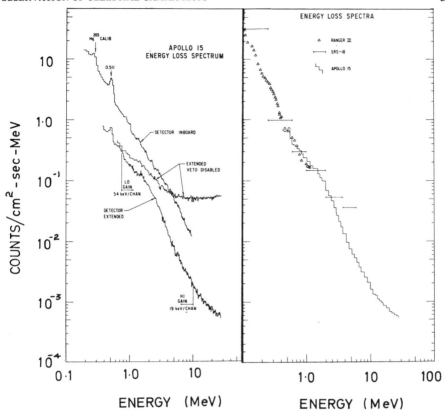

Fig. 14: Uncorrected energy loss spectra of the Apollo 15 γ-ray experiments.

Fig. 15: Corrected energy loss spectra of the Apollo 15 γ-ray experiments.

The same authors also attempted to allow for counts originating from radioactive spallation nuclei. They have taken the spallation spectrum proposed by Dyer and Morfill (1971) and Fishman (1972), multiplied by an arbitrary normalization factor. This factor was then determined by the requirement that the spectrum be smooth even across the 0.6-3 MeV range where the spallation gamma rays are. In fact, the normalization factor found in this way leads to approximately only half the intensity of spallation photons compared to the expected value according to Dyer and Morfill, and Fishman.

Type of Reaction	Energy Interval			
	1-3 MeV	3-5 MeV	5-10 MeV	1-10 MeV
(n,n'γ)	7,4%	18,1%	4,6%	8,8%
(n, x γ)	6,9%	35,8%	42,0%	15,2%
total	14,3%	53,9%	46,6%	24,0%

TABLE II

It can be seen that severe problems remain regarding background corrections in the low energy gamma ray domain. This is the reason why the double Compton scattering technique attracts interest in spite of its low efficiency. Here, the forward/backward ratio is $\gtrsim 10^4$, and the gamma ray thus must come from above, have its first interaction in the upper, its second interaction in the lower counter in order to be counted (Schönfelder et al., 1973). Local gamma ray production thus can make no contribution.

The only type of background here appears to be reactions caused in the upper counter by neutrons, where a gamma ray is also emitted. This may then interact in the lower counter and give rise to a false count. Table II gives an estimate of the background generated in this way (Schönfelder et al., 1974).

If one uses pulse shape discrimination in the upper counter, one may even reduce this type of background.

At higher energies, the spark chamber image provides a very powerful tool to discriminate gamma pair events from background. However, locally produced gamma rays, and low energy electrons scattering in the spark chamber may fake true gamma rays and it is therefore desirable to compare measurements carried out at different geomagnetic cut-offs or to use satellite experiments.

III. OBSERVATIONAL RESULTS

1. Lines

Line emission has been discovered from the sun in the August 4, 1972, event by Chupp et al. (1973) (see also Fig. 8). However, we are not concerned here with gamma rays from the sun.

A line feature at 473 keV from the direction to the galactic centre has been reported by Johnson et al. (1973) with a total photon flux of 1.8×10^{-3} photons/cm^2 sec. The interpretation of this line is uncertain at present (see Fishman et al., 1972; Ramaty et al., 1972, Clayton, 1973).

Gamma ray lines may serve as a tool to study nucleosynthesis (Clayton et al., 1969) and the structure of supernova shells (Clayton, 1974).

As Clayton (1973) points out, the most abundant species having a radioactive progenitor is Fe^{56}. Its progenitor is Ni^{56} which has γ-decays in the range from 0.84 to 3.26 MeV. In a simple model, Clayton computes that 1.7×10^9 supernovae would have produced 2.3×10^8 $M_\odot$ of Fe^{56}, the correct mass fraction of the Solar abundance (1.3×10^{-3}) before the birth

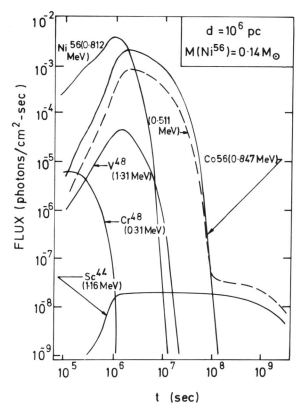

Fig. 16: Prominent γ-ray line fluxes as a function of time. The calculation assumed that about 0.5 $M_\odot$ of silicon-burning shells containing 0.14$M_\odot$ of ^{56}Ni have been ejected from a supernova arbitrarily placed at a distance of 10^6 parsecs.

of the sun of the galactic mass of 1.8×10^{11} $M_\odot$.

Then, one can compute that 3×10^{54} Fe^{56} atoms are produced per week. The time dependence of γ-ray line flux expected in this way is shown in Fig. 16.

Can this radiation be seen as a universal background? The average iron number density in the universe is $1.3 \times 10^{-3} \times 1.7 \times 10^{-31} = 2.3 \times 10^{-12}$ cm^{-3} or $2.2 \times 10^{-34} g/cm^3$. The flux of these γ-rays is

$$\frac{dF}{d\Omega} = \frac{c}{4\pi} q_\gamma \, n(Fe^{56}) = \frac{3 \times 10^{10}}{12.5} \times 2.8 \times 2.3 \times 10^{-12} \text{ photons/cm}^2 \text{ sec ster}$$

$$= 2.7 \times 10^{-2} \text{ photons/cm}^2 \text{ s ster}$$

Clayton then tries to determine the photon energy spectrum. The calculations then become model dependent; his highest flux value would be just beyond 1 MeV at an intensity of 1.3×10^{-2} MeV/cm^2 s ster MeV. One may compare this with Fig. 17 for the diffuse flux and see how interesting the result is. Of course the argument depends how easily the γ-rays can escape from the source.

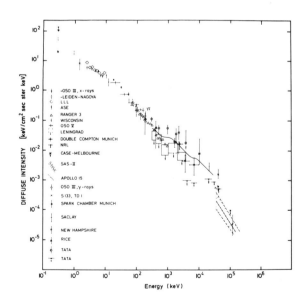

Fig. 17: Compilation of results on the diffuse γ-ray background.

2. The Diffuse Flux

In the X-ray regions above $\sim$ 1 keV, a diffuse flux of hard photons is measured that is isotropic to within a few percent. The plane of the galaxy is not visible at these energies. At about 100 MeV, a diffuse flux is also visible, the isotropy of which is not known to a very high degree. Between these two bracketing measurements, a number of observations have been carried out that substantiate the existence of a diffuse gamma ray flux, but that leave open the following questions (Fig. 17):
(a) Previous measurements have indicated a sharp steepening in the spectrum between 20-30 keV. Is this a reality?
(b) Measurements between 1 and 50 MeV seemed to show higher flux values than one would have expected from a power-law interpolation between the flux values < 1 MeV and at 100 MeV.
(c) At what energy does the galactic disc begin to appear above background?

Fig. 17 shows a compilation of results on the diffuse flux.

(a) Arguments have been advanced that the sharp break is a feature mainly caused by background. Following Schwartz (1973) only points in the energy region from 1 to 100 keV have been

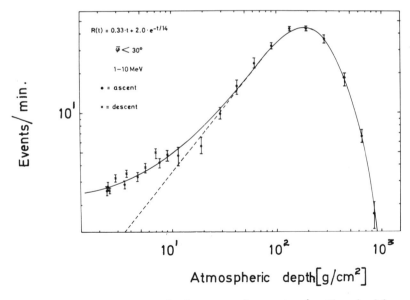

Fig 18: Growth curve of the counting rate in the double-Compton telescope as the instrument reaches the top of the atmosphere.

accepted that utilize direct means to assess effects of electron contamination or have been carried out under various geomagnetic conditions.

In addition, results of OSO-III have been incorporated that have since been published (Dennis et al., 1973). It is clear that there exists only a very gradual change.

(b) A very active debate still ensues about the actual values of the diffuse flux in the 1-50 MeV range. The Munich group first (Mayer-Hasselwander et al., 1972) reported high flux values in this region, they were soon supported by findings from Apollo 15 (Trombka et al., 1973). However, the background problem is very severe in this energy interval, and doubts remain as to whether or not the results were real.

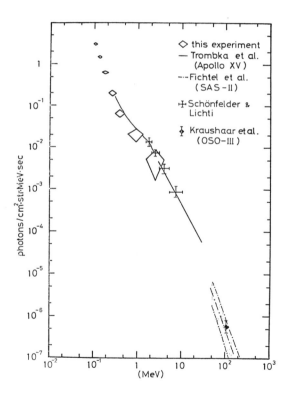

Fig. 19: Recent results of the Nagoya group on the diffuse background.

In the energy range from 1-10 MeV, the experimenters on board of Apollo 15 and 16 have carried out extensive background investigations as was discussed in Section II.4 and have converted their energy loss spectrum into a photon spectrum. Their results are probably reliable in this energy range. At higher energies, they still feel that their corrections are not entirely adequate (Trombka, 1974).

In the same energy range a reliable measurement has been carried out using a double Compton telescope (Schönfelder et al., 1973; Schönfelder and Lichti, 1974). The growth curve of these authors is shown in Fig. 18, but corrections have to be applied due to neutron interactions as shown in the Table, Section II.4.

The findings in this energy range have recently been supported by experiments from the Nagoya group, using a shutter technique (Fig. 19) (Tanaka, 1974). It would thus appear that, at $\sim$ 2 MeV, the diffuse flux lies about a factor of 3 above the power law interpolation between the values at 100 keV and 100 MeV, while in flux region from 10-30 MeV the actual spectrum is still unclear.

Above 50 MeV, the measurements of the diffuse flux by SAS-II have indicated a very steep spectrum (Fichtel et al., 1973) so that their results would appear to join the high flux values at lower energies.

In the author's opinion, the results claimed in the energy interval from > 10 to $\sim$ 100 MeV remain somewhat in doubt so that the confirmation of their reality is urgently required.

It is also interesting to realise that this additional component, if found to be isotropic, contains an energy density of one to several x 10^{-5} eV/cm^3.

Regarding the nature of this excess, the quality of the spectral measurements is still too poor to allow any conclusions to be drawn. However, it should be remembered that gamma rays are practically not absorbed even at very large redshifts. Thus, one cannot avoid considering cosmological effects on the diffuse gamma ray spectrum.

One other result is interesting. Above 100 MeV, 10^{-26} photons/sec ster are produced per hydrogen atom by the cosmic radiation as measured near the sun. The integral diffuse flux is 3 x 10^{-5} photons/cm^2 sec ster. We thus obtain the value L x n = 1(kpc cm^{-3}) for the product of length (kpc) and density n(cm^{-3}) at which the cosmic radiation could exist throughout the intergalactic medium. Thus, at a density of 10^{-5} cm^{-3}, the cosmic radiation could exist at galactic value out to distances of 100 Mpc (Thompson, 1974; Fichtel, 1974).

(c) It is difficult to assess the increasing anisotropy of the galactic disc as it appears above the diffuse flux as a function of energy. The reason is that at higher energies statistics is too poor, as is angular resolution at lower energies.

If one considers the energy spectrum of gamma rays received from a strip of $|b^{II}| < 10°$ around the galactic centre, this will disappear in the diffuse background at around 10 MeV. In agreement with this is that observations with the double Compton telescope between 1 and 10 MeV did not find an enhancement from the galactic plane (Fig. 20). The half-angle of the cone of sensitivity of this instrument is 15° (Schönfelder and Lichti, 1974).

Theoretical interpretation of the observations of the diffuse flux will be given by Stecker (this institute).

3. Emission from the Galactic Plane

After a lengthy discussion in the literature, most scientists seemed to agree that the results of OSO-III could be explained by π^0-gamma rays from cosmic ray collisions with the

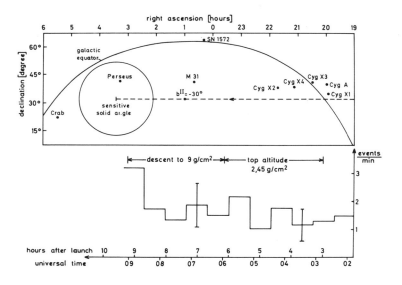

Fig. 20: Experimental results of the double Compton telescope showing that the Galactic plane is not seen by this instrument.

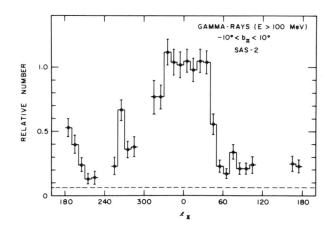

Fig. 21: The distribution of high energy Galactic γ-rays along the Galactic plane. Results from SAS-2.

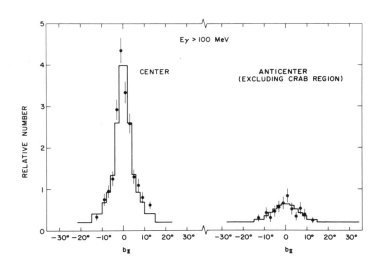

Fig. 22: The distribution of γ-rays as a function of b^{II} from the centre (a) and anticentre (b) directions. Histograms are Gaussian distributions. The curve (a) has two Gaussians fitted, the sharper one with the instrumental resolution.

interstellar gas, if the galactic centre region was excluded. There, an additional component was apparently present (Kraushaar et al., 1973; Stecker, 1971).

More recently, the results from SAS-II (Kniffen et al., 1973) have added statistical weight to the OSO-III observations, and they have also indicated that the feature towards the galactic centre exhibits the shape of a broad ridge rather than a distribution peaked at $1^{II} = 0°$ (Fichtel et al., 1974; Bignami et al., 1974). Fig. 21 shows the γ-ray distribution as a function of galactic latitude towards the galactic centre and anticentre respectively.

It seems to be interesting to combine the OSO-III and SAS-II data for all galactic longitudes because both cover about the same width in latitude. However, it may be that some of the sharper features of the SAS-II results are somewhat smeared out by this procedure. Fig. 23 shows the result.

Turning now to the energy spectrum observed from the centre region (Kniffen et al., 1973), this is shown in Fig. 24. One immediately recognizes that this is not a pure $\pi°$-spectrum, but rather appears to have a soft component. Of one takes into account the points obtained by the Imperial College Group recently

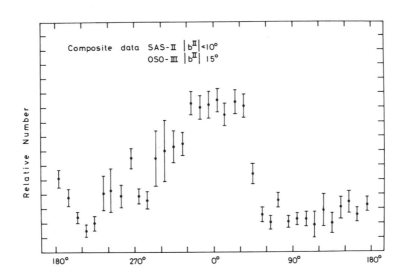

Fig. 23: Galactic longitude distribution compiled from the OSO-3 and SAS-2 data.

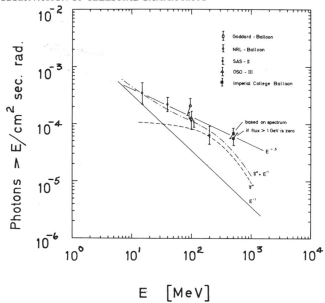

Fig. 24: Energy spectrum of γ-rays received from the Galactic centre.

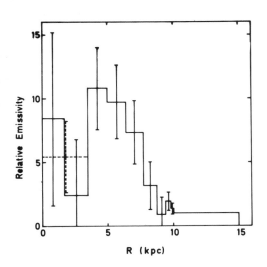

Fig. 25: Results of an unfolding of the Galactic γ-ray distribution of Fig. 21. This indicates that the emissivity must have a toroidal structure.

(Sood et al., 1974), one gets a very hard spectrum at high energies.

Strong (1974 and private communication) has unfolded the distribution in galactic longitude of the SAS-II results of Fig. 21 and has shown in general terms that this can be understood in terms of a toroidal structure of galactic gamma ray emissivity (Fig. 25). A similar result has been obtained by Puget and Stecker (1974). This general picture has been outlined in more detail by Stecker et al. (1974) and Bignami and Fichtel (1974), who consider enhanced π^0-production from cosmic rays either in a ring or at about 4-5 kpc distance from the centre, or in the spiral arms. However, Cowsik and Voges (1974) and Beuermann (1974) have pointed out that the electrons present in the cosmic radiation must be considered also. Indeed, because some of them are secondaries from the same collisions that produce the gamma rays in the first place, they should then also be enhanced. Electrons, however, will produce gamma rays by Compton collisions on starlight. Since starlight is strongest in the galactic centre and shows no toroidal structure, the ensuing gamma-rays will in turn exhibit a distribution with galactic longitude that is not flat but rather shows a maximum at $l^{II} = 0°$. It is in this context that the mixed distribution of the OSO-III and SAS-II results shown in Fig. 23 is interesting.

Recently Solomon and Stecker (1974) have proposed a molecular arm feature to be the origin of the toroidal structure of gamma ray luminosity. This will be discussed in more detail by Stecker (this institute), as will be other theories of galactic gamma ray emission.

4. Localized sources

Recently, SAS-II has also added to the wealth of data existing on the Crab, and the new compilation of results is shown in Fig. 26. It appears that most of the flux is pulsed, the values > 100 MeV are: $(3.2 \pm 9) \times 10^{-6}$ photons/cm^2 s for the total flux, and $(2.2 \pm .7) \times 10^{-6}$ photons/cm^2 s for the pulsed flux.

Another feature has emerged from the SAS-data that had been anticipated by the present author several years ago (Pinkau, 1970). This is the observation of the Vela supernova remnant (Thompson et al., 1974) (Fig. 27). The centre of the gamma ray enhancement is within 1° of Vela-X and PSR 0833-45. The observed excess is 5×10^{-6} photons/cm^2 sec > 100 MeV, detected at 8.5 σ above background.

Thompson et al. (1974) suggest that this source may be due

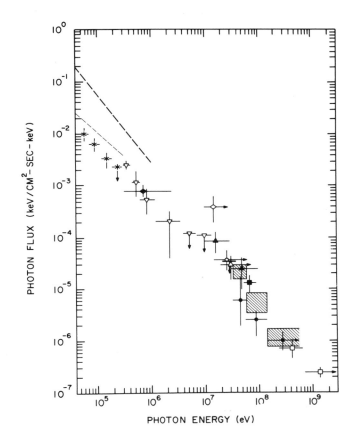

Fig. 26: Compilation of results on the γ-ray emission from the Crab.

to either a galactic-arm-segment feature at a distance > 2 kpc, or due to the Vela supernova remnant. Since the latter interpretation is more interesting with respect to origin of cosmic rays, we will consider it further in the next section.

IV   DISCUSSION OF THE VELA SUPERNOVA REMNANT

One of the methods to make progress in answering the question of galactic vs. extragalactic origin of the nucleonic component of cosmic rays would be to measure gamma ray sources, just as the non-thermal radio sources are an indication for cosmic electron production. Since enhanced gamma radiation has been observed from the direction of Vela it is interesting

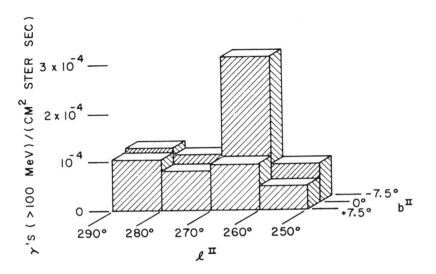

Fig. 27: Count data on high energy γ-rays received from the Vela region.

to study this source in detail.

Following Woltjer (1972) who in turn uses Sedov's (1959) work, one can distinguish three phases in a supernova remnant.

I. $M_o \gg \frac{4\pi}{3} \rho_o R^3$. Here the mass M_o ejected from the supernova at time 0 with velocity v_o and total energy ε_o is larger than the swept-up material from interstellar space with local density ρ_o.

II. $M_o \ll \frac{4\pi}{c} \rho_o R^3$, $\int |\frac{d\varepsilon}{dt}|_{rad} \, dt \ll \varepsilon_o$. The remnant's behaviour is now dominated by the swept-up matter. That this is so, can be seen more simply by considering Shklovsky's (1968) treatment:

Conservation of momentum:

$$(M_o + \frac{4}{3} \pi R^3 \rho_o) V = M_o V_o \qquad (4.1)$$

$$V = \frac{dR}{dt}$$

$$\int (M_o + \frac{4}{3} \pi R^3 \rho_o) \, dR = \int M_o V_o \, dt$$

$$M_o R + \frac{1}{3} \pi R^3 \rho_o = M_o V_o t \qquad (4.2)$$

If $\frac{4}{3} \pi R^3 \rho_o \gg M_o$, then

$$V = \frac{3 M_o V_o}{4 \pi R^3 \rho_o} \qquad (4.3)$$

$$R = \left(\frac{3 M_o V_o t}{\pi \rho_o}\right)^{1/4} \qquad (4.4)$$

and thus

$$R = 4 V t \qquad (4.5)$$

The theory of an adiabatically expanding shock front (Sedov, 1959) leads to slightly different solutions (see Milne, 1970); (Wallerstein and Silk, 1971):

$$R(pc) = 2 \cdot 10^{-11} \left[\frac{E_o(\text{erg})}{n_H(\text{cm}^{-3})}\right]^{1/5} |t(\text{yr})|^{2/5} \qquad (4.6)$$

$$V_{shock} = \frac{2}{5} \frac{R}{t} \qquad (4.7)$$

$$V_{obs} = \frac{3}{4} V_{shock} \qquad (4.8)$$

$$T = \frac{3}{16} \frac{\mu}{k} V_{shock}^2 \qquad (4.9)$$

$$T(^\circ K) = 1.45 \cdot 10^{-9} \left(V_{shock} \left|\frac{cm}{s}\right|\right)^2 \qquad (4.10)$$

III. $\int \left|\frac{d\varepsilon}{dt}\right|_{rad} dt \sim \varepsilon_o$. Radiative cooling becomes important. Then, the shell moves at constant radial momentum

$$\frac{4}{3} \pi R^3 \rho_o V = \text{constant}$$

Against these considerations one can now look at the observations. Seward et al. (1971) have obtained X-ray data from Vela and conclude that the temperature was $\sim 4 \times 10^6$ °K if the X-rays are thermal. Since cooling is important only below 10^5 °K (Cox and Tucker, 1969), the remnant is in phase II.

From eq. (4.10) we derive a shock velocity V_{shock} = 530 km/s. Optical observations (Wallerstein and Silk, 1971) lead to a value V_{obs} = 240 km/s which, with eq. (4.8) results in V_{shock} = 320 km/s.

From the pulsar data, the age of Vela is estimated to be 1.1×10^4 years, while the sling effect (Sewart et al. 1971) places this age to be more like 10^5 years.

The distance of Vela (Brandt et al., 1971) is thought to be r = 470 pc, and the total energy required for ionization is $E_o = 5 \times 10^{51}$ ergs. Its angular diameter in radio, optical and X-ray observations is $\sim 5°6$, giving R = 23 pc.

The local interstellar medium is thin, about 0.4 atoms/cm^3. Then, the swept-up mass is $\sim$ 400 $M_\odot$.

Conclusions from observations:

Shock velocity $V_{shock} \sim 300 \ldots 500$ km/s

Age $t \sim 1.1 \times 10^4 \ldots 10^5$ years

Radius $R \sim 23$ pc

Distance $r \lesssim 470$ pc

Density $n_H = 0.4$ cm^{-3}

From eq. (4.7) one obtains for the age a range of between 18000 and 30000 years. From eq. (4.6), this leads to a range of initial energy release of between 9×10^{50} ergs and 2.5×10^{51} ergs. It thus appears possible to form a consistent picture, choosing, for example, a shock velocity of 500 km/s, radius 23 pc, age 18000 years, and initial energy 2.5×10^{51} ergs.

Woltjer (1972) estimates the magnetic field to be approximately 40 μ Gauss.

From a compilation of optical observations (Brandt et al. 1971) and the magnitudes of γ^2 Vel and ζ Pup one estimates the additional starlight density to be about 0.5 eV/cm^3 which gives about 1 eV/cm^3 together with the general background starlight. This should be at a temperature of an O5 star, i.e. about 50000 °K.

The fate of relativistic particles can now be considered. Using the formalism of Felten and Morrison (1966), we can estimate the expected Compton gamma ray flux on starlight from the observed radio spectrum. The latter is shown in Fig. 28.

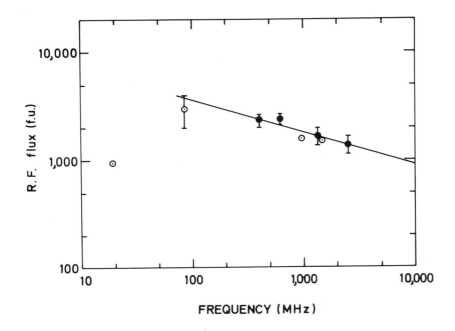

Fig. 28: Radio spectrum from the Vela supernovae remnant.

The electromagnetic radiation expected from electron interactions is:

Radiosignal:

$$I_\nu \left(\frac{\text{Watts}}{\text{m}^2 \text{ HZ ster}}\right) = 4.8 \times 10^{-20} \ (4.9 \times 10^2)^{3-m} \ n_o \ R \ H^{\frac{1+m}{2}} \ \nu^{\frac{1-m}{2}} \ ;$$

Compton γ-rays:

$$I_k^c \left(\frac{\text{MeV}}{\text{MeV cm}^2 \text{ s ster}}\right) = 10^3 \ (56.9)^{3-m} \ n_o \ R \ \rho \ T^{\frac{m-3}{2}} \ k^{\frac{1-m}{2}} \qquad (4.11)$$

Bremsstrahlung γ-rays:

$$I_k^{BS} \left(\frac{MeV}{MeV\ cm^2\ s\ ster}\right) = 80\ \frac{1}{m-1} \left(\frac{k}{m_e c^2}\right)^{1-m} n_o\ n_H\ R$$

Here, $n_o\ \gamma^{-m}\ d\gamma\ (cm^{-3})$ is the electron spectrum, R (light-years) the length of the line of sight, H (μ Gauss) the magnetic field, ν (MHz) the radio frequency, ρ (eV/cm³) the starlight energy density, T(°K) its temperature, k(eV) the photon energy), $m_e c^2$ the electron rest energy in the same units as k, and n_H the nuclear density (cm^{-3}) on which the electrons radiate Bremsstrahlung.

From Fig. 28, we have $\frac{1-m}{2} = -0.3$, T = 50000 °K, k = 10^8eV, H = 40 μ Gauss, and ν(MHz) is correlated to k by the fact that the same electrons must produce the radio waves and the 100 MeV gamma rays, i.e.

$$\nu = 0.013\ \frac{kH}{T} = 1083\ MHz \qquad (4.12)$$

For this frequency, I_ν = 2000 f.u. = $2 \times 10^{-23} \left(\frac{Watts}{m^2\ Hz\ ster}\right)$.

Assuming that the radio source has the same angular dimension as the gamma ray source we obtain from eq. (4.11)

$$I_k^{Compton} = 2.8 \times 10^{-9} \left(\frac{MeV}{MeV\ cm^2\ s}\right),$$

$$I_k^{BS} = 2.6 \times 10^{-8} \times n_H \left(\frac{MeV}{MeV\ cm^2\ s}\right).$$

This corresponds to a Compton photon flux of $1.4 \times 10^{-8}\ \frac{photons > 100\ MeV}{cm^2\ s}$ if one integrates the diverging spectrum up to 1 GeV. Compton collisions therefore cannot have caused the gamma rays.

In the case of Bremsstrahlung the flux > 100 MeV is $4.3 \times 10^{-8} \times n_H$.

This leaves nuclear interactions and gamma ray production via π^o-decay as a possible mechanism. Pinkau (1970) has estimated a photon flux of 4.5×10^{-6} photons/cm²sec to be produced if the product of the total energy input into cosmic rays W(erg) times local matter density, $n(cm^{-3})$ is 7×10^{50} erg/cm³.

Since the local matter density was assumed to be 0.4 cm^{-3}, a total energy of 1.7×10^{51} ergs would have to be present today in the Vela supernova in order to produce the observed gamma rays by nuclear interaction.

However, the adiabatic expansion energy loss has to be considered.

The energy loss by adiabatic expansion can be written as

$$-\frac{dE}{dt} = \frac{V}{R} E \qquad (4.13)$$

which, with eq. (4.7) gives

$$-\frac{dE}{dt} = \frac{2}{5} \frac{E}{t}, \text{ and } E = E_o \left(\frac{t_o}{t}\right)^{2/5} \qquad (4.14)$$

One may assume that adiabatic energy loss sets in if the remnant comes into phase II, i.e. if the swept-up mass is about 1 $M_\odot$. This occurs at R = 3 pc or t_o = 110 years. Then, for E = 300 MeV (threshold for pion production), E_o = 2.3 GeV. There is a factor of 3 more energy above 300 MeV in a cosmic ray energy spectrum (slope -2.6) than above 2.3 GeV. Thus the supernova would have had to release about 5×10^{51} ergs in the original event in cosmic rays. This is quite a large requirement, but it does not appear impossible.

The conditions could be relaxed somewhat if expansion is assumed to occur in a shell, or if the local matter density was actually larger than 0.4 cm^{-3}.

V. CONCLUSIONS

In this presentation it has been shown that gamma ray observations begin to yield information that can be used to understand the physics of galactic structure and of the distribution of cosmic rays and matter. The investigation of individual sources like supernova remnants allows us to possibly detect cosmic ray sources in the galaxy, and Vela is a prime candidate for this. The diffuse flux measurements allow us to investigate the intergalactic medium and possibly very early epochs in the universe.

These results must be seen against the large effort still required to improve instrumentation with respect to energy- and angular resolution and background rejection. The human eye is vastly superior to gamma ray astronomy instruments! Only determined efforts, and a few bright ideas, can help to overcome this deficiency.

REFERENCES

Berger, M.J. and Seltzer, S.M., Nucl. Instr. Meth., 104, 317, 1972.

Beuermann, K.P., Proc. 8th ESLAB-Symp., Frascati, 1974.

Bignami, G.F. and Fichtel, C.E., Astrophys. J., 189, L65, 1974.

Brandt, J.C., Stecher, T.P., Crawford, D.L. and Maran, S.P., Astrophys. J., 163, L99, 1971.

Caravane Collaboration of Cos-B, Proc. 8th ESLAB-Symp., Frascati, 1974.

Chupp, E.L., Forrest, D.J., Higbie, P.R., Suri, A.N., Tsai, C. and Dunphy, P.P., Nature, 241, 333, 1973.

Clayton, D.D. and Fowler, W.A., Comments on Astrophys. and Space Phys. I, 147, 1969.

Clayton, D.D., Astrophys. J., 188, 155, 1974.

Clayton, D.D., Workshop on γ-ray Astrophysics, NASA-SP-339, 1973.

Cowsik, R. and Voges, W., Proc. 8th ESLAB-Symp., Frascati, 1974.

Cox, D.P. and Tucker, W.H., Astrophys. J., 157, 1157, 1969.

Dennis, B.R., Suri, A.N. and Frost, K.J., Astrophys. J., 186, 97, 1973.

Dyer, C.S. and Morfill, G., Astrophys. and Space Sci., 14, 243, 1971.

Evans, R.D. "The Atomic Nucleus", McGraw-Hill, 1955.

Felten, J.E. and Morrison, P., Astrophys. J., 146, 686, 1966.

Fichtel, C.E., Kniffen, D.A. and Hartman, R.C., Astrophys. J., 186, L99, 1973.

Fichtel, C.E., Goddard Preprint X-662-74-57, 1974.

Fishman, G.J., Astrophys. J., 171, 163, 1972.

Higbie, P.R., Chupp, E.L., Forrest, D.J. and Gleske, I.U., Preprint U. of New Hampshire, UNH-71-24, 1971.

Johnson, W.N., Harnden, F.R. and Haymes, R.C., Astrophys. J., 172, L1, 1972.

Kniffen, D.A., Hartman, R.C., Thompson, D.J. and Fichtel, C.E., Astrophys. J., 186, L105, 1973.

Kraushaar, W.L., Clark, G.W., Garmire, G.P., Borken, R., Higbie, P., Leong, V. and Thorsos, T., Astrophys. J., 177, 341, 1973.

Mayer-Hasselwander, H.A., Pfeffermann, E., Pinkau, K., Rothermel, H., Sommer, M., Astrophys. J., 175, L23, 1972.

Milne, D.K., Aust. J. Phys., 21, 201, 1968.

Milne, D.K., Aust. J. Phys., 23, 425, 1970.

Morfill, G. and Pieper, G.F., GSFC Preprint X-600-73-239, 1973.

Neuert, H., "Kernphysikalische Meβverfahren", G. Braun, Karlsruhe, 1966.

Pinkau, K., Phil. Mag., 2, 1389, 1957.

Pinkau, K., Il Nuovo Cimento, 33, 221, 1964.

Pinkau, K., Phys. Rev. Lett., 25, 603, 1970.

Pinkau, K., Nucl. Instr. Meth., 104, 517, 1972.

Pinkau, K., Proc. 13th Int. Cosmic Ray Conf., Denver, Vol. 5, 3501, 1973.

Pinkau, K., Proc. COSPAR Meeting, Sao Paulo, III.A.1.3, 1974.

Puget, J.L. and Stecker, F.W., Ap. J., 191, 323, 1974.

Ramaty, R., Borner, G. and Cohen, J.M., Preprint Goddard Space Flight Center, Greenbelt, 1972.

Samini, J., Share, G.H. and Kinzer, R.L., Proc. 8th ESLAB-Symp., Frascati, 1974.

Schönfelder, V., Hirner, A. and Schneider, K., Nucl. Instr. Meth., 107, 385, 1973.

Schönfelder, V. and Lichti, G., Proc. 8th ESLAB-Symp., Frascati, 1974.

Schömfelder, V. and Lichti, G., Astrophys. J., in press, 1974.

Schwartz, D., see discussion in: "Gamma Ray Astrophysics", NASA SP-339, Washington, 1973.

Sedov, L.I., "Similarity and Dimensional Methods in Mechanics", Acad. Press, New York, 1959.

Seward, F.D., Burginyon, G.A., Grader, R.J., Hill, R.W., Palmieri, T.M. and Stoering, J.P., Astrophys. J., 169, 515, 1971.

Share, G.H., Kinzer, R.L. and Seeman, N., Astrophys. J., 187, 511, 1974.

Shklovsky, I.S., "Supernovae", John Wiley and Sons, London, 1968.

Solomon, P.M. and Stecker, F.W., Proc. 8th ESLAB-Symp., Frascati, 1974.

Sood, R.K., Bennett, K., Clayton, P.G. and Rochester, G.K., Proc. 8th ESLAB-Symp., Frascati, 1974.

Stecker, F.W., "Cosmic Gamma Rays", NASA SP-249, 1971.

Stecker, F.W., Puget, J.L., Strong, A.W. and Bredekamp, J.H., Astrophys. J., 188, L59, 1974.

Strong, A.W., Proc. 8th ESLAB-Symp., Frascati, 1974.

Tanaka, Y., Proc. 8th ESLAB-Symp., Frascati, 1974.

Thompson, D.J., Proc. 8th ESLAB-Symp., Frascati, 1974.

Thompson, D.J., Bignami, G.F., Fichtel, C.E. and Kniffen, D.A., Astrophys. J., 190, L51, 1974.

Trombka, J.I. and Schmadebeck, R., NASA SP-3044, 1968.

Trombka, J.I., Senftle, F. and Schmadebeck, R., Nucl. Instr. Meth., 87, 37, 1970.

Trombka, J.I., Metzger, A.E., Arnold, J.R., Matteson, J.L., Reedy, R.C. and Peterson, L.E., Astrophys. J., 181, 737, 1973.

Trombka, J.I., Proc. 8th ESLAB-Symp., Frascati, 1974.

Wallerstein, G. and Silk, J., Astrophys. J., 170, 289, 1971.

Woltjer, L., Ann. Rev. Astron. Astrophys., 10, 129, 1972.

Womack, E.A. and Overbeck, J.W., J. Geoph. Res., 75, 1811, 1970.

COLLAPSED STARS, PULSARS AND THE ORIGIN OF COSMIC RAYS

F. Pacini
Laboratorio Astrofisica Spaziale, Frascati, (Rome)

I   INTRODUCTION

In the last two or three decades it has become clear that the origin of cosmic rays is related to major astrophysical phenomena, such as the activity in supernovae and their remnants, pulsars and violent events in extragalactic systems. These phenomena are characterized by the presence of large amounts of non-thermal energy in the form of magnetic fields and relativistic particles.

During this Advanced Study Institute Dr. Burbidge will discuss the extragalactic phenomena possibly related to cosmic rays and Dr. Colgate will do the same for the explosion of supernovae. My task is to present the relevance of collapsed stars and, in particular, of pulsars.

In recent years the discovery of pulsars has been regarded as a decisive step toward understanding the origin of cosmic rays, since it is beyond doubt that these objects produce relativistic particles. As we shall see later, however, the detailed acceleration mechanisms remain unclear and it is not even clear whether the pulsars can be regarded as the dominant source of cosmic rays in our Galaxy.

In my lectures I am planning to begin by presenting the various equilibrium configurations which are possible at the end of stellar evolution. I shall then describe the pulsar phenomenon from an observational point of view and finally introduce the basic electrodynamics of pulsars and the mechanisms which have been proposed for the acceleration of particles.

II COLLAPSED STARS

1. Generalities

It is well known that a normal star shines by virtue of the continuous conversion of simple into more complex nuclei and that this process can at most release energy with an efficiency of order 1%. This upper limit is set by the nature of nuclear forces and corresponds to the maximum binding energy of a proton in an atomic nucleus, the nucleus Fe^{56}. In a normal star the gravitational energy that binds an individual proton to the rest of the star is much less than the nuclear energy that binds a proton into a nucleus. Accordingly, the gravitational energy of a normal star contributes little to the overall energy budget.

When a star has reached the end point of evolution and the nuclear reactions cannot provide more energy, its luminosity carries away the internal thermal energy and the gas pressure decreases (unless the gas is already degenerate). This leads to a catastrophic collapse during which the gravitational binding increases and the nuclear binding remains unchanged.

It is probably appropriate to define a <u>collapsed star</u> as a star whose mass M and radius R satisfy the inequality

$$G \frac{M^2}{R} \gtrsim 0.01 \, Mc^2 \quad \text{i.e.} \quad R \lesssim 10^2 \, R_s \tag{2.1}$$

($R_s = \frac{2GM}{c^2}$ is the well known Schwarzschild radius). For such an object the gravitational binding dominates any other form of internal binding. We note that this definition is somewhat arbitrary: it does not include white dwarfs among collapsed stars but it does include neutron stars, some supermassive objects and, of course, black holes. Newly formed collapsed objects can be a powerful stockpile of energy since they are likely to rotate very fast due to the conservation of angular momentum.

The rotational energy can be roughly as high as the gravitational energy and therefore can exceed the nuclear energy available to the original star. As pointed out by P. Morrison, there is also a basic difference in the quality of the energy available. In a normal star, say $\sim$ one solar mass ($\sim 10^{57}$ particles), the bulk of the stellar energy is distributed between $\sim 10^{57}$ degrees of freedom, each of these with an energy $kT \sim$ a few KeV. If the same star collapses to a neutron star size ($\sim$ 10km.), the bulk of the energy is stored in just one degree of freedom, rotation, and the corresponding kT can exceed 10^{53} ergs. Even if a normal star has a fairly large kT in the rotational mode (or in some other macroscopic mode), this is usually a small fraction of the total stellar energy. From the point of view of thermodynamics it is obvious that only a small fraction of the energy released by ordinary stars should appear in the form of high energy particles and that the opposite is true for collapsed stars.

2. Degenerate Stars

If we exclude supermassive stars from our considerations, collapsed stars are high-density objects where the matter is in a degenerate state. We shall therefore review briefly the possible equilibrium configurations for degenerate stars. (For more details, see Zel'dovich and Novikov, 1971).

If M and R are the stellar mass and radius, P and ρ the typical internal pressure and density, hydrostatic equilibrium requires that the gravitational potential energy per unit mass should be of the same order as the kinetic energy per unit mass

$$G \frac{M}{R} \sim \frac{P}{\rho} \sim V_s^2 \qquad (2.2)$$

(V_s is the sound velocity inside the star). This implies

$$P \sim \left(\frac{4}{3}\pi\right)^{1/3} G M^{2/3} \rho^{4/3} \qquad (2.3)$$

If the pressure is determined by highly degenerate particles the equation of state is a function of the density only. According to eq. (2.3), the stellar mass is then also a unique function of the density. The pressure is just the density of momentum flux and therefore in a degenerate gas with Fermi momentum

$$p_f = (3\pi^2)^{1/3} \hbar n^{1/3}$$

the pressure $P \sim n v p_f$ is given by

$$P \propto n^{5/3} \quad \text{(if } p_f \ll mc\text{)} \qquad (2.4)$$

$$P \propto n^{4/3} \quad \text{(if } p_f \gg mc\text{)} \qquad (2.5)$$

In the case of a free gas (non-interacting particles) of nuclei (Z, A) and electrons, the electron degeneracy is non-relativistic up to densities

$$\rho \simeq \frac{A}{Z} \times 10^6 \text{ gr. cm}^{-3}.$$

Beyond this limit the electron energy exceeds the rest-mass energy.

Stars where the atomic nuclei determine the mass density (and therefore the stellar gravity) while the degenerate electrons provide the supporting pressure have been known for a long time: they are called <u>white dwarfs</u>. The typical parameters of white dwarfs are a mass $M \sim 0.1 - 1\ M_\odot$, a density 10^5-10^6 gr. cm^{-3} and a radius $R \sim 10^9$ cm. These parameters can be roughly inferred by inserting the proper equation of state $P(\rho)$ in the equilibrium equation (2.3). It is easy to see that in the non-relativistic regime where $P \propto \rho^{5/3}$ the mass of a degenerate configuration is $M \propto \rho^{1/2}$. On the other hand, in the relativistic region (that is above $\sim 10^6$ gr. cm^{-3}) $P \propto \rho^{4/3}$: the density disappears in eq. (2.3) and the mass becomes constant. The limiting value for the mass is of the order of $1.4\ M_\odot$.

In real life an increase of the matter density much above 10^6 gr. cm^{-3} is accompanied by basic changes in the equation of state. Indeed, the Fermi energy of the electrons becomes sufficiently high to transform the nuclei into neutron rich nuclei because of inverse beta-decay $(Z, A) + e^- = (Z-1, A) + $ neutrinos. The opposite reaction, ordinary beta-decay, is impossible because all the cells in the phase space are occupied and electrons cannot be produced. The density at which neutronization begins depends on the composition of the stellar matter. For a non-interacting gas of electrons and protons this happens around 2×10^6 gr. cm^{-3} but a larger density is required in the case of heavier nuclei.

The effect of neutronization is that of decreasing the number density of electrons which provide the supporting pressure for the star. This softens the equation of state and leads to an adiabatic exponent $\gamma \equiv \frac{d \ln P}{d \ln \rho} < \frac{4}{3}$. As we shall see later, an adiabatic exponent less than $\frac{4}{3}$ implies instability. We cannot expect the existence in nature of configurations with a density in the region where neutronization is taking place.

Above $10^{11}-10^{12}$ gr. cm^{-3} pressure and density are determined by the degenerate neutron gas. If the neutrons were non-interacting the adiabatic exponent would be $\gamma = \frac{5}{3}$ up to densities $\sim 10^{15}$ gr. cm^{-3}. Above this value the neutrons are relativistic and again $\gamma = \frac{4}{3}$. In real life nuclear interactions cannot be neglected and they are attractive below $\sim 10^{14}$ gr. cm^{-3}, repulsive above. The effect of nuclear interactions is to extend the instability region where $\gamma < \frac{4}{3}$ up to densities $\sim 10^{13}$ gr. cm^{-3}. Above this density stable equilibrium is possible again and, for obvious reasons, one can talk of <u>neutron stars</u>.

The minimum mass of a neutron star is poorly known because of uncertainties in the equation of state but should be around 5×10^{-2} $M_\odot$. The maximum mass is also poorly known but should be around 2 - 3 $M_\odot$, with a density $\rho \sim 10^{15}$ gr. cm^{-3} and a radius $R \sim 10$ km.

A proper investigation of neutron stars requires the adoption of the general relativistic equations. Also, beyond the nuclear density 3×10^{14} gr. cm^{-3} the neutrons start to be converted into heavier particles. Various kinds of hyperons become stable and complicate further the equation of state of superdense matter. (For a review of various problems concerning the equation of state at ultra-high densities, see Canuto, 1974).

3. Stability of Degenerate Stars

So far we have discussed equilibrium configurations but we have not considered whether this equilibrium is stable or unstable. It is easy to see that a star can be stable only if $\gamma > \frac{4}{3}$. Indeed, if we squeeze a star, stability implies that the pressure increases faster than the gravitational pull. Vice-versa, if we force a star to expand, the pressure should decrease faster than the gravity. From equ. (2.3) one finds immediately that the left-hand side increases faster than the right-hand side if $\frac{dP}{d\rho} > \frac{4}{3} \frac{P}{\rho}$, that is if $\frac{d\ln P}{d\ln \rho} > \frac{4}{3}$. This is the basic physical reason why there cannot be stable stars intermediate between white dwarfs and neutron stars.

4. Astrophysical Problems Related to Neutron Stars

Until 1968 the only type of degenerate stars observationally known to exist were the white dwarfs. Neutron stars were studied intensively from a theoretical point of view and it was widely believed that some supernovae explosions would leave behind a neutron star as well as an expanding shell. In particular, various arguments were suggesting the possible presence of a neutron star inside the Crab Nebula, the remnant of a stellar explosion which occurred roughly 900 years ago. As is well known, the Crab Nebula emits at all frequencies from radiowaves up to γ-ray photons: the bulk of the emission is caused by relativistic electrons moving in a magnetic field around 10^{-3} gauss. The more energetic electrons have a lifetime against radiation losses shorter than the age of the Nebula. Indeed, the lifetime of electrons radiating in a field B at a frequency ν is given by $T_{\frac{1}{2}} \simeq 10^{12} \nu^{-\frac{1}{2}} B^{-3/2}$ years. It is immediately seen that all electrons radiating at or beyond the optical region have a lifetime $\lesssim$ a few years. The present-day nebular electrons cannot have been produced in the initial explosion and therefore in the Nebula there should be a continuous injection of particles at a rate matching the nebular electromagnetic losses, about 10^{38} ergs sec^{-1}. A simple analysis of the radiated spectrum reveals that the electrons injected into the Nebula are distributed in the energy range $10^8 \lesssim E \lesssim 10^{14}$eV according to power laws. The average energy of the injected electrons is around 10^{10}eV.

A similar problem exists for the origin of the magnetic field. A field around 10^{-3} gauss over the Crab volume ($\simeq$(1pc.)^3) cannot result directly from the presence of a central dipole or from the radial expansion of a pre-existing stellar field. Both explanations would demand sources of unrealistic strength and, furthermore, the theoretical expectations would not match the observed geometry of the field.

The enigma of the origin of the relativistic particles and of the magnetic field led various authors to suggest that a neutron star could be present inside the Crab Nebula and act as an energy stockpile for its activity.

A newly born neutron star is likely to be a very excited object. Apart from its thermal content (which is dissipated very fast), there is also much energy stored in vibrational and rotational forms.

In the case of vibrations, the fundamental period is always of order

$$P \sim \frac{R}{V_s} \sim \frac{1}{\sqrt{G\rho}} \qquad (2.6)$$

Neutron stars with $\rho \sim 10^{15}$ gr. cm^{-3} vibrate about one thousand times per second (in the case of white dwarfs $P \sim$ several seconds). Neutron stars can also be expected to rotate very fast: a slowly spinning star like our sun would be spinning $\gtrsim 10^3$ times per second if it collapsed to a neutron star size. In the case of rotation, the basic limitation is that the rotational energy cannot exceed the gravitational energy. It is then immediately found that a star can rotate at most as fast as it vibrates.

Can the mechanical energy stored in neutron stars play an important role in connection with the activity observed in the Crab Nebula? The answer to this problem came in 1968 with the discovery of pulsars and, in particular, with the discovery of the pulsar NP 0532 in the Crab Nebula.

III OBSERVATIONAL ASPECTS OF PULSARS

Pulsars are characterised primarily by the emission of sharp flashes of radiowaves at almost exactly maintained time invervals. The typical ratio between the time length of a pulse τ and the period P is $\delta \equiv \frac{\tau}{P} \simeq 1-10\%$. More than one hundred pulsars are presently known, a large proportion of them concentrated along the galactic plane. A vast amount of data is available about these sources but in the following we shall only present the basic characteristics.

1. Periods' Distribution and Slowing Down

The periods range from 33 msec. for the Crab Nebula pulsar NP 0532 up to 3.75 sec. for NP 0527. The existence of pulsars with periods << 1 sec. immediately rules out white dwarfs as candidates for this phenomenon. As we have seen earlier, the typical mechanical periods of white dwarfs are $\sim$ few seconds. Since there are no configurations intermediate between white dwarfs and neutron stars, the latter remain the only possibility. Vibrations can be ruled out because they are in the millisecond range. Rotating neutron stars are the obvious candidate since they could very well span the periods' range between 33 msec. and a few seconds. Furthermore, in full agreement with the notion of rotating neutron stars, those pulsars whose period has been monitored with sufficient precision show a tendency toward lengthening the period. It is found that the apparent lifetime $\frac{P}{\dot{P}}$ generally increases with the period and ranges between $\sim 10^3$ years for NP 0532 up to about 10^8 years for some slow objects. It is interesting to notice that the quantity $\frac{P}{\dot{P}}$ for NP 0532 roughly matches the known age of the Crab Nebula.

Most pulsars have periods in the range 0.5 - 1 sec. Since they spend only a small fraction of their life with a very short period, it is not surprising that relatively few objects have very short periods. The lack of pulsars with very long periods can only be explained in terms of a fairly steep dependency of the pulsar luminosity on the period. Two pulsars (NP 0532 and PSR 0833) have shown abrupt period changes with speeding-ups $\frac{\Delta \dot{P}}{P}$ in the range $10^{-6}-10^{-9}$. More recently, similar but smaller irregularities have been detected also in slower pulsars. (PSR 1508 + 55 and PSR 0329 + 54).

2. Distances

The same pulse appears systematically first at high and later at low radiofrequencies. This is due to the dispersive properties of the interstellar medium. The delay depends on the number of electrons along the line of sight $\int n_e \, d\ell$. Since n_e is roughly known from independent methods, one can estimate the distance of pulsars. The typical distances range between a few hundred up to a few thousand parsecs.

3. Intrinsic Power, Spectra, Brightness Temperatures

The radio power emitted from pulsars ranges between 10^{27} ergs sec^{-1} up to 10^{31} ergs sec^{-1}. A basic uncertainty on the emitted power stems from our ignorance of the pulsar emission diagram. The radio spectra emphasize the low frequencies (say, around 100 MHz) and generally decrease rapidly toward the high frequencies. Several pulsars show also the presence of intrinsic cut-off's below 100 MHz.

The sharpness of the pulses indicates that the emitting region is not larger than $c\tau$ (τ is the pulse time length): this corresponds to emitting regions of the order of a few Kms up to a maximum $\sim$ 100Km. The brightness temperatures of the emitting regions are found to range between $\sim 10^{20}$ °K up to 10^{28} °K for the Crab pulsar (occasionally even higher). This entails that the electric field radiated by NP 0532 can reach and exceed 10^{10} volts m^{-1} at the source. Also, the power flux through the emitting surface can exceed 10^{12} watts cm^{-2}. (F. Drake has pointed out that this is an amount of power per square centimeter equivalent to the electrical power produced by all plants on the earth). There is no doubt that such brightness temperatures demand an extremely coherent radiation process at radiofrequencies.

4. Optical and X-Ray Emission

Only the fastest pulsar NP 0532 has been detected also at optical and X-ray frequencies (radiopower $\sim 10^{31}$ ergs sec^{-1}; optical power $\sim 10^{34}$ ergs sec^{-1}; X-ray power $\sim 10^{37}$ ergs sec^{-1}). The lack of optical emission from the second fastest pulsar PSR 0833 (which has a period only $\sim$ 3 times that of NP 0532) entails a period dependency of the optical luminosity $L \alpha P^{-n}$ with $n \gtrsim 8.4$. The brightness temperature of the Crab Pulsar at optical and X-ray frequencies is $\sim 10^{10}$ °K, perfectly compatible with an incoherent (but non-thermal) process. The optical and X-ray pulses are found to be simultaneous with the radiopulses when the effect of interstellar dispersion has been taken into account.

5. Pulse Shapes and Intensity Fluctuations

The pulse shapes are rather complicated, often with separate components or "sub-pulses". In some cases the sub-pulses are persistent features; in others, especially for small structures, no coherence exists from pulse to pulse. The pulses also show considerable intensity fluctuations, partly intrinsic, partly due to interstellar scintillation. No fluctuations exist for the optical, X-ray pulses from NP 0532.

6. Drifting Sub-pulses

In several sources, the sub-pulses reappear in consecutive pulses. However, as first noted by Drake and Craft, a given sub-pulse can arrive somewhat earlier in each successive pulse. This phenomenon, called "the marching sub-pulses", is probably fundamental for the understanding of the radiation process and seems to indicate the existence of organized motions in the emitting region.

7. Polarization

The polarization characteristics are different from pulsar to pulsar but are well marked in any individual source. It is interesting that the radiation in some cases can reach a degree of polarization close to 100% and that the polarization can vary in a simple way across the pulse. Also, in the presence of marching sub-pulses, each sub-pulse maintains its polarization characteristics.

8. Association with SN Remnants

Only the two fastest pulsars NP 0532 and PSR 0833 have been found inside SN Remnants. The absence of extended SN Remnants around slow pulsars is easy to understand since they have ages largely exceeding the lifetime of a Supernova Remnant. The lack of observed pulsars in some young Remnants such as Cas A can imply that not all explosions leave behind a neutron star but could also be explained in terms of the unknown pulsar emission diagram. In addition, because of the high radioluminosity of Cas A, it is unlikely that even a source like NP 0532 would have been discovered in it against the strong background. As of today, it is very difficult to determine directly or indirectly (for instance, by comparing the number of pulsars in the galaxy with the frequency of stellar explosions) what is the fraction of Supernovae which leave behind a neutron star.

9. Pulsars in Binary Systems

No evidence exists for the presence of pulsars in binary systems. This can be due to either of the following (or to a combination of them): a) most binary systems are disrupted when one of the stars becomes a supernova, b) the lifetime of pulsars in binary systems is short and the energy output from the pulsar is soon replaced by accretion from the companion (this would suffocate the pulsar).

10. The Energy Balance of Pulsars

A measurement of the slowing down of pulsars determines the loss of rotational energy $I\Omega\dot\Omega$ (I is the moment of inertia, Ω is the rotation frequency). For most sources the energy radiated away as pulses is just a fraction 1-10% of the total energy loss (note however that both these quantities cannot be determined with great precision, since we ignore the geometry of the pulsar beam and the exact value of the stellar moment of inertia). Where does the remaining energy go? The case of the Crab Nebula pulsar NP 0532 has shown that the bulk of the stellar rotational energy is transformed into relativistic particles. Indeed, immediately after its discovery, it was realised that the loss of rotational energy of the pulsar is $\sim 10^{38}$ ergs sec^{-1}, the amount needed to compensate for the radiation losses of the nebular particles. There can be little doubt that NP 0532 is the energy source for the nebular activity and one can reasonably suspect that also older pulsars, on a smaller scale, transform their rotational energy into energy of fast particles. Pulsars are basically machines where the energy set free in the gravitational collapse is transformed into rotational energy and then into a relativistic form.

IV   PULSAR ELECTRODYNAMICS

1. Generalities

The electrodynamics of a rotating neutron star closely resembles that of a rotating magnetized conductor. The differences between the two cases are largely due to the fact that the magnetic fields of neutron stars can be enormous. At least two processes can lead to huge fields: a) differential rotation - if any - in the collapsing star. If the stellar core spins at a different rate than the envelope, a magnetic linking between the two regions results in the twisting of the field lines and the gradual building-up of large internal toroidal fields which in principle could reach values of order $10^{15}-10^{17}$ gauss, b) flux conservation. Since the conductivity of the stellar material is extremely high, during the collapse the flux will be conserved. An ordinary star like our sun with radius $R \sim 10^{11}$ cm would amplify the field by a factor $\sim 10^{10}$. Depending upon the initial condition, this could lead to strengths in the range $10^{10}-10^{15}$ gauss or so.

Note that the processes a) and b) could have different

implications for the stellar stability. If the field increases because of twisting, one can reach a situation where the magnetic energy exceeds the gravitational energy. Indeed, if the magnetic energy derives from the energy stored as differential rotation and can be as high as $\frac{1}{2} I \Omega^2 \lesssim G \frac{M^2}{R}$. If the angular momentum is conserved, $\frac{1}{2} I \Omega^2 \alpha R^{-2}$ and the rotational energy increases faster than the gravitational energy. On the other hand, if the flux is conserved, the ratio between magnetic energy and gravitational energy does not depend on the radius.

2. Electrodynamics of a Sphere Rotating Around the Magnetic Axis

If a magnetized sphere rotates and one connects the pole to the equator through a nonrotating circuit, an electromotive force arises and the circuit is traversed by a current. (Faraday experiment). In a laboratory experiment, say with a sphere of size $\sim$ 10 cm, field $\sim 10^4$ gauss, spinning frequency $\sim 10^3$ sec^{-1}, the difference of potential between poles and equator is just a few volts. In the case of neutron stars (size $\sim 10^6$ cm, field $\sim 10^{12}$ gauss, rotating $\sim$ 1000 times a second) the difference of potential can exceed 10^{16} volts.

The case of a magnetized neutron star, rotating about an axis parallel to the rotation axis, has been first investigated by Goldreich and Julian (1969).

Since the star is a very good electrical conductor, the internal charges redistribute themselves within the sphere under the influence of a magnetic force $\frac{q(\underline{\Omega} \times \underline{r}) \times \underline{B}}{c}$. The star material becomes polarized and there is an electric field $\underline{E}$

$$\underline{E} = - \frac{(\underline{\Omega} \times \underline{r}) \times \underline{B}}{c} \tag{4.1}$$

If the field results from a central dipole $\underline{\mu} = \frac{1}{2} \underline{B}_o R^3$ the components in the polar coordinates (r, δ) are

$$B_r = \frac{2\mu}{r^3} \cos \delta \qquad B_\delta = \frac{\mu}{r^3} \sin \delta \tag{4.2}$$

$$E_r = \frac{\mu \Omega}{cr^2} \sin^2 \delta \qquad E_\delta = - \frac{2\mu\Omega}{cr^2} \sin \delta \cos \delta \tag{4.3}$$

One finds
$$\text{div } \underline{E} = - \frac{2\mu\Omega}{cr^3} (2 \cos^2 \delta - \sin^2 \delta) \tag{4.4}$$

The electric potential ϕ is given by

$$\underline{E} = -\text{grad } \phi \qquad (4.5)$$

$$\phi = \frac{\mu\Omega}{rc} \sin^2 \delta + \text{const} \qquad (4.6)$$

A similar result is obtained for a uniform magnetic field $\underline{B} \parallel \underline{\Omega}$.

Assume for a moment that the star is surrounded by a vacuum. At first sight this assumption appears legitimate because of the very strong gravitational field (even at a temperature $T \sim 10^8$ °K the scale height of the atmosphere is $\sim 10^{-4} R$). The electric field outside the star is then obtained by solving the Laplace's equation, matching the potentials (or the tangential electric field) across the surface.

Inside the star the condition of perfect conductivity implies $\underline{E} \cdot \underline{B} = 0$. Outside the star, however, if there is a vacuum one finds that

$$\underline{E} \cdot \underline{B} = -\frac{\Omega R}{c} \left(\frac{R}{r}\right)^7 B_0^2 \cos^3 \delta \qquad (4.7)$$

For a pulsar rotating about once per second, the parallel electric field $E_{\shortparallel}$ is $\sim 10^{10}$ volts cm^{-1} and the electric force on a charge largely exceeds the gravitational force. The outer parts of the stellar surface cannot be in equilibrium and the particles are shot out along the magnetic field lines.

The charges drawn off the surface form a magnetosphere around the neutron star. When equilibrium is established, in this magnetosphere $\underline{E} \cdot \underline{B} = 0$ (at least as long as one can neglect inertial and gravity effects). The magnetic field lines are equipotentials.

The pulsar magnetosphere can be divided into two regions. The first region (corotating magnetosphere) contains the field lines that close before a critical distance $R_c = \frac{c}{\Omega}$. In this region the particles can only slide along the rigidly rotating field lines. The charge density is given by the Poisson equation

$$n_- - n_+ = \frac{2 \underline{\Omega} \cdot \underline{B}}{4\pi e c} \qquad (4.8)$$

(The density of particles could be much larger if the charge separation is small).

The rotation of the space charge $(n_- - n_+)$ is equivalent to a toroidal current $j = (n_- - n_+) e \Omega r$. The magnetic field genera-

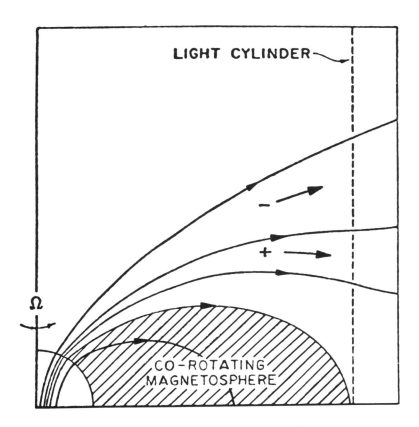

Fig. 1: Pulsar magnetosphere and light cylinder, according to the Goldreich and Julian model.

---

ted by this current is negligible close to the star but becomes very important at larger distances: at $r \sim R_C$ it becomes comparable to the field produced by the currents inside the star. This entails the necessity of a self-consistent solution where the space charge given by the field is the appropriate field source. The basic equations of the problem are well established but they have not been solved in realistic cases.

The corotating magnetosphere cannot extend beyond the critical distance R_c because otherwise the velocity Ωr would exceed the speed of light. The lines of force which close within the light cylinder on the stellar surface have a polar distance from the rotation axis $\delta \gtrsim \delta_o = \left(\frac{\Omega r}{c}\right)^{1/2}$ (typically $\delta \gtrsim 10^{-2} - 10^{-1}$ radians).

The lines of force which pass beyond the critical distance R_c define the so-called open magnetosphere: in this region there cannot be pure corotation and the plasma escapes freely under the influence of electromagnetic effects.

The potential difference across the <u>polar cap</u> $0 \leq \delta \leq \delta_o$ is

$$\Delta \phi \sim \frac{1}{2} \left(\frac{\Omega R}{c}\right)^2 R B_o \qquad (4.9)$$

This potential difference is available for an electrostatic particle's acceleration up to energies of order

$$E_{max} \sim \frac{1}{2} e \Delta \phi \sim 3 \times 10^{12} \frac{R_6^3}{P^2} B_{12} \text{ eV} \qquad (4.10)$$

(The period P is in seconds, B_{12} is in units of 10^{12} gauss, R_6 is in units of 10^6 cm.).

If we assume that the rotation and magnetic axes are parallel, the poles are at lower potential than infinity: the negative charges flow out primarily from the poles and the positive charges from lower latitudes. These poloidal currents lead to a toroidal component of the magnetic field B_t. Inside the speed of light cylinder $B_t \lesssim B_p$ (B_p is the poloidal field). At $r \gg \frac{c}{\Omega}$ however B_t becomes the dominant component. We shall return later on to the relation between B_t and the magnetic field in Supernovae remnants.

In the original Goldreich-Julian model, it is assumed that $\underline{E} \cdot \underline{B} = 0$ along the open field lines up to infinity. The field lines act as conductive wires: the potential drop between the axis of rotation and the first open line is frozen into the escaping plasma and the acceleration of particles takes place very far from the pulsar where the model breaks down. In real life, the assumption $\underline{E} \cdot \underline{B} = 0$ is unlikely to hold in the open magnetosphere. As a result the particles are probably accelerated relatively close to the star.

3. The Oblique Rotator Model

If the magnetic field is not symmetric with respect to the rotation axis, the problem entails time - dependency. In a laboratory experiment, a dipole rotating about an axis different from the magnetic axis radiates magnetic dipole waves at the

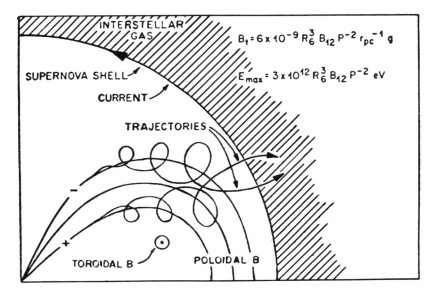

Fig. 2: Schematic diagram showing the supernova cavity, shell and interstellar gas (from Goldreich and Julian, 1969).

---

basic rotation frequency. In the case of a neutron star, the near-zone ($r \ll \frac{c}{\Omega}$) is similar to the one discussed for an aligned rotor, with the extra complication of time dependency (see e.g. Mestel 1973). At $r \gg \frac{c}{\Omega}$, however, the electromagnetic field becomes a wave field. The energy loss is of order

$$I\Omega\dot{\Omega} \simeq \frac{B_c^2}{8\pi} \ c \ 4\pi \left(\frac{c}{\Omega}\right)^2 \simeq B_c^2 \ c \ R_c^2 \qquad (4.11)$$

where

$$B_c = \frac{B_0 R^3}{R_c^3}$$

(This expression is almost identical to the one which can be obtained in the Goldreich-Julian model). Slowing down measurements determine the strength of the magnetic field in the proximity of the light cylinder. This ranges between 10^6 gauss for the Crab Nebula pulsar (where $R_c \sim 10^{10}$ cm $\sim 10^2$ stellar radii) down to one gauss or less for some slow pulsars (where $R_c \sim 10^{10}$ cm $\sim 10^4$ stellar radii). If the field obeys a dipole law in the near zone $r \ll R_c$, one obtains

$$I\Omega\dot{\Omega} \sim \frac{B_o^2 R^6 \Omega^4}{c^3} \qquad (4.12)$$

i.e. $\dot{\Omega} \alpha \Omega^3$. If the plasma outflow were dominating, the field would fall off with the radial law $B \alpha r^{-2}$ and similar arguments would lead to a lower expected value for the braking index n in the relation $\dot{\Omega} \alpha \Omega^n$.

There is observational support for a largely dipole structure of the magnetic field in the near magnetosphere of NP 0532: accurate timing of this pulsar has revealed that $n \sim 2.5$. The small difference between the dipolar and the observed values could indicate a slight radial distortion of the field lines under the pressure of the outflowing plasma.

With a dipole geometry, the above mentioned values for the magnetic strength at $r \sim R_c$ correspond to surface fields around 10^{12} gauss. (Note that the real surface fields could be much stronger if higher multipole components dominated the dipole moment close to the star).

Beyond the speed of light cylinder the electromagnetic energy flux should be constant and the magnetic field (by now predominantly toroidal) falls off as $1/r$.

In the oblique rotator model, the large-scale magnetic field produced by a pulsar is the magnetic component of a low frequency wave.

It has been shown by various authors that low frequency electromagnetic waves with $f \equiv \frac{eB}{mc\Omega} \gg 1$ accelerate particles very efficiently. The reason for this high efficiency is quite simple. Since the gyrofrequency $\frac{eB}{mc}$ is much larger than the wave frequency Ω, the particles move in a strong, nearly static, crossed electric and magnetic field. In a very short time ($\ll \Omega^{-1}$) they reach relativistic velocities along the wave propagation and then they ride the wave at constant phase. In the case of a plane wave, a particle exposed to it acquires a Lorentz factor $\gamma = f^2$; in a spherical wave $\gamma = f^{2/3}$. In the Crab Nebula, at the beginning of the wave zone $f \sim 10^{11}$ and therefore the electrons could acquire an energy $\sim 10^{13}$ eV. It is easy to see that a young pulsar could accelerate particles up to the highest energies found in cosmic rays.

In this theory, the particles produced at any given instant are monoenergetic but a power law can be obtained if one considers that the properties of the pulsar and of the SN Remnant change with time. Kulsrud et al. (1972) have indeed argued that pulsar-powered SN Remnants could produce a power-law cosmic ray spectrum.

It is unclear whether calculations based upon the scheme of vacuum waves interacting with test particles are relevant to the description of the pulsar environment. The answer depends on the density of the circumstellar plasma. First, one wonders whether these low frequency waves can really propagate away from the pulsar. In principle one expects very little - if any - thermal plasma because the waves would exert enough pressure to expel this material (there is indeed observational evidence that the circum-pulsar region in the Crab Nebula has been cleared of ambient plasma). On the other hand, the pulsar itself continuously ejects particles along the open field lines and one should worry at least about this amount of plasma. It is then found that the usual propagation condition $\Omega > \omega_p$ (ω_p is the rest plasma frequency) has to be modified in order to account for the increased mass of the relativistic electrons: the new condition is found to be $\Omega > \frac{\omega_p}{<\gamma>}$. Whether this condition is satisfied around the Crab pulsar depends on the assumed rate of particle outflow. If the number of escaping charges matches the charge density given by the Poisson equation, then the waves are probably able to propagate. If however the charge separation is small and the plasma density by far exceeds the charge density (as some observational evidence seems to indicate) then the reality of low frequency waves becomes an open question.

The simple picture of accelerating particles in vacuum by means of pure electromagnetic waves could also fail on other grounds: for instance it is likely that a static field is superimposed on the wave field and this modifies the particle's dynamics (Pacini and Salvati, 1973). Also, even a small amount of plasma would introduce an index of refraction and change the ratio E/B in the wave.

In conclusion, the simple picture of low frequency, large amplitude waves accelerating particles deserves great attention and undoubtedly represents a very attractive possibility. The plasma physics of this problem is poorly known, but various groups have now tackled the question and one can reasonably expect some progress in the near future.

4. Theory Versus Real Life

The first part of Table 1 summarizes the main consequences of the unipolar mechanism and of the emission of low frequency magnetic dipole radiation.

Basic Theoretical Expectations

UNIPOLAR INDUCTION
- energy loss $B_{o\parallel}^2 \, \Omega^4$
- particles extraction → magnetosphere
- poloidal currents → extended toroidal field
- electrostatic acceleration

DIPOLE RADIATION
- energy loss $B_{o\perp}^2 \, \Omega^4$
- large amplitude low frequency waves
- extended magnetic wave field
- acceleration in the wave

Theory Versus Real Life

Theory	Observation
$\dot{\Omega} \propto \Omega^3$	$\dot{\Omega} \propto \Omega^{2.5}$
$I\dot{\Omega}\Omega \sim 10^{38}$ ergs sec^{-1}	$L_{Crab} \sim 10^{38}$ ergs sec^{-1}
charges' outflow $\sim 10^{33}$ sec^{-1}	particles' outflow $\sim 10^{40}$ sec^{-1}
monochromatic energy spectrum	power law energy spectrum
$B_{static} \sim 10^{-4}$ gauss	$B_{static}/B_{wave} \gtrsim 10$
$B_{wave} \sim 10^{-4}$ gauss	$B \sim 3 \times 10^{-4} - 10^{-3}$ gauss

TABLE I

One can compare the predictions of these models with real life as this manifests itself in the Crab Nebula. This is done in the second part of Table 1 and one can see some good arguments as well as some puzzling discrepancies. Various explanations are possible for some of the existing discrepancies. For instance, the difference between the predicted braking index and the observed value can be due to several causes, including a decreasing moment of inertia for the star or maybe a disalignment of the rotation and magnetic axes. As mentioned earlier, the discrepancy could also simply be caused by the outflowing plasma which perturbs radially the geometry of the field lines.

A more serious problem exists when one compares the expected charges' outflow $\sim 10^{33}$ sec^{-1} with the number of particles which are known to be continuously accelerated in the Crab Nebula, roughly 10^{40} sec^{-1} (average energy 10^{10} eV). There is no doubt that the pulsar is the source of energy for this acceleration but one wonders whether the acceleration can take place far from the pulsar, for instance when the low frequency waves interact with the thermal filaments. This possibility has been suggested by various authors but there are arguments which tend to make it unlikely. A reasonable alternative would be a very small charge separation in the outflowing plasma (one part in 10^7). In this case, as we have mentioned before, the simple vacuum picture for the acceleration mechanisms is likely to fail. Maybe this would automatically remove the other discrepancy, that is the expectation of a monochromatic spectrum. Power laws could possibly be obtained also by injecting particles at various distances from the star, either in the near zone or in the wave region.

We shall return later to the subject of the discrepancy between the predicted nebular field $\sim 10^{-4}$ gauss and the estimated, somewhat higher strength: for the moment we simply remark that this could be a consequence of the existence of a conductive shell at a finite distance from the pulsar which prevents the electromagnetic energy from propagating at the speed of light, thus making possible some storage of magnetic energy.

5. The Origin of Pulsar Radiation

The pulsar radiation could in principle provide some direct information about the neutron star magnetosphere. Its origin, however, is extremely controversial, and there is no general agreement on the nature of the radiation process and on whether the pulses arise close to the stellar surface or in proximity to the speed of light cylinder.

A fully deductive theory for the origin of the pulses in the neutron star magnetosphere is not yet available. In any case, the basic requirement for producing pulses is that the radiation of individual particles is anisotropic. This is realized if the particles move relativistically, so that the cone of emission covers an angle of order γ^{-1} around the instantaneous velocity. Also, all particles seen by a given observer should be moving in the same direction within an angle of order γ^{-1}. A duty cycle $\sim$ a few percent typically implies $\gamma > 10^2$.

The high brightness temperature of the radio pulses can only be achieved by an extremely coherent radiation mechanism. Two general classes of coherent mechanisms have been considered and they involve either bunches of particles moving in phase ("antenna mechanisms") or negative absorption ("maser mechanisms"). In the case of bunches, the usual thermodynamic limitation $kT_b \lesssim$ (energy per radiating charge) refers to individual bunches and not to the single particles in the bunch: the brightness temperatures which can be achieved are therefore increased by a factor equal to the number of charges per bunch. If the radiation is due to electrons with energy, say, 100 MeV($\gamma \sim 10^2$), then $T_b \sim 10^{26}$ °K corresponds to about 10^{14} electrons per bunch. Also, in order to have coherent radiation at wavelengths $\lambda \sim 10-100$ cm, the size of the bunch along the visual line has to be less than one wavelength. In real life, one would certainly expect bunches of different size and the emitted spectrum would reflect essentially the distribution of sizes (as is well known, the spectra of incoherent processes essentially reflect the energy distribution of the particles).

As we have already mentioned, only the fastest pulsar NP 0532 has been detected at optical and X-ray frequencies. The optical and X-ray pulses are simultaneous to the radio pulses but are remarkably steady. This, together with the relatively low brightness temperature at high frequency (10^{10} °K in the optical band, 10^5 °K in the X-ray band), suggests that the emission is normal incoherent radiation, probably by the same particles which radiate coherently at radiowavelengths.

The simplest and historically first model of pulsar radiation was proposed by T. Gold immediately after the discovery of these objects. In this model, streams of plasma are ejected from selected hot spots on the star surface and corotate with the star up to the speed of light cylinder. At $r \sim R_c$, the plasma is relativistic and gives rise to an emission beamed in the direction of motion. The emitted spectrum is given by the usual theory of emission from particles in circular orbits. The critical emitted frequency is $\sim \Omega \gamma^3$ and therefore the Lorentz factor would have to be $\sim 10^2-10^3$ in order to produce radio emission.

The spectral characteristics of Gold's process remain unchanged if the radiation does not arise in the corotating magnetosphere but derives from relativistic motions along the curved field lines (at $r \sim R_c$ the field lines have a radius of curvature $\rho \sim R_c = \frac{c}{\Omega}$). The problem here is that the open lines diverge as they approach the critical distance and it is not clear what would define a preferred sector with small angular extent. The short duty cycle could be understood more easily if the emission arises when the particles slide along the open lines close to the star (we recall that in a dipole field the bundle of the open lines close to the surface subtends a rather small angle). In this type of model the pulses are considered a by-product of the extraction and acceleration of particles. If in the proximity of the star $\underline{E} \cdot \underline{B}$ is close to the vacuum value, the particles would reach extremely high energies. For the Crab Nebula these energies are around 10^{16} eV and the corresponding radiation is peaked at a photon energy $h\nu \sim 10^{12}$. The gamma-ray photons would be moving with a finite angle with respect to the strong magnetic field and would annihilate into e^+e^- pairs if the condition $\varepsilon_\gamma B_\perp > 4 \times 10^{18}$ is satisfied (ε_γ in eV, $B_\perp$ in gauss). It is assumed that the positrons will be turned around by the electric field and flow back to the surface of the star while the electrons would come out in bunches rather than in a steady flow. Since the electrons are produced with a small (but finite) pitch angle, in their motion there would be a gyration as well as a relativistic sliding along the curved field lines. Their emission would result from two separate processes, the normal synchrotron radiation and in addition the curvature radiation. Since the sizes of the bunches are much larger than the Larmor radius, the synchrotron radiation is incoherent, but coherence effects dominate the curvature radiation. An analysis of this combination of curvature radiation and synchrotron process has been carried out some time ago (Pacini and Rees, 1970): we briefly recall the main arguments and results.

If the radio emission at frequency ν is attributed to bunches of particles moving along the field lines with velocity corresponding to a Lorentz factor γ_{bunch}, then the following condition should be satisfied

$$\frac{1}{2\pi} \frac{c}{\rho} \gamma_{bunch}^3 \gtrsim \nu \qquad (4.13)$$

The effective Lorentz factor of the bunch is related to the velocity v of the particles and to the pitch angle β by the relation

$$\gamma_{bunch} \lesssim \left(1 - \frac{v^2}{c^2} \cos^2 \beta\right)^{-1/2} \qquad (4.14)$$

Close to the surface, say at 5 stellar radii, the radius of curvature would be $\rho \sim 2 \times 10^7$ cm and therefore $\gamma_{bunch} \gtrsim 200$. The brightness temperature of NP 0532 with an emitting region close to the star could reach 10^{28} °K; considerations similar to those made earlier then require 10^{16} electrons or 10^{13} protons per bunch.

The interpretation of the pulsar radiation in terms of curvature plus synchrotron processes near the star is incompatible with the idea that electrons are involved. The incompatibility lies in the position of the low frequency cut-off of the optical radiation from NP 0532. If this cut-off is due to the synchrotron reabsorption, its position and the corresponding spectral flux determine the strength of the magnetic field in the emitting region, provided one knows the mass of the radiating particles. For electrons, one finds that $B_\perp \sim 10^4$ gauss, while for protons one obtains $B_\perp \sim 10^9$ gauss. Since the pitch angles are only a few degrees, the emission could arise close to the star only if it were due to protons with no contribution by electrons. On the other hand, if the observed optical and X-ray emission is due to electrons, the position of the cut-off indicates that the pulses are emitted in proximity to the speed of light cylinder where one expects $B \sim 10^4$ gauss. The hypothesis that the pulses arise in a region surrounding R_c (either just inside or a few basic wavelengths away on the outside) would offer an immediate explanation of why only the Crab Nebula pulsar emits optical and X-ray radiation. Indeed, if the high frequency radiation is due to the synchrotron process, the total output should roughly scale like the product of the energy outflow in relativistic particles (proportional to P^{-4}) multiplied by the square of the magnetic field in the emitting regions B_c^2 (in a dipole field $B_c \propto P^{-3}$ and therefore $B_c^2 \propto P^{-6}$). One would therefore expect a scaling-law $L_{syncro} \propto P^{-10}$: this does not conflict with the present limit on the optical emission from the Vela pulsar.

The above considerations can be summarised by saying that the characteristics of the incoherent high frequency radiation would favour an emission process taking place close to the light cylinder. On the other hand, if this were the case, the small duty cycle would have to be the consequence of a poorly understood pattern of the electromagnetic field and of the particles' motion in it.

V. PULSARS AS SOURCES OF COSMIC RAYS AND MAGNETIC FIELD

1. Generalities

As we have seen in the previous lectures, there can be no doubt that pulsars produce relativistic particles. The evidence for this is overwhelming and stems from the relation between NP 0532 and the Crab Nebula, as well as from the anisotropic character of the pulsar radiation. The two basic mechanisms of pulsar electrodynamics lead to the expectation of a large scale magnetic field around rotating neutron stars. In the following we shall comment briefly on the question of whether pulsars can be an important source of cosmic rays and also possibly of the magnetic field inside our Galaxy.

2. Energy Requirements

From an energetic point of view, pulsars are equivalent to SN explosions. The presence of a galactic cosmic ray background with energy density around 1 eV cm^{-3} requires the continuous production of cosmic rays at a rate $\sim 10^{40}$ ergs sec^{-1}. If the frequency of SN explosions is roughly one every 50 years, each SN should yield 10^{49} ergs in the form of relativistic particles. The energy which has been released by the pulsar NP 0532 into the Crab Nebula is roughly $10^{48}-10^{49}$ ergs. Pulsars may therefore account for the energy budget of cosmic rays in our Galaxy. A problem could however arise if cosmic rays are trapped inside the Remnant and decrease their energy because of adiabatic expansion.

3. Composition

The surface of neutron stars is likely to be made essentially of iron nuclei. If cosmic rays are accelerated close to the neutron star (or from material drawn off the star), the particles should primarily be iron: this would contradict the evidence available concerning the composition of cosmic rays. If pulsars have to account for the origin of cosmic rays, it may be necessary to invent a mechanism to accelerate material different from that of the neutron star surface. One possibility is that low frequency electromagnetic waves accelerate particles drawn off the expanding debris of the SN shell, (Kulsrud et al., 1972). Another equivalent possibility is that the number of pulsars born in binary systems is not negligible and that the pulsars accelerate particles expelled from the companion. (Since most of the energy is released very early in the pulsar life, this would be compatible with the notion of very short lifetimes for pulsars in binary systems.

On a different line, we note that the bulk of the rotational energy could be released by a pulsar in the first few months or years (for the Crab Nebula, however, various lines of evidence indicate a figure $\sim$ 300 years). If so, the cosmic ray particles would be degraded by various absorption processes, including fragmentation and photodisintegration. It is unrealistic to investigate at present the details of these processes since we know relatively little about the early aftermath of a SN explosion and about how soon a pulsar starts to produce fast particles.

4. Spectrum

As noted earlier the theories presently available lead to the expectation of a monochromatic acceleration for the particles in the pulsar neighbourhood. Probably this can be removed when more realistic theories become available. In the meantime one should consider the figures quoted earlier just as an indication of the efficiency of the mechanisms under consideration. We are still far from the possibility of accounting for the cosmic ray spectrum on the basis of our present knowledge of pulsar electrodynamics. A thorough discussion of various points related to the acceleration of cosmic rays from pulsars has been presented recently by Salvati (1974).

5. Production of Magnetic Field Inside SN Remnants

As discussed earlier, both the aligned unipolar inductor and the oblique rotator model lead to the expectation of a large scale toroidal magnetic field, decreasing like $\frac{1}{r}$ beyond the critical distance. In the first case the field is generated by the poloidal currents escaping from the open magnetosphere and the nebular particles radiate in it via the normal synchrotron process in a static field. In the second case, the magnetic field is a component of the low frequency waves emitted directly by the pulsar at the basic rotation frequency. A relativistic electron would radiate in this field in a way similar to synchrotron radiation in an equivalent static field, but there would be a substantial amount of circular polarization (Rees, 1971). For the Crab Nebula, a model where the large scale magnetic field is an oscillating wave field leads to the prediction of a few percent circular polarization. No evidence for it has been found to a level of order 0.1% (Landstreet and Angel, 1971), but these measurements do not rule out the simultaneous presence of low frequency waves plus a static field stronger by at least a factor of 10.

The field strength predicted by both mechanisms is possibly one order of magnitude less than the lower limit inferred from the upper limit to the γ-ray flux (Fazio et al., 1972). This discrepancy can only be understood if the toroidal field generated by the currents keeps accumulating inside the Remnant. In essence, one has to suppose that new magnetic lines are continuously produced by the pulsar and propagate outwards with the speed of light up to a certain critical distance r_w. The spacing of the field lines is $\frac{c}{\Omega}$ in the internal region (this region can be described in the way sketched by Goldreich and Julian, 1969). Beyond r_w the situation changes due to the existence of a conductive supernova shell. The field lines can only propagate outwards at a velocity less than the velocity of the shell, which leads to a continuous increase in the nebular magnetic energy. (See Fig. 3). The critical distance r_w is the distance at which the ambient energy density is of the same order as the energy density in the outflowing relativistic wind.

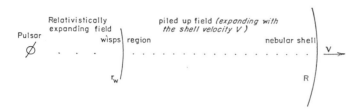

Fig. 3: A schematic drawing of the magnetic field behaviour in a SN Remnant.

If this conjecture is correct, a certain fraction of the stellar rotational energy is gradually converted into magnetic energy of the nebula. Assuming that the magnetic energy increases at a rate given by the pulsar input minus the adiabatic losses (due to the expansion of the shell), one obtains the time evolution of the magnetic field inside a SN Remnant. Figure 4 depicts the evolution in the case of the Crab Nebula, assuming an initial pulsar period P $\sim$ 16 msec (this results from an extrapolation back in time of the present period) and an expansion velocity $\sim$ 1000 km sec^{-1}. One can see that the nebular magnetic field did increase during the first few hours and reached a maximum strength of several hundred gauss. Only after about 3 years the field dropped below 1 gauss. The present day expected value would be close to 10^{-3} gauss, assuming that the bulk of the pulsar energy goes into magnetic fields.

If this model is correct, equipartition of energy between electrons and magnetic field should not exist in SN Remnants, even if they are produced in the same amount. The reason is that magnetic fields lose energy because of expansion losses while electrons lose energy both because of expansion and radiation losses. The proton component, however, would lose energy mainly because of expansion: the ratio (proton energy/

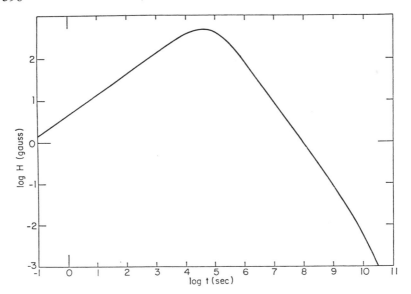

Fig. 4: Time evolution of the magnetic field in the Crab Nebula based upon a uniform expansion velocity.

magnetic energy) would remain constant until the SN Remnant merges with the interstellar medium. This leads to the speculation that - if pulsars account for the production of cosmic rays in the galaxy - they may also account for the energy density of galactic magnetic fields.

VI. CONCLUSION

Many important aspects of the relation between pulsars and cosmic rays are either controversial or unclear. However, the basic process operating in pulsars, that is the gradual conversion of macroscopic rotational energy into magnetic fields and relativistic particles, seems to offer a good starting point for understanding the production of fast particles (and possibly also magnetic fields) in the Galaxy.

REFERENCES

Canuto, V., "Matter at very high densities," NORDITA, Preprint 1974.
Fazio, G.G., Helmken, H.F., O'Mongain, E. and Weekes, T.C., Ap. J. (Letters), 175, L117, 1972.

Goldreich, P. and Julian, W., Ap. J., 157, 869, 1969.

Kulsrud, R., Ostriker, J. and Gunn, J., Phys. Rev. Letters, 28, 636, 1972.

Landstreet, J. and Angel, J.R., Nature, 230, 103, 1971.

Pacini, F. and Rees, M.J., Nature, 226, 622, 1970.

Pacini, F. and Salvati, M., Ap. Letters, 15, 39, 1973.

Rees, M., "The Crab Nebula", I.A.U. Symposium, No. 46, D. Reidel Publishing Company, 1971.

Salvati, M., "Production of fast particles from pulsars", Preprint, 1974.

Zel'dovich, Ya. B. and Novikov, I.D., "Relativistic Astrophysics", Vol. 1, University of Chicago Press, 1971.

Additional References

Section II

The basic properties of pulsars are also reviewed in:

Smith, F.G., Reports Prog. Phys., 35, 399, 1972.

Section III

The basic theory of pulsars is summarised in an excellent review article:

Ruderman, M., Ann. Rev. Astronomy and Astrophysics, 10, 427, 1972.

Some original references for the electrodynamics are:

Gunn, J.E. and Ostriker, J.P., Ap. J., 165, 523, 1971.

Mestel, L., Nature, 233, 149, 1971.

Pacini, F., Nature, 216, 567, 1967.

Pacini, F., Nature, 219, 145, 1968.

Some original references for the radiation theory are:

Gold, T., Nature, 218, 731, 1968.

Sturrock, P.A., Ap. J., 164, 529, 1971.

HISTORICAL SEARCHES FOR SUPERNOVAE

F.R. Stephenson

School of Physics, University of Newcastle, Newcastle upon Tyne, England.

I. INTRODUCTION

The object of this paper is to discuss historical records of supernovae which have appeared in our own galaxy other than the well known supernovae of 1054, 1572 and 1604.

Much of the observational material which I want to examine is very fragmentary and leaves a lot to be desired. However, we are in the unfortunate position of having no telescopic observations of supernovae in our galaxy, since the last was seen in 1604. That is not to say that none may have occurred more recently. The supernovae of 1572 (seen by Tycho Brahe) and 1604 (seen by Kepler) have left non-thermal radio sources which are comparable in intensity with other sources in various catalogues of probable SNR's (e.g. Milne, 1970; Downes, 1971; Ilovaisky and Lequeux, 1972a). Scarcely any of these latter sources have ever been identified, even tentatively, with a historically recorded 'new star', and the possibility exists that several may be the remnants of supernovae which have occurred in relatively recent times but have escaped detection (optically) on account of interstellar absorption. This is very probably true for 3C 144 (Cas. A). From the rate of expansion of the optical filaments of this object, van den Bergh and Dodd (1970) estimate the age as 306 $\pm$ 8 years, while Gull (1973) prefers an age of only 200 $\pm$ 40 years. The apparent gap since 1604 may thus not be real.

Lacking any telescopic observations of galactic supernovae, it is important to analyse as carefully as possible the historical records of those events which we do have access to.

II. OBJECTIVES

There are basically four fundamental objectives in the study of historical records of supernovae. Let us examine these in turn.

The radio sources 3C 144 (SN 1054) - the famous Crab Nebula, 3C 10 (SN 1572) and 3C 358 (SN 1604) are of precisely known age. We thus know just how long the remnants have been developing, although their evolutionary history is complicated by varying densities of surrounding interstellar matter swept up by the expanding shells. Even so, I am of the opinion that much more would be known about the evolution of SNR's if the ages of a few further remnants could be established with certainty. The author (Stephenson, 1971a) has given detailed arguments in favour of regarding 3C 58, a well established SNR, as the remnant of a supernova which appeared in 1181, and Gardner and Milne (1965), following the careful historical analysis of Goldstein (1965) and Goldstein and Ho (1965) have proposed MSH 14-4$\underline{15}$ (PKS 1459-41) as the relic of the brilliant new star of 1006. Other suggested associations are more tenuous. Valuable reviews are given by Minkowski (1968) and Shklovsky (1968).

Baade (1943 and 1945) showed conclusively that the light curves of SN 1572 and SN 1604, as determined from the original observations, were of Type I. It follows that both supernovae were of low mass (roughly one solar mass). A similar demonstration has not been satisfactorily given for any other historical supernova. This remark applies particularly to SN 1054, which from the meagre data available concerning the light variation might have been of either type (regarding Type II as including Types III and IV for the purposes of the present study). The author in his 1971a paper gave somewhat tentative arguments in favour of SN 1181 as of Type I, but again the observational data is very meagre. There is, however, hope that data on the maximum brightness and duration of visibility for other historical supernovae may lead to a distinction between Types I and II in certain cases, but there is a pressing need for more data on Type II light curves.

Several records allow a fairly reliable estimate of the apparent magnitude at maximum. If the colour of the star is recorded or satisfactory allowance can be made for interstellar extinction in some other way, it may be possible to derive a reasonably good value for the parallax. On the assumption that the supernovae of 1572 and 1604 had intrinsic brightness typical of Type I (absolute magnitude - 19.0), Minkowski (1964) and Woltjer (1964) obtained distances of respectively 2.4 and 3-4 kpc (SN 1572) and 6.7 and 10 kpc (SN 1604). These figures are probably as reliable as any obtained by other means (e.g. 21 cm absorption data). It may prove possible to make useful estimates of distances for other historical supernovae.

A particularly important consideration from the point of view of the present conference is the frequency of occurrence of supernovae in the galaxy. Recent estimates, using various methods of approach, are in reasonable agreement. Tammann (1974) has summarised recent work on this problem and regards the result for τ, the mean interval between supernova outbursts, determined by Ilovaisky and Lequeux (1972b) as typical. This value is $\tau = 50 \pm 25$ years.

From historical records of stars of long duration, recorded mainly in the Far East, the author (Stephenson, 1974) estimated that as many as 12 supernovae could have been observed in the last 2200 years (I would say definitely 7). The full significance of this estimate is difficult to appreciate since we can make no satisfactory allowance for incomplete records, interstellar extinction, etc., but a value for τ at least as small as that deduced by Ilovaisky and Lequeux seems indicated.

III. SOURCES OF HISTORICAL RECORDS OF POSSIBLE SUPERNOVAE

Usable historical records of new stars appear to be found in only four principal sources: Medieval European monastic chronicles, Arabic chronicles, astrological works etc., Far Eastern histories and diaries and post renaissance European scientific writings. I am informed by Professor A.J. Sachs of Brown University, U.S.A. (personal communication) that the extensive series of Late Babylonian astronomical texts which he has studied contains no reference to either comets or new stars. It should however be remarked that these texts are in a very fragmentary state, but the situation is still very surprising. Let us consider the four principal sources in turn:

1. Medieval European Monastic Chronicles

These are in the main concerned with affairs of the monastery and local events. However, occasional reference is made to earthquakes and the more remarkable astronomical events, especially eclipses and comets. Frequently an eclipse or comet is regarded as the major event of the year. Occasionally annalists show a considerable interest in astronomical matters, recording aurorae, meteors and even occultations, as well as the more striking phenomena, but such interest is rare. As far as the appearance of new stars is concerned almost no interest seems to have been expressed. Either the chroniclers did not have adequate knowledge of the constellations to recognise a new starlike object (I think this is very important) or they were unconcerned about such matters. To give an example of the former attitude, Simeon of Durham, a 12th century English chronicler describes an occultation of a certain "bright star" by the totally eclipsed Moon which occurred on a date which corresponds to AD 775 November 24 (ref. Arnold,

1885). The source of Simeon's account is unknown, but the description is clearly that of an eyewitness. Computation (to be published elsewhere) shows that the bright star was in fact the planet Jupiter, but it would appear that the observer was ignorant of this: to him it was just a star. We might imagine that had a bright star suddenly appeared in the sky, an observer of this calibre would not have recognised the fact.

On the other hand, the latter attitude has been neatly crystallised by Sarton in an unidentified passage quoted by Needham (1959, p. 428). He writes, "The failure of medieval Europeans and Arabs to recognise such phenomena was due, not to any difficulty in seeing them, but to prejudice and spiritual inertia connected with the groundless belief in celestial perfection."

Whatever the explanation, the paucity of medieval European records of novae and supernovae is disappointing in the extreme. By comparison the number of eclipse and cometary observations is enormous. Probably the best list of new stars in European chronicles is that of Newton (1972, pp. 102-114), which is compiled mainly from the chronicles published by Pertz (1826→). However the observations which he lists are of little astronomical interest and several indicate the poor state of astronomical knowledge which prevailed in Europe at the time. Let me quote one example. Albertus (ref. Pertz, 1826→) in his chronicle of the monastery of Stade (Germany) records that in the year 1245 (Albertus was a contemporary) a star appeared about May 4. We are told that it was toward the south in Capricorn. It was large and bright, but red. It could not be Jupiter, because Jupiter was then in Virgo. Many claimed that it was Mars because of its colour. After July 25 it was no longer bright and it continued to diminish day by day. From Tuckerman's (1964) tables it is evident that the star was, in fact, Mars. The planet was almost stationary in Capricorn from about May until September and was in opposition on July 21. Newton (p. 112) thought that the star was a nova and cited the gradual loss in brightness as evidence of this, although he emphasised the need for a check on the planetary positions before a definite judgement was reached. However, the fading of the "star" was in fact due to Mars receding from the Sun after opposition, although the indicated date seems a little early for noticeable loss of brightness.

It gives me little surprise that observers of the calibre of Albertus and his contemporaries should fail to notice the famous supernova of 1054, which is so well documented in Chinese annals. This was about as bright as Venus at maximum and so could have been mistaken for a planet. On the other hand I find it almost astonishing that the new star of 1006, which was compared with the Moon in intensity by the Arabs (Goldstein, 1965 - see also below), should have aroused virtually no interest in Europe. There are only two or three observations of this, although it far outshone all the other stars and planets in the sky.

To summarise, European records of new stars are a disappointment and with the exception of the few observations made in 1006 of historical interest only. Further research into the many unpublished chronicles is likely to repay few dividends.

2. Arabic Writings

I have studied Medieval European and Far Eastern chronicles for several years (mainly for the eclipse reports which they contain), but I must confess to being anything but a specialist on Arabic writings. The best I can do here is to emphasise just what the patient work of Goldstein (1965) was able to accomplish for the supernova of 1006 among scattered Arabic sources and express the hope that a competent Arabist may be persuaded to search for possible references to the new star of 1054, (it would be particularly valuable if details on the light variation could be found). Sarton's remark, quoted above, embraced Arabic astronomers as well as European, but we have ample evidence that they were extremely competent observers (ref. Newton 1970). Goldstein found references to the star of 1006 in several Arabic writings (astrological treatises and chronicles) and they provide valuable information which supplements the Far Eastern observations. Further details will be given below. To conclude this brief survey, I would consider that Arabic writings are potentially valuable sources of observations of novae and supernovae if sufficient interest can be aroused among historians. As yet, the only work on Arabic records seems to be that of Goldstein, already cited. Johnson (1958) has touched briefly on the subject.

3. Far Eastern Observations

The reason why we possess so many observations of novae and supernovae from China, Japan and Korea is mainly astrological. I have discussed this at some length in my 1971a paper (see also Muller and Stephenson, 1974), and thus only want to summarise the position here.

From very early in Chinese history, astronomers were appointed by the ruler to maintain a constant watch of the sky both day and night. The motive was purely astrological; celestial events were regarded as precursors of terrestrial events. A quotation from Chapter II of the Chin Shu ("History of the Chin Dynasty" written AD 635) relating to the "guest star" of AD 369 (a possible supernova) illustrates this concept. The following translation is by Ho (1966, p. 242):

"During the second month of the 4th year of the T'ai-ho reign period of Hai Hsi (Kung) (24th March to 22nd April AD 369) a guest star appeared at the western wall of the "Purple Palace" and went out of sight only at the seventh month (19th August to 17th September). According to the standard prognostication one may expect the Emperor to be assassinated by his subordinates whenever a guest star guards the Purple Palace. During the 6th year, Huan Wen dethroned the Emperor and gave him the title "Duke of Hai Hsi".

The "Purple Palace" (Tzu-wei) is the circle of perpetual visibility at the latitude of the Chinese capital (declination roughly 65 deg.).

Prognostications frequently referred to the Emperor, but this was by no means necessarily the case. For example the guest star of AD 393, also a possible supernova, "foreboded war and death at (the region of) Yen", (Ho. p. 243). In order to detect these warnings from heaven and take counter measures as quickly as possible court astrologers were appointed to maintain a continual watch of the sky. It is to this tradition that we owe the splendid series of guest star and other astronomical observations (e.g. eclipses, comets and conjunctions) which stretches almost uninterrupted from about 200 BC until modern times in China. Much the same astronomical/astrological system was adopted by the Koreans and Japanese at a later date (there are fragmentary guest star records from Korea from about 100 BC and very detailed ones after 900 AD, while Japanese observations commence about 700 AD).

Since the substance of a particular prognostication depended very much on which asterism the guest star appeared in, Far Eastern records are usually careful to give a fairly good location. Normally only the asterism is named which might mean that the star appeared anywhere within an area of say 100 square degrees of sky, but occasionally positions are much more accurate than this - e.g. the position of Kepler's supernova was measured by the Koreans to the nearest degree in right ascension and declination.

Practically all of the useful records of guest stars have been published in the catalogues of Hsi (1955), Ho (1962) and Hsi and Po (1965). However, as translations vary so much, the author has found it necessary to consult original sources.

4. Post Renaissance European Observations

What I have in mind here are the observations of the supernovae of 1572 and 1604 made by Tycho Brahe, Kepler and others. These are much superior to even the best of the oriental observations since the positioning is so precise (to about the nearest minute of arc) and detailed light curves can be drawn from the various comparisons in brightness with stars and planets.

From our point of view, it is unfortunate that true "scientific astronomy" began so late in the West, but this is a subject where we are entirely at the mercy of what our predecessors chose to record and what proportion of their writings has survived to the present day.

IV. LIGHT CURVES OF NOVAE AND SUPERNOVAE

Light curves of Nova Aquilae (1918) and Nova Pictoris (1925) are shown in Figs. 1 and 2. These are taken from the work of Spencer Jones (1961) and Payne-Gaposchkin (1957, fig. 4.16). Nova Aquilae was a fairly typical "fast" nova but by virtue of its great brilliance at maximum (mag. -1.1, the brightest discovered since the invention of the telescope), it remained brighter than +5.5, probably the lower limit for unaided eye visibility, for some 160 days. The galactic latitude was only - 1 deg. so that there is a definite possibility that a similar nova occurring in ancient times might have been mistaken for a supernova by present day astronomers investigating the records. However, it is difficult to estimate the frequency of very bright nova in low galactic latitudes. From the list of galactic novae compiled by Payne-Gaposchkin (1957, pp. 2-8) excluding the rather special cases of η Car (1843) which was possibly a subluminous supernova (Zwicky's type V), and the recurrent nova T CrB (1866) no nova brighter than mag. +3 was discovered before the present century, although there are many fainter ones.

"Slow" novae follow no particular pattern of light curve and the light variations of Nova Pictoris were extremely irregular.

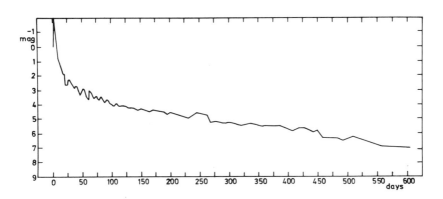

Fig. 1: Light curve of Nova Aquilae (1918).

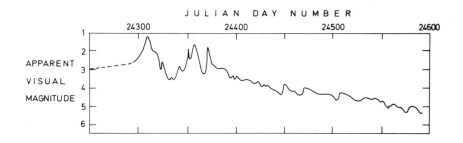

Fig. 2: Light curve of Nova Pictoris (1925).

However the star remained brighter than +5 for 250 days. The galactic latitude in this case was rather high (-25 deg.) so that confusion with a supernova would be out of the question.

Supernovae because of their great distance (several kpc) tend to hug the galactic equator. However, because of their nearness to us (distances of the order of 100 pc) one would expect the distribution of the very brightest novae to be fairly isotropic. This and the fact that the smooth light curve of a supernova of either type is more likely to lead to extended observation are my main points in explaining why nearly all early guest stars of long duration lie in low galactic latitude (i.e. they are to be interpreted as supernova, see Table I). Presumably bright novae which remain visible for a lengthy period are very rare, otherwise the distribution of guest stars of extended visibility would be much more isotropic.

Apart from the slow fall off and smoothness of the supernova light curve, my main concern is the paucity of Type II observations more than 100 days after maximum. This is mainly the result of supernovae of this type being intrinsically fainter and also less frequent than Type I, but as a result if only the direction of visibility of a star is recorded it is impossible to tell whether it is of Type I or II. Figs. 3 and 4 show the composite light curves of Type I (Barbon et al. 1974a) and Type II (Barbon et al. 1974b). The Type I curve is very well defined for several hundred days after maximum and indeed Barbon et al. (1974a) have recognised two sub groups with only slightly differing rates of decay, but we just cannot say precisely what form the Type II curve takes even after as little as 100 days. Personally I should like to see more work done on Type II light curves with large aperture telescopes.

HISTORICAL SEARCHES FOR SUPERNOVAE

Historically observed new stars of long duration

Year (AD)	Constellation	Duration	b^{II}
185	Centaurus	8 months	$0° \pm 2°$
369	Draconis, etc.	5 months	(?)
386	Sagittarius	4 months	$- 9 \pm 6$
393	Scorpius	8 months	0 ± 5
902	Cassiopeia	1 year (?)	$+ 13 \pm 3$
1006	Lupus	several years	$+ 12 \pm 2$
1054	Taurus	22 months	$- 6$
1181	Cassiopeia	6 months	$+ 2 \pm 2$
1572	Cassiopeia	18 months	$+ 1$
1592 A	Cetus	15 months	$- 56$
1592 B	Cassiopeia	4 months	$- 3 \pm 1$
1592 C	Cassiopeia	3 months	$- 3 \pm 1$
1604	Ophiuchus	1 year	$+ 7$

Table I

Notes to Table I

(a) This table is a revised version of Table 1 of Stephenson (1974).

(b) Galactic latitudes and uncertainties are estimated except for the supernovae of 1054, 1572 and 1604, and the star 1592A, which was probably Mira.

(c) The position of the star of 369 is only vaguely described (see text).

(d) The 902 star may have been a comet of short duration (see text).

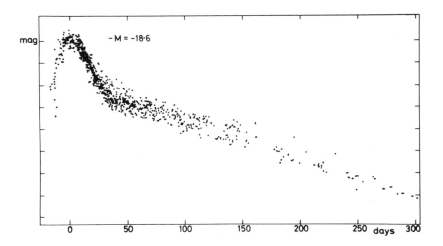

Fig. 3: Composite blue light curve obtained by the fitting of the observations of 38 type I supernovae. One magnitude intervals are marked on the ordinates.

V SUPERNOVA REMNANTS

Unlike novae, supernovae leave as a remnant a powerful source of synchrotron radiation. Most observations of SNR's have concentrated on the short wave radio range, and although catalogues of X-ray sources are available (cs.Gorenstein, 1974) these are far from complete. I propose to restrict my attention here to radio sources.

Minkowski (1968) has expressed the view that the only non-thermal sources in the galaxy are supernova remnants. I think it would be difficult to argue with this point of view. Weiler and Seielstad (1971) in discussing 3C 58, have outlined the basic reasons for regarding a particular discrete radio source as a galactic SNR. We may list these here.

(a) Optical filamentary structure observed. This may not be detected if there is heavy interstellar obscuration (as is true for 3C 58);

(b) Large angular size (several arcmin). Most extragalactic sources are effectively point sources (e.g. quasars);

(c) Fairly simple shape (in the radio region). The few extended extragalactic sources are of rather complex form;

(d) Low galactic latitude (less than about 5 deg. in all but a few cases);

(e) Negative spectral index (using the convention $I_v \propto v^{+\alpha}$. Thermal sources (the only galactic discrete thermal sources are probably HII regions) have a positive and variable spectral index in different frequency ranges. A synchrotron emitter has an almost constant value of α over a wide range of frequencies.

(f) Radio polarisation studies reveal the presence of a considerable magnetic field.

Catalogues of probable SNR's have been compiled recently by Milne (1970) Downes (1971) and Ilovaisky and Lequeux (1972a). It is difficult to judge how reliable or complete such catalogues are, but I would imagine even their authors cannot feel too confident on either point. On the other hand there is little we can do at present in the investigation of historical records of supernovae but to make use of them. Catalogues of general discrete sources such as 4C (Gower et al. 1967) give little or no information about individual sources and contain an overwhelming proportion of extragalactic objects. Perhaps the best I can do is to point out those probable SNR's in the various catalogues which agree reasonable well with the positions described in the historical records and then ask for the co-operation of radio astronomers.

VI HISTORICAL OBSERVATIONS OF SUPERNOVAE

Historical observations of new stars take many forms. A study like the present one has to attempt to answer the question, "How may early observations of supernovae be confidentaly identified among the records of other phenomena such as ordinary novae and comets?" The author (Stephenson, 1971a) considered that the most satisfactory approach was to examine only the records of those new stars which were seen for several months. Far Eastern astronomers in particular had considerable difficulty in classifying stars correctly into the various categories which they used. In principle the term "guest star" (k'o hsing) was applied to starlike objects which did not move (this term was used in describing the SN of 1054, 1572 and 1604). But there are frequent records of sightings of moving guest stars. There were separate terms for comets - "broom stars" (hui hsing) and "rayed stars" (po hsing), the latter probably referring to comets which did not approach close enough to the Sun to develop a detectable tail. A comet has a star-like nucleus and hence confusion with a nova is possible unless a sizeable tail is apparent. The following observation recorded in the Sung Shih ("History of the Sung Dynasty" - written 1345) is typical and illustrates the care with which oriental positional observations could be made.

"On a ping-ch'en day in the 4th month of the fourth year of the T'ien-hsi reign period (25 May 1021) a guest star appeared at the NW of the front star of Hsien-yuan. It was as large as a plum. It moved rapidly past the large star of Hsien-yuan (Regulus), entered the T'ai-wei enclosure, concealed the star Yu Chih-fa, trespassed against Tzu-chiang and passed P'ing-hsing from the NW. After 75 days it entered the horizon and went out of sight". (trans. Ho, 1961, p. 183).

This object was quite clearly a comet, but it never seems to have formed a tail.

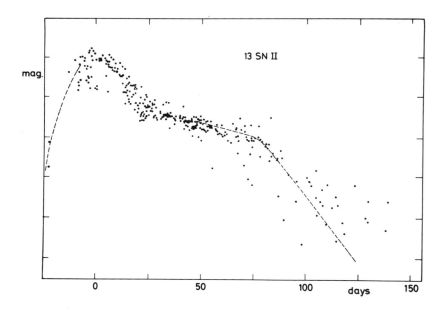

Fig. 4: Composite light curve obtained from 13 type II supernovae.

When a star was visible for several months a careful description of any motion is almost invariably given (see Stephenson, 1971a). If the record is silent on the question of motion and the period of visibility is lengthy there can be little doubt that we are dealing with an observation of a nova or supernova.

On the other hand, discussions of new stars which were seen for only a few days or weeks serves little purpose. The chances are that such objects were ordinary novae (if not comets).

New stars of very long duration (at least six months) for which there is no record of any motion are recorded in the following years: AD 185, 393, 1006, 1054, 1181. The maximum recorded duration of visibility for a moving star is six months - a tailess comet in A.D. 277 (Ho, 1962, p. 157). Much has been written on the stars of 1054, 1181, 1572 and 1604 (cf. Stephenson, 1974). For most of the remainder of my talk I want to discuss the stars which appeared in 185, 393 and 1006 and examine possible associations with non-thermal galactic radio sources. Brief comments are also made on the somewhat disappointing objects which appeared in 369 and 902.

Recently the author (Stephenson, 1971b) analysed observations of guest stars of unrecorded duration which were in close conjunction with the Moon on precisely recorded dates and was thus able to locate the positions with an accuracy of about 1 square degree. He compared the positions with those of discrete radio sources but none of his proposed identifications can be regarded very confidently. The main attraction of the records is the high accuracy with which the stars could be located. It is hoped that the paper might encourage research on post novae in the restricted areas of sky indicated.

As most new stars were seen only in the Far East, I should make a few brief comments about oriental dates. These present fewer problems than might be imagined. Anything closer than the year is, of course, of little interest to astrophysicists (except where the duration of visibility is involved), but it is desirable to have direct access to all of the available material. Oriental dates are normally expressed as follows:

(a) Year of a particular reign period of the ruler concerned;

(b) Lunar month (there are 12 months in most years, but an intercalary month is frequently inserted so that the seasons occur at approximately the same time of year);

(c) Cyclical day. A 60-day cycle independent of months and years is used.

Conversion of dates to the Julian or Gregorian calendar can be rapidly effected using the tables of Hsüeh and Ou-yang (1956). The use of a 60-day 'week', which has been in continuous operation since well before 1000 BC, enables a ready check on date conversion by a simple computer program. I might add I have made numerous calculations of observations involving the Moon (eclipses and occultations) and find that the recorded dates are almost invariably exact.

1. The New Star of AD 185

This star, which seems to have been recorded only by the Chinese, appeared in a very southerly declination but was still observed for many months. All that we know about the star is contained in the following entry in chapter 22 (part of the astronomical treatise) of the Hou Han Shu ("History of the Later Han Dynasty" - written AD 450, but based on original records). We read (translations are by the author except where otherwise indicated):

"On the day kuei-hai in the tenth month of the second year of the Chung-p'ing reign period (AD 185 December 7) a guest star appeared within (chung) Nan-mên; it was as big as half a mat; it was multi-coloured (literally: "it showed the five colours") and it was fluctuating. It gradually became smaller and disappeared in the sixth month of the following year (186 July 5 to August 2)".

The period of visibility of this star was 7 or 8 months so that the probability of it being a supernova is high. As will be shown below, the meridian altitude at Lo-yang, the Chinese capital of the time, was no more than about $5°$ and may have been considerably less than this. Atmospheric dispersion would thus explain the colour effects described. Hsi and Po (1965) have pointed out that the character yen ('mat') is very similar to ting (a type of bamboo rule used by astronomers), but in any case the reference to the apparent size of the object is spurious due to distortion in the atmosphere at such a low altitude.

It is possible to make a reasonable estimate of the apparent brightness of the star at maximum on the assumption that it was a supernova of Type 1 - supernovae of Type II appear to decay more rapidly after the maximum is passed. In 8 months the brightness would have fallen by at least 6 magnitudes. Allowing a further 3 mag. for atmospheric extinction and assuming that the star was no longer seen when its magnitude had reached +5, the apparent magnitude at maximum was brighter than -4. However, from the description it seems to have been excessively brilliant and I would consider -6 a conservative estimate.

The determination of the location of the new star requires careful consideration of the constituents of the asterism Nan-mên, for there are several catalogued SNR's in this area (in Centaurus). Hill (1967) supposed that Nan-mên was "apparently associated with the stars α, β and ε Centauri", while Ho (1966, p. 105) was of the opinion that the constituents were β and ε Cen. Early oriental star charts of which I possess copies (the originals vary in date from 900 to 1750 AD; cf. Stephenson, 1971a) are all agreed that Nan-mên consisted of only two stars. For example in the Chin Shu we read, "The two stars of Nan-mên ("Southern Gate"), situated south of K'u-lou, form the outer gate of the heavens and govern garrison troops" (trans. Ho, 1966, p. 105). Although most early oriental star charts are crudely drawn, estimates of R.A. and dec. are sufficient to prove that only α, β and ε Cen. could have been members of the asterism. These stars, especially the first two, are particularly bright (mags. -0.3, +0.6 and +2.3) and no other star in the vicinity is brighter than mag. +4.

Both α and β Cen. lay close to the horizon at Lo-yang in AD 185 and their meridian altitudes there (latitude 34.7 deg. N) were respectively $3°$ and $5°$. Owing to the effects of precession, they are now only visible south of about latitude $30°N$. Some time around 700 AD they ceased to be visible from the Chinese capital (then Ch'ang-an) although they could still be seen from the southern half of the empire. The greater brightness of α Cen as compared with β would tend to compensate for the increased atmospheric extinction due to its slightly more southerly declination so that from a particular latitude in Central China (say $35°N$) the stars would be roughly equally bright and furthermore would tend to enter the circle of perpetual invisibility at about the same epoch. The evidence so far is thus against ε Cen. being regarded as one of the components of Nan-mên since α and β Cen. form a natural pair of objects of similar brightness and visibility.

Figure 5 is copied from a star map in the Ku-chin T'u-shu Chi-ch'êng, a Chinese encyclopedia dating from 1725, but containing star charts more than a century earlier. I have omitted the Chinese characters identifying the various asterisms and substituted the Romanisation in the case of Nan-mên. The region of sky covered is roughly from about 10^h to 14^h R.A. and $-60°$ to $+50°$ dec. This chart is unusual among oriental star charts in that it makes some attempt to distinguish between stars of different brightness. The three bright stars are undoubtedly α Vir. (near the centre of the chart) and α and β Cen. (near the lower edge). It is evident that the map is not drawn to scale, but the forms of the various asterisms are very similar to those shown on other oriental charts. In particular, there can be no question that Nan-mên was understood to consist of only α and β Cen.

The other star maps which I have access to do not differentiate between stars of different brightness. In order to decide between α, β and ε Cen. as the constituents of Nan-mên, the most satisfactory procedure is to measure on each chart the angle between the line representing the great circle joining the two stars and that representing a circle of declination passing between them. In Fig. 5 this angle is very close to zero, assuming

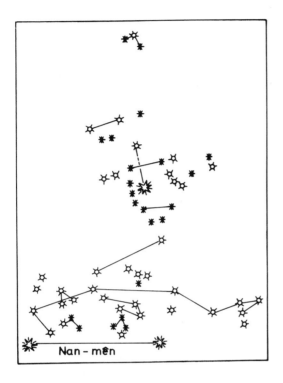

Fig. 5: Copy of a star chart in the "Ku-chin T'u-shu Chi-ch'êng" showing Nan-mên.

that the edges of the chart represent meridians and declination circles. Before discussing the results obtained from other early maps, it is necessary to examine the real situation.

Figure 6 is based on figure 4 of Hill (1967) and shows the positions of the three stars α, β and ε Cen. (represented by asterisks) at the epoch AD 185, together with catalogued SNR's in the area (represented by squares). I have corrected a slight error made by Hill in the R.A. of α Cen. The angle defined in the last paragraph above is about $+10°$ (measured from the west) for α and β Cen. and $+70°$ for β and ε Cen. (at the present time the angles are scarcely any different). On the oriental star charts this angle varies from zero to $+35°$, with a mean of $+15°$.

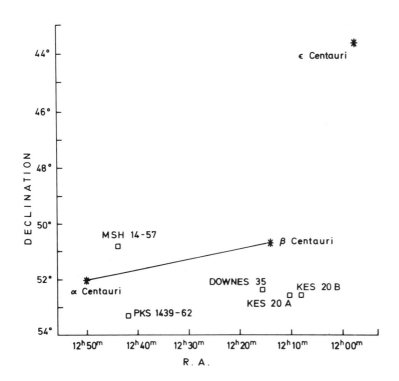

Fig. 6: The relative positions of the stars α, β and ε Centauri and SNR's near α and β Centauri for epoch 185 A.D.

The evidence thus appears to be conclusive that the constituents of Nan-mên are and were α and βCen. and there is no evidence that ε Cen. was ever regarded as a member of the asterism. The use of the term chung ("within") in the record of the guest star now becomes critical. The normal expression to denote the location of a new star in an asterism is the rather vague term yu ("at" or "in the vicinity of"). However, chung, which is only rarely used, is much more specific and in the present context can be understood to mean that the guest star appeared roughly between α and β Cen. Other examples of the use of chung are in describing the occurrence of sunspots within the Sun and in the name for China itself - Chung-kuo ("The Middle Kingdom"). We can note that the galactic latitude of the new star was extremely low ($0° \pm 2°$), which is characteristic of a supernova.

The radio sources shown in figure 6 are probable SNR's listed in the catalogues of Milne (1970), Downes (1971) and Ilovaisky and Lequeux (1972a) which lie close to α and β Cen. It is clear that only three sources answer the Chinese description at all satisfactorily – Downes 35, PKS 1439-62 and MSH 14-57. Hill (1967) regarded PKS 1439-62 as the probable remnant of the guest star, mainly because it showed optical filamentary structure, and this conclusion is supported by Westerlund (1969). Personally I would prefer this identification since a star so close to the horizon (meridian altitude at Lo-yang about $1\frac{1}{2}°$) would be likely to suffer the severe dispersion and distortion described in the text. However, discussion of the suitability of Downes 35 and MSH 14-57 would be invaluable.

Owing to an error, I recently (Stephenson, 1974) thought that several catalogued SNR's "could all fit the Chinese description satisfactorily". This was based on the misconception that the Hou Han Shu text read yu instead of chung. I now wish to withdraw this remark. I regard the supernova of AD 185 as one of the most promising objects for detailed study by radio astronomers.

2.   AD 369

The position of the supernova of 185 is very carefully described, but in comparison the new star which appeared in 369 can only be located very roughly. Once again we have only a single record, again from China. This is contained in chapter 13 of the Chin Shu and is quoted in full above (section III). For convenience I will repeat the astronomical details of the text.

"During the second month of the fourth year of the T'ai-ho reign period of Hai-hsi a guest star was seen at the western wall of the Tzu-wei ("Purple Palace") until the seventh month when it disappeared."

The star was seen from about April to September in AD 369. Using similar reasoning to that for the AD 185 object, if the star was a supernova, as seems likely from the duration of visibility, it was certainly brighter than mag. +1 at maximum, but we cannot say how much brighter.

Unfortunately the celestial co-ordinates are very uncertain. The only way in which we can even deduce an approximate location is to assume that the galactic latitude was close to zero, for the western wall of Tzu-wei was very extensive, covering almost 12^h of R.A. near dec. +70°. It would seem that any attempt to locate the remnant of the star seems doomed to failure.

HISTORICAL SEARCHES FOR SUPERNOVAE 417

3. AD 393

The report of this star in chapter 13 of the Chin Shu is extremely brief. "During the second month of the 18th year of the T'ai-tuan reign period, a guest star was within (chung) Wei up to the ninth month when it disappeared".

The period of visibility of the star was from about March to November in AD 393, making about 8 months in all (there was an intercalary seventh month in the Chinese calendar). When the star disappeared, it may well have been close to the Sun (i.e. heliacal setting) so that it could have still been bright. From the period of visibility, it was certainly brighter than -1 at maximum, and may have been considerably brighter than this. Wei ("Tail") is a bowl-shaped asterism (the tail of the Scorpion in Western Astronomy) lying across the galactic equator so that the period of visibility and low galactic latitude are strongly indicative of a supernova. Wei is very well defined, consisting of ϵ, μ, ζ, η, θ, ι, κ, λ and ν Sco. (cf. Ho, 1966, p. 95). Unfortunately, four catalogued SNR's fit the Chinese description very well. These are shown in figure 7, along with the stars in the vicinity of Wei. The epoch is 1950 (proper motion of the stars is negligible). The location, although covering a rather large area, is well defined in view of the special shape of the asterism, and the use of the term chung, already discussed under the guest star of AD 185. I have only included on the chart those catalogued SNR's which lie within the "bowl", since the expression chung would not apply to stars outside this area. As it happens, no other catalogued SNR lies close to this region. I would welcome a discussion of the four sources RCW 120, CTB 37A, CTB 37B and No. 51 in Milne's 1970 catalogue as possible remnants of this supernova, but it may not be possible to make a firm decision.

4. AD 902

The record of this star is very vague concerning its "fixed" nature and duration. There seems a strong possibility that the object was a comet. Observational details are to be found in chapter 22 of the Hsin T'ang Shu ("New Book of the T'ang Dynasty" - written in 1061).

"During the first month of the second year of the T'ien-fu reign period, (11 February to 12 March 902) a guest star like a peach was at Tzu-wei beneath (i.e. to the north of) Hua-kai. It gradually moved and reached Yü-nu. On the day ting-mao (2 March) a meteor (liu-hsing) left Wen-chǎng and reached the guest star; the guest star did not move. On the day chi-szu (4 March) the guest star was at K'ang and guarded it. In the following year (ming nien) it still had not faded away".

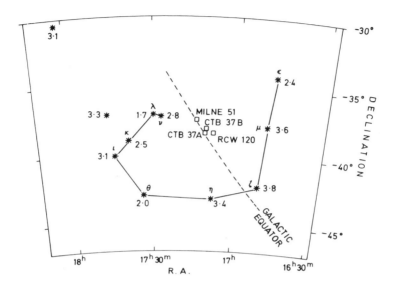

Fig. 7:   Catalogued SNR's within the asterism Wei (SN 393).

The whole record is most difficult to interpret. We are told quite specifically that the star moved, so that a tailless comet seems to be the obvious interpretation. Further, all the details relate to the first month and only at the end do we find that it was still visible the following year. This is much too long a duration for a comet. I feel that we must read jih ("day") for nien ("year"), although the characters are very different, so that the last sentence commences, "On the following day". All the events then occurred in the same month and the latter part of the entry could be merely interpreted as a commentary to the effect that although the meteor seemed to come in contact with the guest star it did not disturb it in any way.

On the other hand we may be dealing with two independent guest stars, but it is impossible to be definite. It should be borne in mind that the record may be several times removed from the original text and is almost certainly a very condensed version of it, with consequent loss of accuracy. This remark is probably true of virtually all Far Eastern astronomical records.

Regretfully we must abandon the 902 star as of uncertain nature.

5. AD 1006

Records of this star from Europe, the Near East and the Far East have been published in detail by Goldstein (1965) and Goldstein and Ho (1965). They are far too numerous and detailed to reproduce here in full and I merely want to summarise what can be learnt about the star, with occasional quotations where necessary.

From the Arabic sources, the star may have been as bright as the Moon, one quarter illuminated (mag. -8), and so was by far the brightest new star on record. It appeared about 1 May 1006 and remained visible for about $3\frac{1}{2}$ months (one text allows the date of disappearance to be deduced as 13 August).

From the European sources (there are probably only two - Benevent and St. Gallen) we learn that the star was very brilliant and was seen for 3 months - in agreement with the Arabic reports.

Far Eastern (Chinese and Japanese) sources state that the star shone so brightly that objects could be seen by its light, that it appeared on 1st May and that it disappeared during the lunar month corresponding to 27 August to 24 September. A Japanese history says that on a day corresponding to 14th September offerings were made to the various shrines because of the guest star, so either it was still visible or had only just disappeared. Chapter 56 of the Sung Shih ("History of the Sung Dynasty" - written 1345) gives additional information and is worth quoting in full:

"On a wu-yin day in the fourth month of the third year of the Ching-tê reign period (6 May 1006) a chou-po star appeared to the south of the Ti lunar mansion and one degree to the west of Ch'i-kuan. Its shape was like that of a half-moon and it shone so brightly that objects could be seen (by its light). It appeared at the east of K'u Lou. During the eighth month (27 August to 24 September) it went below the horizon following the rotation of the heavens. During the eleventh month (24 November to 22 December) it was again sighted at the Ti lunar mansion. Thereafter its heliacal rising took place during the eleventh month in the morning at the east, and its heliacal setting during the eighth month in the southwest" (trans. Ho in Goldstein and Ho, 1965).

The star was thus observed by the Chinese for several years. It is difficult to understand why it was seen in the Far East for a full month longer than in the West, unless the occidental astronomers ceased to be interested in the spectacle once it began to fade significantly. In $3\frac{1}{2}$ months a Type 1 supernova diminishes in brightness by only about 4 mag. so that if the star was of this type it would still be about as bright as Venus. On the other hand, the somewhat unusual supernova S And., which appeared in M 31 in 1885, faded by some $7\frac{1}{2}$ mag. in a similar time, (see Payne-Gaposchkin, 1957, fig. 9.3). We can be reasonably confident in concluding that the star was not a supernova of Type I, but beyond this it is difficult to classify it (as is true for S. And.).

Most of the sources are very vague as to the position of the star. Surprisingly, one of the most useful observations is that made at the monastery of St. Gallen in Switzerland. In the chronicle of St. Gallen, which is mainly concerned with affairs of the monastery (see Stephenson, 1974) the star was described as "seen for three months in the inmost extremities of the south, beyond all the constellations which are seen in the sky". The latitude of St. Gallen is $47.4°N$, and Goldstein believed that for the star to be seen there at all, its declination at the time must have been no further south than $-41\frac{1}{2}°$. However, the region around St. Gallen is extremely mountainous and I (Stephenson 1974) pointed out that making the necessary terrain correction (based on a map of the area), a star of declination more southerly than $-37°.0 \pm 0.°5$ would not rise above the visible horizon. Far Eastern sources are somewhat vague regarding declination, but several mention that it was within Chi-kuan in Lupus - see figure 8 which is taken from figure 1 of Goldstein and Ho. At the epoch 1006 none of the stars of Chi-kuan were further north than $-36°N$, which sets fairly tight limits on the declination.

The Chinese astronomers measured the R.A. of the star to the nearest degree: "It was found in the third degree of the Ti (lunar mansion)", which implies between 2 and 3 deg. E of α Lib., the determinant star of the lunar mansion Ti. Goldstein and Ho give the equivalent R.A. as between 211 and 212° in 1006. Correcting for precession, the 1950 location is between R.A. $15^h 04^m$ and $15^h 08^m$, and dec. $-41°.5$ to $-39°$. The only catalogued SNR in this region is PKS 1459-41 (1950 co-ordinates: $15^h 00^m$, $-41°45'$). The Lupus Loop lies in this region, but the general diffuse appearance of this SNR and its low surface brightness are characteristic of a very old supernova. The sharpness, brightness and symmetry of PKS 1459-41 (MSH 14-415) are suggestive of a very young event (Milne, 1971). The location of the star is in excellent agreement with that of the supernova PKS 1459-41, and I have no hesitation in supporting the identification, which was first made by Gardner and Milne (1965) and confirmed by Milne (1971). The galactic latitude is very high for a SNR ($b^{II} = +15°$), which led Gardner and Milne to suggest a fairly small distance from us. This is supported

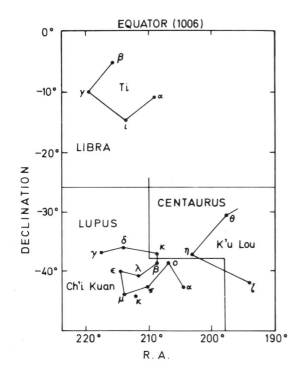

Fig. 8: Chart showing area of sky in which SN 1006 appeared.

by the extreme optical brilliance of the supernova at maximum. Milne (1970) estimated the distance to be as small as 0.4 kpc. Gardner and Milne made the further important point that the rarity of supernovae at such a high galactic latitude favours the identification.

VII CONCLUSION

Apart from the well known stars of 1054, 1572 and 1604, there are sound reasons for regarding the new stars which appeared in AD 185, 393, 1006 and 1181 as supernovae. Of these four, there is solid evidence that the non-thermal galactic radio source PKS 1459-41 is the remnant of SN 1006. The position of the radio source 3C 58, a well established SNR, agrees extremely well with that of the new star of 1181, and apart from HB 3, which from its low surface brightness and large angular extent is probably very ancient, and 3C 10, the remnant of SN 1572, there is no other catalogued SNR in the area. Dr. A.S. Wilson of Sterrewacht te Leiden, who was a participant at the conference, informs me that a detailed study of 3C 58 is in progress and will be submitted for publication in the near future.

The situation for the two remaining stars (185 and 393), although less promising, holds definite hope for the future. Three catalogued SNR's fit the observed position of the 185 star almost equally well and any one of four probable SNR's could be the remnant of the 393 object. It may prove possible to eliminate some of these radio sources, e.g. by demonstrating that they are extragalactic or, if in fact SNR's, that they are much too old (or too young) to qualify as remnants of supernovae which appeared some 2000 years ago, but the author is not qualified to judge. I would like to end by requesting the co-operation of radio astronomers in an attempt to isolate the remnants of both stars. No other historically recorded new star can be confidently regarded as a supernova with an identifiable remnant.

ACKNOWLEDGMENT

It is a pleasure to express my gratitude to Mr. A.C. Barnes, School of Oriental Studies, University of Durham, for much valuable help, advice and discussion.

REFERENCES

Arnold, T. (ed.), "Simeonis monachi Dunelmensis Historia Regum Anglorum et Dacorum",H.M. Stationary Office, London, 1885.

Baade, W., Ap. J., 97, 119-129, 1943.

Baade, W., Ap. J., 102, 309-317, 1945.

Barbon, R., Ciatti, F. and Rosino, L., in C.B. Cosmovici (ed), "Supernovae and Supernova Remnants", D. Reidel Publishing Company, Dordrecht-Holland, pp. 99-102, 1974a.

Barbon, R., Ciatti, F. and Rosino, L., in C.B. Cosmovici (ed.), "Supernovae and Supernova Remnants", D. Reidel Publishing Company, Dordrecht-Holland, pp. 115-118, 1974b.

Bergh, S. van den and Dodd, W.W., Ap. J., 162, 485-493, 1970.

Downes, D., Astron. J., 76, 305-316, 1971.

Gardner, F.F. and Milne, D.K., Astron, J., 70, 754, 1965.

Goldstein, B.R., Astron. J., 70, 105-114, 1965.

Goldstein, B.R. and Ho Peng Yoke, Astron. J., 70, 748-753, 1965.

Gorenstein, P., in C.B. Cosmovici (ed.), "Supernovae and Supernova Remnants", D. Reidel Publishing Company, Dordrecht-Holland, pp. 223-242, 1974.

Gower, J.F.R., Scott, P.F. and Wills, F., Mem. Roy. Astron. Soc., 71, 49-142, 1967.

Gull, S.F., Mon. Not. Roy. Astron. Soc., 162, 135-142, 1973.

Hill, E.R., Austr. J. Phys., 30, 297-307, 1967.

Hsi Tsê-tsung, Acta. Astr. Sinica, 3, 183, 1955. Trans. in Smithson. Contr. Astrophys. 2, 109-130, 1958.

Hsi Tsê-tsung (Xi Ze-zong) and Po Shu-jen (Bo Shu-ren), Acta. Astr. Sinica, 13, 1-21, 1965. Trans. NASA Tech. Trans. TTF-388. Abridged trans. by K.S. Yang in Science, 154, 597-603, 1966.

Ho Peng Yoke, in A. Beer (ed.), "Vistas in Astronomy", vol. 5, Pergamon Press Ltd., Oxford, pp. 127-225, 1962.

Ho Peng Yoke, "The Astronomical Chapters of the Chin Shu", Mouton, Paris, 1966.

Hsüeh Chung-san and Ou-yang I, "A Sino-Western Calendar for Two Thousand Years", AD 1-2000, Peking, 1956.

Ilovaisky, S.A. and Lequeux, J., Astron. Astrophys., 18, 169-185, 1972a.

Ilovaisky, S.A. and Lequeux, J., Astron. Astrophys., 20, 347-356, 1972b.

Johnson, H.M., Astron. Soc. Pac., Leaflet No. 353, 1958.

Milne, D.K., Austr. J. Phys., 23, 425-444, 1970.

Milne, D.K., Austr. J. Phys., 24, 757-767, 1971.

Minkowski, R., Ann. Rev. Astron, Astrophys., 2, 247-266, 1964.

Minkowski, R., in B.M. Middlehurst and L.H. Aller (eds.), "Nebulae and Interstellar Matter" (Stars and Stellar Systems, vol. VII), Chapter 11, University of Chicago Press, 1968.

Muller, P.M. and Stephenson, F.R., in G.D. Rosenberg and S.K. Runcorn (eds.), "Growth Rhythms and History of the Earth's Rotation", John Wiley and Sons Ltd., London, 1974.

Needham, J., "Science and Civilisation in China", vol. 3, Cambridge University Press, 1959.

Newton, R.R., "Ancient Astronomical Observations and the Accelerations of the Earth and Moon", Johns Hopkins Press, Baltimore, 1970.

Newton, R.R., "Medieval Chronicles and the Rotation of the Earth". Johns Hopkins Press, Baltimore, 1972.

Payne-Gaposchkin, C., "The Galactic Novae", North Holland Publishing Co., Amsterdam, 1957.

Pertz, G.H. (ed.), "Monumenta Germaniae Historica, Scriptores", 32 vols., Hahn, Hanover, 1826 →.

Shklovsky, I.S., "Supernovae", John Wiley and Sons Ltd., London, 1968.

Spencer Jones, H., "General Astronomy" (Fourth Edition), Edward Arnold Ltd., London, Fig. 112, 1961.

Stephenson, F.R., Quart. J. Roy. Astron. Soc., $\underline{12}$, 10-38, 1971a.

Stephenson, F.R., Astrophys. Lett., $\underline{9}$, 81-84, 1971b.

Stephenson, F.R., in C.B. Cosmovici (ed.), "Supernovae and Supernova Remnants", D. Reidel Publishing Company, Dordrecht-Holland, pp. 75-85, 1974.

Tammann, G.A., in C.B. Cosmovici (ed.), "Supernovae and Supernova Remnants", D. Reidel Publishing Company, Dordrecht-Holland, pp. 155-185, 1974.

Tuckerman, B., Mem. Am. Phil. Soc., $\underline{59}$, 1964.

Weiler, K.W. and Seielstad, G.A., Ap. J., $\underline{163}$, 455-478, 1971.

Westerlund, B.E., Astron. J., $\underline{74}$, 879-881, 1969.

Woltjer, L., Ap. J., $\underline{140}$, 1309-1313, 1964.

SUPERNOVAE AND THE ORIGIN OF COSMIC RAYS (I)

S.A. Colgate

New Mexico Institute of Mining and Technology
Socorro, New Mexico, 87801

I. INTRODUCTION

There have been few papers on the theoretical origin of cosmic rays in the last decade that have not in some way or other invoked the phenomenology of a supernova. Some of these have been particularly vague about the details of the supernova itself only regarding supernovae as in some fashion making violence in our galaxy, and out of it emerges the cosmic ray spectrum. On the other hand, other theoretical interpretations of cosmic rays, in particular, the shock origin model, have almost monotonously dwelt upon the morphology of supernovae themselves. Cosmic rays are then only a small effect in the whole range of explosive phenomena.

I have been asked in this first of two talks to describe the various models of supernovae. Somehow or other a supernova occurs, and this rather singular occurrence is marked by one single characteristic observation which has been unique in the astronomical history of supernovae. The observation is that in some dim, distant, fuzzy object of space, presumably a galaxy, some non-central part of that fuzziness brightens up and in a matter of a week to a month; reaches some form of a maximum, decays away, and has a color temperature not much different from the average star. It is only the fantastic hierarchy of observation and deduction that has cast that dim fuzzy object in the heavens as a distant galaxy as awesome as our own composed of some 10^{11} to 10^{12} stars, some part of which brightens up to a luminosity that is as bright as the whole galaxy together. We do not even know that this sudden flaring up at some point in the galaxy is indeed a star, but the presumption was made originally because of the association in the minds of the observers of supernovae with nova, but there is no a priori reason other than the logic of subsequent arguments to say that this flaring up phenomenon, called a supernova, has its origin as a star rather than some larger or smaller assemblage of galactic matter.

Let us review briefly the history. The sudden flaring up of a star has been recognized many times by astronomers and has been called a nova, but when Hubble recognized galaxies as associations of 10^{11} stars or more then when a star brightened up in a galaxy to the luminosity of the whole galaxy this phenomenon was called a supernova. The two observations that the color temperature was comparable to a normal star and that the light curve rose to a maximum in something like a week are sufficient alone to deduce that something like a solar mass of matter must have been expanded at a velocity corresponding to several MeV/nucleon. Exotic explanations have been put forward for this major release of energy, but none were considered so wild at the time as Fritz Zwicky's suggestion that supernovae may be the result of the formation of a neutron star. As you know, this concept has had significant substantiation not only from the theoretical work that I will be describing, but also from the fact that neutron stars interpreted as pulsars have been seen in the remnants of both the Crab and the Vela. Past astrophysicists had to grasp at almost science fiction to explain the supernova, but now it appears feasible to transport the energy released in the binding of a neutron star to the outer layers of the imploding star by a large neutrino flux, thereby pushing and ejecting the outer layers. Conversely, it is equally plausible that most stars evolving into a carbon-oxygen core will thermonuclearly detonate and destroy themselves. The neutrino transport theory of supernovae has recently been bolstered by the new theoretical considerations concerning the weak interaction force and how this relates to resonant neutrino scattering of large nuclei. The thermonuclear origin of supernovae, on the other hand, has resisted all attempts at finding some way to avoid the detonation before implosion to a neutron star. The observational difficulty with the purely thermonuclear origin is that since the star completely destroys itself, that is, ejects all the thermonuclearly exploded matter into the galaxy, the enrichment of the galaxy in heavy elements would be many fold, at least two orders of magnitude above current observations. The implication is that with the initiation of a thermonuclear explosion somehow or other the conditions for initiation are delayed until the star has reached a sufficiently high density such that beta decay reactions can go fast enough to allow the further collapse to a neutron star before the outer layers are ejected.

I will be concerned with this problem of the self-consistency of the theoretical ejection from supernovae and the observed element distribution in the galaxy as well as the implied delay in the formation of a pulsar on which the offending matter is imploding. This delay has a significant bearing on the likelihood of early pulsar spin-up origin for supernovae and cosmic rays.

Let us now make more detailed calculations of the total mass and kinetic energy of the ejecta of the most energetic supernova. The key question to many astrophysicists is, "What is the source of the energy of the explosion?", and of course, the key point in such an explanation is the amount of energy that must be produced usually measured in MeV per nucleon in order to eject matter at the required velocities.

The brightest supernovae are approximately 25 magnitudes brighter at maximum than the sun. This corresponds to a luminosity 10^{10} times that of the sun, or about 4×10^{43} ergs/sec. The light curve of Type 1 supernovae will typically have a half-width of about one week, or approximately 10^6 sec. Consequently, the optical energy radiated during the light curve maximum is about 4×10^{49} ergs. If the supernova surface were at the same temperature as the sun, then the luminosity would be proportional to R^2, so that the radius of the supernova at maximum would have to be 10^5 times the solar radius, or 7×10^{15} cm. On the other hand, if the supernova temperature were 30,000°K or 5 times greater than the sun, the optical output would increase only by a factor of 5 or 6, compared to a solar temperature of 6000°K. (This is a consequence of the shape of the Planck spectrum, and the energy radiated in the invisible ultraviolet would be very large (10^{52} ergs)). Therefore, it is more reasonable to assume a temperature of about 10,000°K, the temperature which is also inferred from the spectra. For this temperature, the radius will be about 3×10^{15} cm; if we assume that the supernovae reaches maximum in 10^6 sec by the radial expansion of the ejected matter, it follows that the velocity of the surface is 1.5 to 2×10^9 cm/sec, which corresponds to about 1 to 2 MeV/nucleon. It is also reasonable to assume that the matter that is made luminous is already out at the large radius 3×10^{15} cm and that its distribution is associated with some initial (several years prior) ejection of matter from the pre-supernova star. The mass distribution in this prior ejected low density matter becomes critical in determining the shape of the light curve. In particular, unless the outer regions of the ejected matter have a scale height which is comparable to or greater than the radius, then the rise of the light as has been shown in the calculations of Falk and Arnett (1974), Colgate and White (1966), and Grasberg, et al. (1970), will rise very rapidly in a matter of less than a day to a maximum and decay thereafter. It is difficult to contrive the hydrodynamics that would give rise to such a mass distribution. I feel that it is a more contrived explanation of the light curve of the supernova than one based upon the expansion of the outer layers of a previously compact star. Naturally, if the matter is already out there, there is no question of talking of the origin of cosmic rays from the shock traversing the envelope of such a star. It is also difficult to use such a model for any early acceleration mechanisms because of the very large amount of matter distributed ahead of the shock wave and ahead of the bulk of the dynamical effects.

We can make a rough calculation based on diffusion theory arguments of the total mass involved. The density of the expanding material is such that bound-bound transitions are negligible and the only opacity is provided by Compton scattering. If we assume that less heat energy is being generated at the time of maximum than earlier, that is, that the energy source has been turned off somewhat early, then diffusion theory predicts that the maximum luminosity will occur at approximately the time when a diffusion wave has penetrated to about one third the radius of the sphere. Therefore, the diffusion velocity should be one third of the expansion velocity of the surface:

$$v_{diff} = \frac{1}{3} u_s. \qquad (1.1)$$

From diffusion theory, the skin depth ∂ is given by $\partial^2 = Dt$, where D is the diffusivity. We substitute $D = \lambda c/3$, where λ = the photon mean free path and c the velocity of light so that

$$\frac{\partial}{\lambda} = \frac{t}{\partial} \frac{c}{3}. \qquad (1.2)$$

If we consider ∂/t to be the diffusion velocity then from Eqs. (1) and (2) we obtain $\partial/\lambda = c/u_s = 10$ (since we previously calculated u_s to be 3×10^9 cm/sec). We interpret this result to mean that the medium is 10 mean free paths thick. This then requires that $\rho r/3 = 10/K$ where K is the Compton opacity; taking $K = (1/5)$ cm^2/g gives $\rho r = 150$ g/cm^2. A computer calculation, Moore (1973), using a more complicated diffusion theory, including expansion and a variable velocity distribution gives a value of $\rho r = 50$g/cm^2. The total mass will then be $4\pi r^2 (\rho r)/3$; using the more conservative value of $\rho r = 50$ gives a mass of 2×10^{33} gms, which is one solar mass. Therefore, from the very simplest observations, namely, that a supernova occurs in distant assemblages of stars and has a brightness corresponding to 10^{10} suns and rises to this maximum in a week and has a black body spectrum corresponding to a temperature of 10,000 degrees, we can infer that something like a solar mass of matter is expanding at the extraordinary energy of 1 to 2 MeV/nucleon. It should be noted that the inferred kinetic energy is at least 100 times the energy radiated in the optical. By way of substantiating this picture, one can observe in some supernova remnants and in some early line shifts evidence for these high velocities. Minkowski (1968) infers this velocity for the expansion of Tycho's nebula which is consistent with these estimates, and for the cases where the He I and He II spectra have been identified late in time for Type II supernovae, the velocities from the line shifts are again consistent with these estimates. However, the average supernova remnant and the line shifts observed for the hydrogen lines of Type II supernovae, have considerably lower velocities, a factor of 2 or more, either indicative of a slower supernova or that the matter which is observed is a smaller, inner mass fraction moving at lower velocities.

II THE ORIGIN OF SUPERNOVAE

We have referred in the previous paragraph to a Type I and Type II supernova and, indeed, there may be a more complicated morphology as suggested by Zwicky. In general, most investigators would agree that there is major distinction in the light curves and the spectra for two kinds of supernovae. Type I supernovae have a more rapidly rising light curve from our previous discussion, and therefore higher specific energy, and occur in old population stars. The spectra of these supernovae have not been uniquely understood despite the work of many investigators (Minkowski (1968), Kirshner, Oke, Penston and Searle (1973), but there is general agreement that very little hydrogen, less than 10%, if any, contributes to this unexplained spectrum. This is very odd matter indeed; these supernovae have less than 10% hydrogen whereas the normal matter of the universe is at least 90% hydrogen, and so this observation alone is indicative of a star that is probably highly evolved and has consumed most of its original hydrogen prior to the time of explosion.

The second type of supernovae, Type II, on the other hand, rises more slowly and in particular has a spectrum that appears as if it is composed of normal matter with dominantly hydrogen lines appearing and, furthermore, with a velocity shift and spread of these lines that is consistent with a slower optical rise time and therefore indicative of slower expansion velocities. These supernovae occur in spiral galaxies and therefore in younger population stars and even with a high concentration at the star formation edge of each spiral arm as noted by Elliott Moore (1973).

The conceptual origin of supernovae fits into a very broad scheme of stellar evolution. The structure of a star is presumed to be a balance between two forces: (1) the attraction of gravity, and (2) a repulsion from the gradient of pressure which in turn is derived from the gradient of temperature and density of the star. These two forces are in equilibrium and normally a very stable equilibrium at that. A very slow evolution takes place because of radiation, namely, the heat loss from the star (although in the final stages neutrino losses may dominate).

During the bulk of the history of the star, the nuclear energy generation and the adiabatic compressibility are strongly stabilizing factors. As the star evolves by burning the nuclear fuel at its center, both these stabilizing factors at some point in the evolution inevitably reverse in sign. This inevitable evolution toward instability was first recognised by Chandrasekhar (1939) as the limit to the maximum mass of a cold stable star. However, the implications for supernovae were first developed at length in the now famous article by Burbidge, Fowler and Hoyle (1957) on the origin of elements. An integral part of the theory of the origin of elements was the rapid nucleosynthesis that would take place under the extreme conditions of a supernovae explosion. They further recognized that at a critical stage of stellar evolution a dynamical collapse would lead to the conditions necessary for rapid nucleosynthesis.

The presumed structure at the point of dynamical collapse was a star whose inner core was composed mostly of iron, the minimum in the packing fraction curve. The mass of iron (called the Chandrasekhar limit) of 1.3 $M_\odot$ is the limit above which unstable collapse can occur.

Stars whose mass is smaller than 1.3 $M_\odot$ will never reach this instability but instead quietly cool off and become stable by virtue of the electron pressure. The reason most stars reach this fate is because most stars are born less massive than the critical 1.3 $M_\odot$ and, in addition, mass loss occurs that reduces the mass during evolution.

The observational astronomer calls these cold dense stars "white dwarfs". It was the triumph of the original theoretical calculations that this critical mass could be predicted on elementary physical grounds, and that the observations strongly support it. It also leads to the conclusion that only some stars whose mass is greater than 1.3 $M_\odot$ can possibly evolve to supernovae, but we are confronted with a larger dichotomy that the frequency of supernovae, per tens to hundreds of years per galaxy, is still small enough such that only a fraction of those stars born and evolving with initial mass greater than 1.3 $M_\odot$ can possibly become supernovae. Currently, the discrepancy is explained by the mass loss predicted and observed to take place during what is called the "red giant" stage. A few stars sufficiently massive avoid the red giant stage so that sufficient mass accumulates in the core to cause collapse. This number may be sufficient to explain the occurrence of Type II supernova that occur almost exclusively in young population I stars. The other Type I supernovae occur in old population II stars, and therefore require some form of mass accretion to take place after ejecting their envelope in the red giant phase. It has been suggested by Whelan and Iben (1973) that the mass accretion can take place by the exchange of red giant envelopes in a binary pair of stars. Such a mass accretion might even manifest itself as an x-ray source as suggested by A.G.W. Cameron (1974).

III THE MECHANISMS OF SUPERNOVA EXPLOSIONS

Burbidge, Burbidge, Fowler and Hoyle (1957) suggested that a dynamical collapse led inevitably to the triggering of a thermonuclear explosion in the less dense outer layers of carbon and oxygen. It was recognized that the matter, once collapsed as iron, is very difficult to eject from a star because of the ever increasing gravitational binding energy. However, the matter that is still only partially evolved to iron, say, in the carbon-oxygen stage of nucleosynthesis is potentially explosive and releases 0.5 MeV/nucleon of thermonuclear energy, and could very possibly blow itself off the star. The remainder would collapse to a neutron star. The reason that hydrogen or helium, both of which have a greater potential energy available in nucleosynthesis, were not considered as potential thermonuclear fuels is because their reaction rates to synthesize something heavier and release the potential energy of binding are far too small to contribute significantly within the dynamic time scales of such an explosion. A thermonuclear explosion is therefore limited to the energy available in synthesizing or burning carbon and oxygen up to the most stable element - iron. This energy of complete synthesis is only 0.5 MeV/nucleon and is inadequate to explain some Type I supernovae explosions as interpreted on the preceding elementary basis. On the other hand, factors of 2 or 4 are small indeed in the range of astrophysical variables, and it has always been considered possible that this elementary interpretation of the velocity of expansion may have some significant error which can allow thermonuclear burning to explain the origin of supernovae.

Recently, Bodenheimer and Ostriker (1974) have suggested that the major fraction of the energy input to a supernova and the light curve are the result of the radiation from a rapidly rotating and newly formed pulsar. The light curve formed is similar to a Type II supernova - slowly rising and falling over a month. The primary difficulty is that unless the neutron star is composed of all the matter of the original core, then it must eject some matter explosively. This ejection would be the normal shock wave models, and the pulsar energy would be very much later in time. On the other hand, the cores of such stars - out to, say, 10^{12} cm radius are of the order of 2 to 2.5 $M_\odot$ - pushing the upper limit of neutron star stability if it all were to implode. It is possible that such a model might just account for Type II supernovae, but highly unlikely for it to account for Type I supernovae.

IV THE EXPLOSION PROCESS AND NEUTRON STAR FORMATION

It was some years after this major work of Burbidge, Burbidge, Fowler and Hoyle that Richard White and myself (1966) started evaluating such an explosion using the hydrodynamic numerical computer calculations that were then available at Lawrence Laboratory in Livermore for nuclear weapons research. Our initial calculations explored the possibility that dynamical collapse alone would cause the matter to bounce on a hypothetical "hard core", possibly a neutron star. Even when no heat loss due to neutrino emission was included in these initial calculations it was evident that a dynamical collapse of an already tightly bound star would lead to only a minuscule ejection of matter and this minuscule fraction of ejected matter depended upon an extraordinarily sensitive transfer of energy by shock waves in the many, many orders of magnitude variation of density of a star. The phenomenon was governed by the generation of a reflected shock wave of the in-falling matter on the presumed neutron star core, and this reflected shock wave had to propagate out through in-falling matter through 10 to 12 orders of magnitude (10^{10} to 10^{12}) change in density to the outer layers before reaching a strength sufficient to eject only the very outermost surface mass fraction of a star. Under these conditions, the ejected matter was some 10^{-3} to 10^{-5} of the mass of the star and its velocity at least an order of magnitude less than what could explain supernova light curves. At this point we decided that the logical conclusion was that the thermonuclear energy had to be released in the outer layers triggered by this relatively weak reflected shock wave from the neutron star core. We did not fully understand thermonuclear detonation waves in carbon and oxygen and there were still many questions of cross sections involved. We took the extreme view and placed in the envelope of a 10 $M_\odot$ star at the optimum position, three solar masses of carbon and oxygen. We then released 0.5 MeV/nucleon energy, expecting to see the ejection of several solar masses of matter.

Contrary to this naive and even optimistic expectation, the detonated matter with a 0.5 MeV/nucleon internal energy was swallowed into the imploding neutron star just as if it had not been heated at all! The reason for this became more evident in retrospect. Without the pressure of a "center" to push on, the inner boundary of the exploded carbon and oxygen saw the reduced pressure created by imploding iron into the neutron star. Without an inner boundary to push against, the exploded matter simply expanded inward toward the neutron star and was swallowed in a transformation from normal matter to neutron matter. There seemed to be no way out of this dilemma because the gravitational binding energy of the carbon and oxygen at the point where it would by synthesized in such a star was always greater than the potential thermonuclear energy that could be obtained by burning it to the endpoint-iron. Therefore, it appeared to be almost impossible to use the gravitational collapse to a neutron star core as a means of triggering a thermonuclear explosion that would blow apart a massive star.

V BINDING ENERGIES

I would like to digress briefly into some comments about the binding energies of matter in various forms. If we considered a zero point to be separated protons and electrons in the universe, (ionised hydrogen) then we would say that such matter could evolve to a lower energy state; namely, the hydrogen atom with 13.6 electron volts binding energy, and this energy would be radiated in photons corresponding to the various transitions of the electrons of the quantum shells. We can envisage a further decrease in the energy state for an increase in the binding energy of hydrogen if the hydrogen atoms combine into molecules and gain an additional electron volt of energy. If we proceed to the interior stars where the densities and temperature are high enough to allow the hydrogen nucleus to transform and combine to make helium, then some 8 MeV/nucleon becomes available as radiation. This is the radiation from the sun.

Finally, as helium nuclei are combined to carbon, oxygen and ultimately, iron, additional energy is available, and one reaches the state of maximum binding energy of a nucleon in the nucleus of iron of roughly 10 MeV/nucleon. This represents the maximum binding energy available in what we might describe as normal matter, but there is another state of matter that was predicted many years ago; first by Landau (1932) who recognised that if nuclear matter were squeezed hard enough by gravity, a transformation to a neutron state could take place, and that this matter would represent a very much larger binding energy than that occurring in the iron nucleus. This binding energy occurs only in a collective state of matter because it depends upon the interaction of the gravitational force acting over a very large assemblage of nucleons (10^{57}), and so cannot exist in the small units of nuclei of normal matter. Oppenheimer and Volkoff (1939) were the first to construct models of such neutron stars based upon the new theory of general relativity of Einstein and nuclear physics.

These models were considered almost akin to science fiction in their day, although the physics that went into constructing them was believed to be the most likely basis for the foundation of the real world. Yet, I can remember personally when I was a child considering the extraordinary properties of neutron star matter where a matchbox full weighs a billion tons, and believing that this was sheer fantasy, and yet now, the observation of pulsars has made the existence of neutron stars an accepted certainty, and furthermore the discovery of a pulsar in the Crab nebula remnant makes their association with supernovae raised from the level of inevitable to dramatic certainty.

Two scientists who in the early days did not take the neutron star hypothesis as science-fiction were Baade and Zwicky (1934). They were the first to propose that neutron stars might be the origin and the result of supernovae. Yet White and I, in the first numerical calculations, used a perfectly rigid neutron star as a "hard" core to the imploding supernovae and, one might comment, with depressingly modest results. At this stage we did not know how to deal with the binding energy of a neutron star and, as we pointed out above, this is the largest binding energy, the lowest negative energy state, that has been postulated for matter. The exact value of this binding energy has been open to relatively wide uncertainties due to the complicated interaction of the nuclear equation of state of such dense matter with the subtleties of the general theory of relativity, and as a consequence, the maximum binding energy of a neutron star has varied from 50 MeV/nucleon up to 200 MeV/ nucleon, almost an order of magnitude. Yet, even the smallest of these limits is 100 times larger than the available energy from the thermonuclear detonation of oxygen and carbon, the most energetic explosive stellar fuel; so that the question was, "How could we use this 100-fold greater energy source to blow up a star?" We knew that this energy source had to appear as heat in the newly formed neutron star, but the question was: "How could this heat be used or conducted to the outer layers which reside at a lower gravitational potential and so lead to the internal pressure that could blow them off or eject them from the star?"

VI   NEUTRINOS IN SUPERNOVA

It was at this point, that a chance discussion with Robert Christy at Caltech led me to understand how neutrinos when they are in thermal equilibrium acted like a lepton gas that had properties similar to a relativistic electron gas or even to a photon gas, and so it would exhibit all the properties of pressure and equation of state that were more familiar to us for electrons and photons. If one could conceive of the heat of formation of this neutron star being in thermodynamic equilibrium with neutrinos as well as photons and electrons, then indeed, there was a possibility that this newly formed star would radiate a neutrino flux like a black body characteristic of the implied extraordinary temperatures of formation. One can relatively easily derive these temperatures because one knows the rate of energy release from the in-falling matter to the neutron star state. The total energy available is 0.1 to 0.2 $M_\odot c^2$ and it is released in the time

necessary for the matter to free fall from the initial radius of
the highly evolved core of the initial star. This radius
($\simeq 10^3$km) is determined by the gravitational binding necessary to
hold the matter together at the temperatures at which the final
nucleosynthesis of the star takes place. The resulting free-fall
time of 10 to 20 milliseconds has been characteristic of all calculations performed (Colgate and White (1966), Arnett (1967),
Wilson (1971) and Schwartz (1967)) and determines the temperature
of emission. The binding energy of the neutron star is then
emitted in the free-fall time as a black-body neutrino flux from
the surface of the neutron star. The equivalent radiation
temperature is then determined by equating the Planck emission
rate to the energy release or

$$\frac{7}{4} \frac{c}{4} aT^4 (4\pi R^2) = 0.2 M_o c^2/\tau \qquad (6.1)$$

The inferred temperature is then roughly 30 MeV. The characteristic
energy of a relativistic particle at such a temperature – neutrinos,
electrons, or photons – is 3kT so that particle reactions characteristic of 100 MeV energies must be considered. Most important of
all, such a hot region will emit neutrinos as a black body flux
and these neutrinos will have a mean energy of up to 100 MeV. The
cross section of such neutrinos on the imploding matter is such
that the imploding matter is roughly one to several neutrino
scattering and absorption mean free paths thick, and, as a consequence, White and myself recognised the possibility that this
neutrino flux could represent the heating or energy transport
mechanism that could carry the binding energy of the neutron star
to the outer layers of the still imploding original star. The
heat and energy deposited might be sufficient to result in the
explosion and ejection of the remaining matter. The original
hydrodynamic calculations (Fig. 1) made the relatively crude
assumption of black body emission from the neutron star surface
and deposition of the heat of the neutrinos nearly proportional
to the cross section in the outer layers. Subsequent calculations,
(Arnett (1967), Wilson (1971), Schwartz (1967)), included a hierarchy of neutrino transport theories and modifications to the cooling history of the imploding matter and although the first calculations showed a relatively straightforward mass ejection, later
calculations by others found limitations to the circumstances in
which enough mass would be ejected to prevent the catastrophe of
collapse to a black hole. This latter question is, of course, at
the heart of much of the speculation about supernovae. Phrased
another way, "Is there a mechanism for transporting the binding
energy of a neutron star to the outer layers of the imploding star
that is sufficiently effective to prevent further collapse? Can
the neutron star always manage to eject sufficient mass such as to
prevent its own mass from accumulating beyond the limit where it,

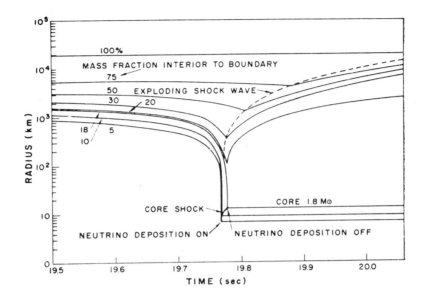

Fig. 1: Radius versus time for 10M$_\odot$ supernova with neutrino deposition. During the initial collapse the neutrino energy is assumed lost from the star, but at the time of formation of a core shock wave (heavy dots) a fraction of the neutrino energy is deposited in the envelope. The deposition ceases when the explosion terminates the imploding shock wave on the core. (From: Ap. J., 143, 660, 1966, Fig. 25, Colgate, S.A. and White, R.H.).

---

too, will start imploding?" This mass is roughly 2 M$_\odot$, and this important question is not yet resolved.

At any rate, some ten years after the first calculations, there has been major concern that the neutrino transport mechanism may not be efficient enough to explain the implied energies of supernovae even though the available energy may be as much as 200 MeV/nucleon, and the energy to be explained or inferred from the observations is only 1 to 2 MeV/nucleon. It is indeed difficult to demonstrate by numerical calculations, the efficiency of a process that depends upon such a complicated hierarchy of phenomena.

Very recently, however, the new universal neutrino theory of Steven Weinberg (1971) has had a major impact upon supernova theory.

As Freedman (1973) has pointed out, a coherence phenomenon occurs in large nuclei that increases the cross section per nucleon for neutrino scattering and absorption is proportional to the atomic weight; so that if the configuration at the start of implosion is composed of heavy nuclei, then the transformation to neutron matter as it falls to the neutron star surface is ideal for allowing the free emission of neutrinos from the forming neutron star surface and then a disproportionately larger scattering cross section on the iron or heavier elements that have not yet imploded. James Wilson (1974) has just completed a calculation using the new neutrino cross sections and the more sophisticated transport theory, which seems to imply that indeed the heat of formation of a neutron star is an efficient and effective mechanism for causing the ejection and explosion of stars (see Fig. 2). The ejection

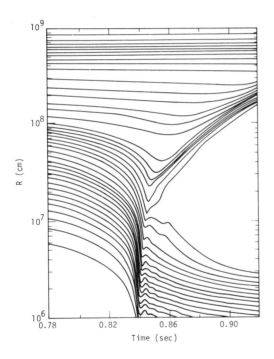

Fig. 2: Radius versus time of computed mass points in the star for the calculation with $a_0 = 1$. (From: Phys. Rev. Ltrs., 32, 849, 1974, Fig. 2, J.R. Wilson).

energies appear to be reasonably close to those inferred from the observations so that there appears to be significant hope of associating the binding energies of a neutron star with the explosion energy observed in supernovae.

On the other hand, a completely different view has been proposed and developed at great length; first by Hoyle and Fowler (1960) and then in greater detail by Dave Arnett (1968). This is the view that a smaller star, less than 8 $M_\odot$, evolves to a carbon core or carbon and oxygen core in such a fashion that its temperature is relatively low and supported primarily by degeneracy pressure. Under these circumstances, carbon or carbon and oxygen is a potentially explosive thermonuclear fuel.

Then if a detonation of the carbon is initiated, there is present the central support pressure necessary to explode the star. This is opposite to the case where an iron core implodes to a neutron star and an outer shell of carbon and oxygen is detonated. In this latter case, there is nothing for the exploding matter to push against and so it is swallowed into the neutron star. It turns out that during the continuing evolution of a carbon-oxygen core, the matter reaches the state where a very small fluctuation in temperature at an exceedingly high density (10^9 to 10^{10} gm/cm^3) initiates a runaway thermonuclear reaction. Detailed calculations by Arnett (1969) and Wheeler and Hansen (1971) have shown that the suggestions of Hoyle and Fowler (1960) are indeed correct that such a detonation will completely disintegrate the star (Fig. 3). The ejection velocities are relatively modest corresponding to less than 0.5 MeV/nucleon, but again the interpretation of supernovae pheonomena may agree with this. There has been a great deal of effort by Barkat, Buchler and Wheeler (1970) and Bruenn (1971) to attempt to find an evolutionary path for the center of such stars such that they can avoid or circumvent the detonation catastrophe. There is still a distinct possibility that Paczynski's (1970) suggestion of a convective enhanced URCA process might help in leading to such an exception. The reason for the effort to circumvent the thermonuclear state is that the present conclusion would be that all less massive stars evolving to a carbon-oxygen core would necessarily undergo a thermonuclear explosion that would disintegrate the star and leave no remnant neutron star. Yet we observe sufficient numbers of neutron stars and, in addition, a neutron star at the center of the Crab and the Vela supernovae remnants so that we are confronted with the following dichotomy. There are enough neutron stars or pulsars observed such that their number is consistent with the notion that supernovae produce neutron stars as remnants. In the one case where a definite association can be made, the Crab, apparently a neutron star is left after a Type I supernova; namely, the smaller, hotter, older population supernovae. Yet it is just this particular class of supernovae that one would expect might evolve through the carbon-oxygen core thermonuclear detonation. It would be very neat if we could find a mechanism such that a carbon-oxygen core

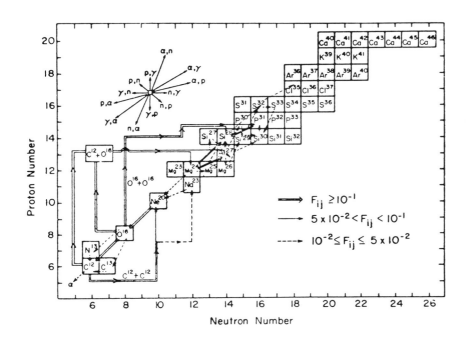

Fig. 3: Primary flows during explosive oxygen burning. Principal flows of the strength and type illustrated are shown at the "standard" explosive-oxygen-burning run (T_{9i} = 3.6, ρ_i = 2.0 x 10^5). Fig. 3 is a "snapshot" at time 7.53 x 10^{-3}s at which point X (^{24}Mg) = 0.18 and X (^{16}O) = 0.49 (From: Ap. J. Suppl. Series, No. 231, 26, 242 Nov. 1973, Fig. 3a, Woosley, Arnett and Clayton).

could evolve stably up to a density of 2 x 10^{10} gm/cm^3.

As Bruenn (1972) has pointed out, the further implosion and detonation of a carbon-oxygen star results in sufficient neutrino energy loss at the time of the detonation such that the star continues to implode to a neutron star state, yet, at the same time the outer layers have been detonated in a carbon-oxygen thermonuclear reaction. Then, as the neutron star is formed by implosion of the inner layers, the Planck neutrino emission from the hot neutron star will be absorbed in the already detonated outer layers augmenting the initial energy from the thermonuclear detonation and ejecting the outer layers at the high energies inferred from the observation. Wheeler, Buchler, and Barkat

(1973) may have found just such a path when the delayed ignition of the initial collapse is considered.

The matter that would be observed later in time would be characteristic of the debris from such a thermonuclear detonation, in particular, a distribution of elements from carbon up to iron some of which would be characteristic of silicon burning.

This picture has the added attraction that the initial star would be highly compact and that the surface layers would correspond to a very small radius, approximately several thousand kilometers, so that the shock wave associated with the detonation and subsequent neutrino heating that would travel through the outer layers, would speed up and form the energy distribution that Johnson and I (1960) have claimed is the origin of cosmic rays. The particular compact structure is ideal for this phenomenon, and the energy imparted to particles greater than a GeV in the shock wave agrees with what is necessary to form cosmic rays in our galaxy.

There is a further theoretical reason to presume that formation of a neutron star occurs with every supernova, and this has to do with the distribution of heavy elements in our galaxy. Arnett and Truran (1970) were the first to point out the rather gross inconsistency of the first supernova calculations based on neutron star formation. The bifraction of the matter of the exploding star at the neutron star surface, as was indicated by the numerical hydrodynamic calculations, would mean that all the matter that had almost fallen into the neutron star and reached densities up to 10^{11} or 10^{12} gm/cm^3 would be ejected with the supernova debris. This matter that has reached densities greater than 2×10^{11} gm/cm^3 is necessarily highly neutron-rich because of the beta decay occurring during implosion and then explosion. This ejection of neutron-rich matter would be several orders of magnitude greater than what is observed in our galaxy. This same argument applies to purely thermonuclear supernovae but not to quite the same degree. As a consequence several years ago, I spent considerable effort trying to investigate whether the numerical hydrodynamics of the explosion would possibly lead to any exception in the ejection of the very neutron-rich matter adjacent to the neutron star. Indeed, it developed that the matter adjacent to the neutron star at the time of the explosion or bifraction of the matter of the star into imploding and exploding fractions is grossly erroneously calculated in the hydrodynamic calculations because one zone has an inner radius of the neutron star boundary and an outer radius that ultimately goes out to infinity. Since the hydrodynamic calculations depend upon averages within the zone, these averages are grossly miscalculated at radii very much larger than the neutron star surface, and phenomena associated with the relatively high densities at the neutron star surface were over-

looked entirely. By means of relatively simple analytical calculations one observes that matter cools so rapidly by neutrino emission that reimplosion occurs and a rarefaction wave progresses outward from the surface of the neutron star and reverses the radial trajectory of the matter previously exploding. The reimplosion of matter back on to the neutron star occurs because the velocity distribution of the ejected matter is such that its velocity is smaller at smaller radius. A certain fraction can, therefore, be captured by the gravitational field of the neutron star. The neutrino emission by the matter falling onto the neutron star is always high enough to ensure that the reimploding matter is cooled sufficiently rapidly that no significant pressure develops. The rate of reimploding matter is governed by a boundary condition, and subsequent free-fall onto the neutron star.
Figure 4 shows the early reimplosion when roughly 50% of the initially ejected matter falls back onto the neutron star, and Figure 5 shows the density distribution as a function of time for the late reimplosion. The later time behaviour is governed by the blow-off from the rarefaction wave as it progresses into the ejected matter. When the ejection velocity becomes sufficiently larger than the escape velocity, the blow-back or rarefaction fan in the interior cannot exceed the excess velocity above escape and reimplosion ceases. This occurs at $t \simeq 10^5$ seconds - near the peak in the light curve. At this point, the reimplosion rate has fallen to a value such that the luminosity at the neutron star surface is

$$L = \rho A \ V \ c^2/10 \simeq 10^{43} \text{ ergs/sec} \tag{6.2}$$

This is about equal to the maximum luminosity of the presumed pulsar

$$L_{max} \simeq (B^2/8\pi) Vol. \omega \tag{6.3}$$

so that up to this point, the stress of the wrapped up magnetic field of the pulsar will be exceeded by the stress of the in-falling matter and no electromagnetic emission can take place. It, therefore, seems unlikely that the pulsars can start emitting electromagnetic radiation until after this matter has been accumulated by the neutron star, and the reimplosion stress decreases below the pulsar limit. The free-fall time is approximately 10^6 sec for the matter emitted at 10^5 seconds so that the pulsar emission should commence only after that time. At 10^6 seconds the expanding matter is still 30 to 100 g/cm^2 thick and so if the pulsar is emitting cosmic rays, they should be spalled to free nucleons and significantly attenuated.

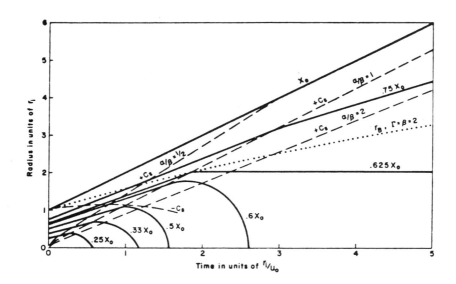

Fig. 4: Radius versus time of the explosion history in linear coordinates and using the reduced variables $X = r/r_i$ and $\tau = t\, U_0/r_i$. Heavy lines, Lagrange coordinates of various mass fractions denoted by the initial radius fraction of the outer boundary X_0. The inner mass fractions reimplode when overtaken by the outgoing rarefaction wave, denoted by $+C_s$. Three such waves (dashed curves) are shown for various ratios of α/β, where α/β is the ratio of internal to kinetic energy. The escape-velocity boundary r_B is shown as a dotted curve for the condition $\Gamma = \beta = 2$. The reimplosion terminates when the rarefaction wave passes the escape-velocity boundary.
(From: Ap. J. 163, 221, Jan. 15, 1971, S.A. Colgate)

We note that pulsars are observed because of a large magnetic field that is undoubtedly frozen into the neutron star at the time of its formation. This magnetic field means that a more modest but still very large (10^8 Gauss) magnetic field must surround the initial star before the explosion. There is a significant probability of observing the electromagnetic pulse from the relativistic ejection of the outer layers occurring within such a magnetic field. Peter Noerdlinger and I (1971) have calculated that it may yet be possible to observe such pulses from red shifts as large as 2 or greater. It would be my long-term hope that we can learn to detect distant supernovae at the time of their formation by the electromagnetic pulse and observe them throughout their

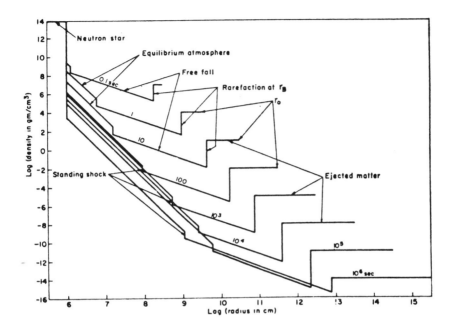

Fig. 5: Density distribution at later times as a function of radius. Outer boundary is $r_o = U_o t$. The rarefaction wave is approximated as a discontinuous decrease in density at the escape velocity boundary r_B. Free fall of the matter takes place to some radius where a standing shock matches on to an equilibrium neutron star atmosphere $\rho \propto r^{-4}$. The rate of mass accumulation is determined by this atmosphere at the surface of the neutron star and in turn is equated to the flux of matter in free fall. (From: Ap. J. 163, 221, Jan. 15, 1971, S.A. Colgate).

initial expansion to luminosity at maximum. If we can do this in large numbers, we have an ideal standard candle for measuring the distance scale of the universe. We need this standard candle to confirm the interpretation of the red shift distance relationship for our universe and ensure that the fantastic construction of observation and deduction which is our present picture of the universe is substantiated in one more major observational way.

REFERENCES

Arnett, W.D., Can. J. Phys., 45, 1621, 1967.

Arnett, W.D., Nature, 219, 1344, 1968.

Arnett, W.D., Astrophys. Space Sci., 5, 180, 1969.

Arnett, W.D., and Truran, J.W., Ap. J., 160, 959, 1970.

Baade, W., and Zwicky, F., Phys. Rev., 45, 138, 1934.

Barkat, Z., Buchler, J.R. and Wheeler, J.C., Ap. J. Letters, 6, 117, 1970.

Bodenheimer, P., and Ostriker, J.P., Ap. J., 191, 465, 1974.

Bruenn, S.W., Ap. J., 168, 203, 1971.

Bruenn, S.W., Ap. J., 177, 459, 1972.

Burbidge, E.M., Burbidge, G.R., Fowler, W.A., and Hoyle, F., Rev. Mod. Phys., 29, 547, 1957.

Cameron, A.G.W., "Hot Vibrating White Dwarf Models of Pulsating X-ray Sources", Harvard College and Smithsonian Observatory Pre-print, 1974.

Chandrasekhar, S., "An Introduction to the Study of Stellar Structure", Chicago: University of Chicago Press, 1939.

Colgate, S.A., and Johnson, M.H., Phys. Rev. Letters, 5, 235, 1960.

Colgate, S.A., and White, R.H., Ap. J., 143, 626, 1966.

Colgate, S.A., and Noerdlinger, P.D., Ap. J., 165, 509, 1971.

Falk, S.W. and Arnett, W.D., Ap. J., 180, L65, 1974.

Freedman, D.Z., National Accelerator Laboratory, Pub.-73/76-THY, Batavia, Illinois, 1973.

Grasberg, E.K., Imshennik, V.S., and Nadezhin, D.K., Ap. and Space Sci., 10, 28, 1970.

Hoyle, F., and Fowler, W.A., Ap. J., 132, 565, 1960.

Kirshner, R.P., Oke, K.B., Penston, M.V., and Searle, L.,
Ap. J., 185, 303, 1973.

Landau, L., Physikalische Zeitschrift Der Sowjetunian, 1, 285, 1932.

Minkowski, R.L., "Stars and Stellar Systems" Vol. 7, ed.
B.M. Middlehurst and L.H. Aller, Chicago: University of Chicago Press, Chap. 11, 1968.

Moore, E.P., "Predetonation Lifetime and Mass of Supernovae from Density-Wave Theory of Galaxies", Astron. Soc. Pacific, 85, 1973.

Oppenheimer, J.R. and Volkoff, G.M.,
Phys. Rev., 55, 374, 1939.

Paczyniski, B., Act. Astr., 20, 47, 1970.

Schwartz, R.A., Ann. Phys., 43, 42, 1967.

Weinberg, S., Phys. Rev. Letters, 27, 1688, 1971.

Wheeler, J.C. and Hansen, C.J.,
Ap. and Space Sci., 11, 373, 1971.

Wheeler, J.C., Buchler, J.R. and Barkat, Z.K.,
Ap. J., 184, 897, 1973.

Whelan, J. and Iben, I., Ap. J., 186, 1007, 1973.

Wilson, J.R., Ap. J., 163, 209, 1971.

Wilson, J.R., UCRL Preprint 75266, Lawrence Livermore Laboratory, to be published in Phys. Rev. Letters, 1974.

SUPERNOVAE AND THE ORIGIN OF COSMIC RAYS (II), A MODEL OF COSMIC RAY PRODUCTION IN SUPERNOVAE

S.A. Colgate

New Mexico Institute of Mining and Technology, Socorro, New Mexico, 87801.

I. INTRODUCTION

In the first of these companion papers of this conference, I have emphasized the mechanisms of supernovae without particular emphasis upon either Type I or II supernovae, but at least outlined the difference between these; namely that the Type I's occur in old population stars, have a rapid rise time, and show no hydrogen in their spectra, and Type II supernovae appear in young Type I population stars, have much slower rise in light curves and slower decay, hydrogen in the spectra, and apparently are considerably more massive both because of the time to evolution and the localization of their occurrence in the leading edge of spiral arms as well as in the population of stars in which they occur.

For the origin of cosmic rays, I would like to emphasize just one type of supernovae; namely, the Type I, that occurs in old population stars, has little or no hydrogen in the ejecta, and where the simple calculation at the beginning of Paper I indicates that an ejection velocity of something like 3×10^9 cm/sec is a minimum velocity self-consistent with a light curve. In that paper, we pointed out that there are several models for the light curves of supernovae other than the excitation of ejected matter. These mechanisms for the production of the light curve other than the ejection from a compact star depend upon a relatively large mass of matter located at a relatively large radius, an extended red giant or ejected planetary nebula distribution at 10^{14} to a few x 10^{15} cm. We emphasize again that Type I supernovae have little or no hydrogen in their spectra, and it is an extreme assumption in stellar evolution to presume that such an extended envelope had no hydrogen. The mechanism of envelope formation depends upon the opacity as a function of temperature which is peculiar to hydrogen and indeed we would not expect an optical light curve that has no hydrogen in the spectrum to be formed from a prior large radius extended envelope. It is the converse of this; namely, the presumed highly compact initial star with no hydrogen in its envelope that we presume to be the type of supernova that gives rise to cosmic rays based upon the theory of shock propagation and shock acceleration.

We will now build up a model of such a supernova, and, indeed, take the extremum of this model to show how one can make a self-consistent theory of the origin of cosmic rays from the shock acceleration of the ejected matter. We have emphasized before that the thermonuclear explosion alone gives a result that would be inconsistent with the enrichment of the galaxy by heavy elements. The frequency of pulsar formation and the fact that the light curve of the supernova in the Crab appears as if it were a Type I is additional evidence that Type I supernovae involve neutron star formation. Therefore, we are not restricted in the theory to presuming velocities self-consistent with the pure thermonuclear origin but instead can choose velocities for the average or bulk of the ejected matter which are self-consistent both with the theory of the rise of a light curve as well as observations, as well as what we might believe reasonably to take place with implosion and explosion on a neutron star core. We already know from the calculations of Arnett and Schramm (1973), Bruenn (1972), Barkat et al. (1972) and Couch and Arnett (1973) that the evolution of a carbon-oxygen core will lead to thermonuclear disruption unless the ignition takes place at the exceedingly high density $\simeq 2 \times 10^{10}$ g cm^{-3}. The reason for this lower limiting density for ignition is that only at central densities greater than this amount will the beta decay of the thermonuclear fusion products of silicon burning to the Ni-Fe group be fast enough to relieve the overpressure of thermonuclear burn such that implosion occurs instead of explosion. You will recall that beta decay rates go roughly as the 5th power of the energy, or in this case, Fermi level of the matter, and since the Fermi level is proportional to $\rho^{1/3}$, the decay rate is approximately $\rho^{5/3}$ versus a free-fall rate of collapse of $\rho^{1/2}$. As a consequence, as one proceeds to higher and higher central density at the time of initiation of thermonuclear burn, one approaches the circumstance where the beta decay rates are faster than free-fall or conversely expansion time, and as a consequence, implosion occurs after detonation or thermonuclear burn and the central core will collapse to a neutron star. The details of the subsequent explosion are still open to considerable controversy. However, the recent results of Wilson (1974) using the neutral current theory for neutrino scattering show in detail that a significant mass fraction 1/4 to 1/3 can be ejected with velocities that correspond to the order of 5 MeV/nucleon. We know from the light curve itself and the arguments in the beginning part of Paper I that something like a solar mass is ejected, and we know from the Doppler shifts that relatively late in time we see matter moving at several MeV/nucleon. Since, as we will discuss in the velocity profiles subsequent to this, the optical observations should strongly emphasize the slower moving matter, we feel that it is self-consistent with the observations and with theory to presume an initial stellar structure with a central density of the order of 2×10^{10} g cm^{-3}, and mean ejection velocity of 3×10^9 cm/sec, and total mass ejected of the order of $1/2$ $M_\odot$. With these numbers we will attempt to describe a self-consistent theory of the origin of cosmic rays based purely on hydrodynamic arguments.

If the energy released from a supernova is generated by the collapse to a neutron star as well as some partial or complete thermonuclear burning, this energy will be generated in a time short compared to the traversal time of sound across the whole star. This is because the central regions collapse in free-fall in a time significantly less than the collapse time of the outer layers, and the thermonuclear detonation will proceed at a velocity significantly greater than the local sound speed. For these reasons, it is axiomatic that a shock wave will be developed in the ejected matter. The criterion for a shock is that a large overpressure is created in a time that is short compared to the traversal time of sound, and when this occurs, a strong shock wave will be the mechanism for distributing that energy over the matter of the star. The shock wave is a mechanism for inelastically transforming matter that is initially at rest into matter in motion. That means for a fluid that the transformation is irreversible and strongly non-adiabatic. As in any inelastic collision, an equal energy is deposited as internal energy as in kinetic energy of motion. This simple equality determines the Hugonial relations that govern the ratios ahead and behind the shock of densities, energies, and pressure. The compression ratio across the shock is determined by the specific heat ratio γ_a, by the relation for the compression $\eta = (\gamma_a+1)/(\gamma_a-1)$ and this compression varies from 4 to 7 for a free particle gas to a relativistic gas like photons, electron pairs, etc. This relatively finite compression ratio is to be compared with a semi-infinite energy or pressure ratio across the shock. The outer layers of the initial pre-supernova star may have temperatures in the range of 10^4 to 10^8 degrees and the temperature behind the shock will be 10^{10} to 10^{11} degrees, and the pressure ratios will be numbers like the 4th power of these temperatures. As a consequence, the conditions behind the shock are in the limit of being highly extreme compared to the conditions ahead and are related only by the finite density ratio corresponding to the above compression ratios and not by the initial temperatures of the matter in the outer layers of the pre-supernova star.

The strength of such a shock wave in the envelope of the pre-supernova star is governed primarily by the density distribution and its initial strength determined by the energy released in the combined detonation and formation of a neutron star. The analytical similarity solutions of shock waves in density gradients were first applied to the envelopes of stars by Ono, Sakashita, and Ohyama (1961) where the methods of Chisnell led to self-similar solutions of the equations of shock propagation. These analytical solutions could not take fully into account the structure of the star coupled with the effects of gravitational binding, and it was only the numerical hydrodynamic calculations of Colgate and White (1966) that could first do this in some detail. These calculations were performed both to try and elucidate the mechanism of supernova explosions, but also with an equal emphasis upon the peculiar properties of the shock wave in the pre-supernova star envelope including the effects of gravitational binding and the energy generation at the neutron star surface. Great care was taken to demonstrate that the numerical hydrodynamics would reproduce accurately the known analytical similarity solutions in power law density gradients where the analytical result was known beforehand. And so it is with some confidence that I feel that the shock velocity distributions produced by a generalized release of energy at the time of the neutron star formation, in the center of a pre-supernova polytrope were calculated. It was these results which have formed the basis for so much of the subsequent consideration of the possible shock origin of cosmic rays.

II. PRE-SUPERNOVA STRUCTURE

The structure of such a pre-supernova star is that of a highly degenerate white dwarf. Because of the extreme density $\gg 2 \times 10^6$ g cm^{-3} the electrons are relativistically degenerate and the equation of state corresponds to an adiabatic $\gamma = 4/3$, independent of the modest interior temperature determined by neutrino emission. Finally, the outer layers that are non-degenerate will have a temperature distribution determined by radiation flow which gives rise to the well-known "radiative zero" solution where $T \propto \rho^{1/3.25}$ or again similar to a gas of $\gamma = 4/3$ (Schwarzschild, 1958). Consequently, the stellar structure should be that of a polytrope of index 3 as discussed extensively by Chandrasekhar (1939). A pure polytrope of index 3 has zero binding energy and so if we endow our model with a binding energy of 10% of the gravitational energy corresponding to the binding of the non-degenerate envelope this corresponds approximately to a polytrope of index 2.5. The ratio of central density to mean density is 23 versus 54 for the polytrope of index 3. We can now calculate a radius for such a pre-supernova star assuming a final central density of 3×10^{10} g cm^{-3}. This assumes a slight compression during the initial implosion of the neutron star and before the shock traverses the envelope. Then

$$\frac{4}{3}\pi R^3 \bar{\rho} = 1.4 \, M_\odot,$$

or (2.1)

$$R = 8 \times 10^7 \text{ cm}.$$

A typical luminosity for such a white dwarf with a central temperature of the order of a few x 10^8 degrees determined by neutrino emission is about ½ the solar luminosity, or 2×10^{33} ergs/sec.

The surface temperature is determined by

$$4\pi R^2 \sigma T^4 = L, \text{ or } T = 1.45 \times 10^5 \text{ deg}. \tag{2.2}$$

The scale height h of the surface layer is determined by the condition

$$kT = h\mu\frac{MG}{AR^2}, \text{ or } h = 500 \text{ cm}, \tag{2.3}$$

where μ = the molecular weight.

The density of the surface layer is determined by the opacity condition

$$K \rho h = \frac{2}{3} \tag{2.4}$$

Using Kramer's approximation to the bound-free opacity results in $\rho \simeq 4 \times 10^{-6}$ g cm^{-3}, but using the more accurate results of the Los Alamos opacity code (Ezer et al., 1963), results in $\rho \simeq 4 \times 10^{-5}$ g cm^{-3}.

This surface condition falls above the radiative zero solution so that as discussed by Schwarzschild (1958), no convective zone will form and instead the density-temperature relation for the outer layers will rapidly converge to the limiting "radiative zero" solution

$$\rho = 10^{-21.1} \, T^{3.25} \, (L_\odot M/LM_\odot)^{\frac{1}{2}} \text{ g. cm}^{-3} \tag{2.5}$$

after a short run of near isothermal outer few layers.

This can be seen readily from Figure 11.1 of Schwarzschild.

The relationship between external mass fraction, scale height, and density for the outer layers with the radiative zero solution, $L = 1/2\ L_\odot$ and $M = 1.4\ M_\odot$ becomes

$$T = 2.75 \times 10^6\ \rho^{1/3.25}\ K$$

$$h = 8 \times 10^6 F^{0.235}\ \text{cm, and} \qquad (2.6)$$

$$\rho = 4.25 \times 10^9 F^{0.765}\ \text{g cm}^{-3}$$

where $F = 4\pi h\rho R^2/M$, and where h is given by equations (2.3) and (2.5)

III. SUPERNOVA MODELS

Figure 1 also shows an estimate of the shock strength and final ejection velocity of the matter of the star. A starting point for this estimate are the calculations of Colgate and White (1966) where the neutrino flux from the forming neutron star core transferred sufficient heat and stress to eject the outer layers with a mean energy of several MeV per nucleon. Despite the intermediate less optimistic calculational results of Arnett (1973) and Wilson (1971) the later work of Wilson (1974) using resonant neutrino scattering as suggested by Freedman (1973) based upon the model of Weinberg (1971) indicates the likelihood of mean ejection energies up to 6 MeV/nucleon. For the likely ejected mass of $1/2\ M_\odot$ this corresponds to 7×10^{51} ergs, close to the value calculated by Wilson (1974). In the calculations of Colgate and White (1966) great care was taken to ensure the accurate numerical representation of the shock wave behaviour in the lower density external mass fractions. We therefore scale these results to the new resonant neutrino scattering models. The behaviour of a shock in the polytropic outer layers of a star can be derived analytically (Ono et al., 1961) and indeed the numerical calculations agreed with this. The inner regions where changing gravity and a scale height comparable to the radius occurs are more complicated to treat analytically and the numerical solutions become necessary. We found that a reasonable representation and approximation of the nonrelativistic behaviour was given by assuming that an inner uniform density sphere of 3/7 of the ejected mass expanded at a surface velocity V_o and tacked on to this is an exterior velocity distribution corresponding to the speed up of the shock wave in the decreasing density of the envelope. As pointed out in Ramaty et al. (1971) and Colgate (1974) this results in the following velocity distribution of ejected matter

$$V = V_o(1 - F)^{1/3}, \qquad 0.75 \geq F \geq F_o = 3/7$$
$$V = 0.67 V_o F^{-1/4}, \qquad F \leq F_o, \qquad (3.1)$$

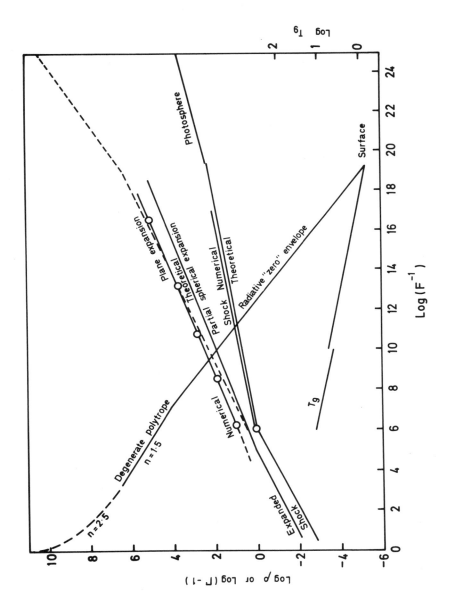

Fig. 1: Dependences of important parameters in the supernova model.

and in a differential mass spectrum:

$$dF = \frac{dM}{M_{ej}} = \frac{6}{7}\frac{d\epsilon}{\epsilon}\begin{cases}(\epsilon/\epsilon_o)^{3/2} & (\epsilon < \epsilon_o)\\(\epsilon_o/\epsilon)^2 & (\epsilon > \epsilon_o)\end{cases} \quad (3.2)$$

where M_{ej} is the total ejected mass and ϵ_o is the transition energy. The total kinetic energy of ejected matter becomes

$$W = M_{ej}\int_{0.75}^{0}\epsilon\,dF = \frac{6}{5}M_{ej}\epsilon_o \quad (3.3)$$

The inner region $F \geq 0.75$ is left indeterminate and was not accurately represented in the numerical hydrodynamics (Colgate, 1971). It is an insignificant addition to the total ejected kinetic energy and is neglected. The exponent $dF \propto \epsilon^{-3}$ for the shock speed up is an approximation that is valid for $F \geq 10^{-4}$ after which $dF \propto \epsilon^{-3.5}$, which is closer to the asymptotic shock solution in plane-parallel geometry with no gravitational effects.

Finally, the shock energy is related to the ejection energy of a given mass fraction in the following peculiar fashion. A strong shock accelerates a particle at rest to a fluid velocity v. Because the collision between the particle and the fluid is inelastic, the particle deposits $1/2\,mv^2$ internal energy in the moving fluid. Therefore, the total energy in the moving fluid is twice the kinetic energy (or internal energy) behind the shock. When the shock fluid expands - adiabatically - during ejection, one might think that the fluid velocity would increase by $\sqrt{2}$, conserving local energy. Instead, the numerical hydrodynamics showed that the velocity was increased by a factor of 2 in the small mass fractions, $F < F_o$, presumably because pressure gradients transfer additional energy from within. Thus, for small mass fractions, the final ejected kinetic energy is 4 times the kinetic or internal energy of the shock fluid, and this factor of 4 must be used to relate the ejected energy distribution of equation (3.2) to the shock conditions at the point of origin.

IV. RELATIVISTIC SHOCK BEHAVIOUR

External mass fraction has been used as a coordinate both because it is a Lagrange coordinate that tags a given fluid sample, and also because the velocity of a nonrelativistic shock depends primarily on the mass fraction and only weakly on the density distribution. Qualitatively, this is because in the nonrelativistic case the sound wave characteristics reach the shock front more readily from fluid far behind the shock, and therefore the mass of the fluid accelerated by the shock becomes more important than its density. The opposite is true in the relativistic case, where the shock behaviour is determined entirely by the density ahead of the shock (Johnson and McKee, 1971). The numerical calculations of the relativistic shock behaviour in a polytropic density distribution by Colgate and McKee (1973) and Colgate, McKee and Blevins (1972) verify this to a high degree of accuracy so that the shock strength is modified accordingly in the region of the stellar envelope where the polytropic index varies from 2.5 to 1.5 and back to 3.25. In Figure 1, we also show the final energy of the ejected fluid based upon the assumption of adiabatic expansion. In the non-relativistic regions, as we previously discussed, this leads to a simple multiplication of x 4. In the relativistic region the partially spherical nature of the expansion has a pronounced effect. Eltgroth (1971) has shown analytically the difference between a spherical and planar expansion, and the curve shown in Figure 1 is the result of the numerical calculations. The limiting behaviour of the two cases is shown in Figure 1 of Colgate, McKee and Blevins.

If we use ε_0 = 6 MeV per nucleon in equations (3.1) and (3.2), then the expanded ejected matter reaches relativistic energies $(\Gamma-1) = 1$ where $\Gamma = 1/\sqrt{1-\beta^2}$ when $F = 1.6 \times 10^{-5}$ and from the previous discussion the shock itself becomes relativistic at a mass fraction corresponding to x 4 this energy or at $F = 10^{-6}$. The numerical calculations showed that to a reasonable degree of accuracy the analytic solution of the relativistic shock could be extended from $(\Gamma-1) = 1$ at $F = 10^{-6}$ to the mass fraction at the breakout of the shock according to the relation

$$(\Gamma-1)_{shock} = (F/F_o)^{-0.178} \qquad (4.1)$$

where in the present case $F_o = 10^{-6}$. This corresponds to the "shock" line in Figure 1 in the relativtic $(\Gamma-1) > 1$ region. The expansion of the fluid corresponds to the upper two curves, the uppermost is the fluid energy factor steeper by the analytical factor 2.7 for a purely plane parallel expansion. The numerical hydrodynamics showed that partially plane parallel-partially spherical expansion of a polytropic envelope results in the middle curve for a more extended white dwarf model $(R = 2 \times 10^8$ cm) than for the present case of the highly condensed one of $R = 8 \times 10^7$ cm, $\rho_c = 3 \times 10^{10}$ g cm^{-3}.

In the latter case, the scale height of the outer layers is so small compared to the radius that almost the entire expansion appears plane parallel, and the energy factor becomes

$$(\Gamma-1)_{expanded} = 16(F/F_o)^{-0.48} \qquad (4.2)$$

We can estimate the temperature immediately behind the shock by knowing the compression factor and, hence, energy density.

The internal energy density E behind a strong relativistic shock measured in the proper frame of the moving fluid is

$$E = 4\rho_i (\Gamma -1)^2 c^2. \quad (4.3)$$

where ρ_i is the rest mass density immediately ahead of the shock (Johnson and McKee, 1971). Since the energy density resides entirely in the photons and pairs, the temperature can be readily calculated as

$$T_9 = 62 \; F^{0.102}, \; F < 10^{-12}, \; kT/mc^2 > 1/3$$
$$48 \; F^{0.102}, \; F > 10^{-12}, \; kT/mc^2 < 1/3$$

V SHOCK STRUCTURE

The equilibrium temperature behind the shock is the primary determining feature of its structure. When the dynamic friction or coupling between the incoming (to the shock frame of reference) rest mass and the equilibrium shocked fluid is small, then, as discussed in Colgate (1974) a high temperature ion equilibrium precursor zone of quasi-thermal ions (rest mass) can form at the front of the shock. Since this precursor zone forms at shock energies as low as 10 to 20 MeV per nucleon, one might expect that in the current case of shock energies up to 100 GeV per nucleon, that an even more extensive nonequilibrium shock precursor zone might form. Instead, the converse is true because as assumed for the Type I supernova models, the density is high enough such that at a given shock strength (energy per nucleon), the temperature is high enough such that electron-positron pairs greatly outnumber the original electrons present in the fluid. The increased lepton density then becomes the origin of the far greater dynamic friction that ensures a quasi-equilibrium shock transition (Colgate, 1970). The range of temperature behind the shock straddles the region $kT \simeq mc^2$, but the major transition to the non-pair dominated specific heat occurs when $kT < mc^2/3$. This occurs for a mass fraction $\simeq 10^{-12}$ which is well within the surface layers of the star. Therefore the important condition of break-out of the shock must concern itself with the non-relativisitic limit of the pair density equation of state. The break-out condition occurs then, when the range of a proton in the pair fluid is equal to the residual surface mass fraction of the star. A canonical magnetic field parallel to the surface will allow the shock to progress further, but since this applies only to a fraction of the surface and may not be present in all pre-supernova stars we will neglect this effect for now and consider it later only for the high energy limit.

The pair density in the nonrelativistic limit is:

$$n_\pm = 2.2 \times 10^{30} \phi^{-3/2} e^{-\phi} \text{ cm}^{-3}. \qquad (5.1)$$

where $\phi = mc^2/kT = 5.95/T_9$.

Using equation (4.4) for T_9 gives

$$n_\pm = 7.4 \times 10^{31} F^{0.15} \exp(-0.097\, F^{-0.1}) \text{ for } F < 10^{-12}$$

The residual thickness of the star measured in the shocked fluid frame is

$$\Delta R' = h_s / [4(\Gamma-1)] = 2.36 \times 10^7 F^{0.413} \text{ cm (by eq. (2.6))} \qquad (5.2)$$

The stopping power of an ionized gas for a relativistic proton when $kT_e/mc^2 < (\Gamma-1)$ is discussed in some detail in Ginzburg and Syrovatskii (1964) as

$$-\frac{dE}{dX} = 2.78 \times 10^{-28} n_\pm z^2 \left[\ln \frac{E}{mc^2} - \ln n_\pm + 73.4 \right] mc^2$$

per cm, (5.3)

or to satisfy the break-out condition

$$(\Gamma-1) = \Delta R' \frac{dE}{dX}$$

$$F = 3.6 \times 10^{-17} \qquad (5.4)$$

This corresponds to a rest mass layer of $\rho h = 1.25$ g cm^{-2} well within the optical surface layer and therefore within the range of validity of the envelope density distribution of equation (2.6). At breakout $(\Gamma-1) \simeq 100$, as indicated in Fig. 1. The temperature at this point becomes $T_9 \simeq 1.1$, and the lepton density $n_\pm$ becomes 3×10^{27} cm^{-3} compared to 2×10^{23} nucleons cm^{-3} for $\rho = 0.33$ g cm^{-3} of rest mass in the proper frame. The photon scattering opacity is therefore 5×10^3 that of normal matter and ensures not only local thermodynamic equilibrium, but also that the radiation remains local to the fluid. One can easily see that since the opacity is of the order 10^4 the nucleon slowing down length, and since the nucleon velocity is c relative to the pair fluid, the photon (Compton) scattering time will be 10^{-4} of the nucleon slowing down time. Since the pair formation and annihilation cross sections are both of the order (1/4) the Compton cross section the pair creation and annihilation will proceed at close to the same rate as Compton scattering, ensuring thermodynamic equilibrium during the nucleon slowing down process.

VI RADIATION DIFFUSION

The "expanded" fluid energy factor $(\Gamma-1)$ for the outer layers shown in Fig. 1 were calculated on the assumption of a purely adiabatic expansion. On the other hand, radiation will diffuse from inner mass fractions to outer mass fractions during this expansion, thereby maintaining a higher temperature or a more nearly isothermal expansion for each successive layer. The result should be a higher final energy for the more transparent outer layers and therefore a flatter slope to the ejected mass fraction and therefore closer to the observed associated cosmic ray energy spectrum.

VII COSMIC RAY ENERGY SPECTRUM

The spectrum of cosmic rays originating from the shock ejection of the outer layers will be just this energy distribution assuming that the ejected matter does not give up a sizable fraction of its kinetic energy in the process of entering the galactic magnetic field.

Let us look at this spectrum assuming for the moment that the shock ejected energies are preserved. In the extreme relativistic limit we have from equation (4.2)

$$(\Gamma-1) \propto F^{-0.48}$$

or that (7.1)

$$F \propto (\Gamma-1)^{-2.08}$$

Since the mass fraction is proportional to the integral number of nuclei external to a given radius and all matter external to this radius will have a greater shock ejection energy, then $F \propto N(>E)$, and the integral energy spectrum becomes

$$N(>E) \propto E^{-2.08} \qquad (7.2)$$

As we have commented in the preceding section, we expect radiation flow to flatten this spectrum, but how much we do not yet know.

In the nonrelativistic region we must realize that equations (3.1) and (3.2) are just a rough approximation to the ejection spectrum, and, in addition, we expect ionization loss during galactic confinement to significantly alter the observed spectrum. In addition, nonrelativistically we must remember that we observe a flux, not a total number and so the spectral exponent will be small by $E^{1/2}$. If we use equations (3.1) and (3.2), then

$$F \propto N(>E) \propto E^{-2} \tag{7.3}$$

and the flux should be

$$\emptyset \propto V\, N(>E) \propto E^{-1.5} \tag{7.4}$$

A better approximation for 100 MeV $\leq E \leq$ 1 GeV, is $F \propto E^{-2.5}$

and $V\, N(>E) \propto E^{-2}$

These exponents are not too far off from the values discussed in this conference. We also note that the upper energy limit to the purely hydrodynamical shock ejection is $(\Gamma-1) \simeq 2.5 \times 10^5$. Let us first, however, examine the hypothesis that the ejected matter can easily enter the galactic magnetic field without significant energy loss.

VIII GALACTIC MAGNETIC INTERACTION

The cosmic rays will expand as a relativistic shell from the supernova. The total energy

$$\begin{aligned} W(E>Mc^2) &= 1/2\, M_\odot F_{Rel} c^2 \\ &\simeq 2 \times 10^{50} \text{ ergs,} \end{aligned} \tag{8.1}$$

which is roughly the canonical value consistent with cosmic ray lifetime, supernova rate, and conservative theories of galactic structure. The hole this energy would blast in the galactic magnetic field of $B \simeq 3 \times 10^{-6}$ Gauss becomes $R = 2 \times 10^{20}$ cm or nearly 100 pc., the thickness of the galaxy.

If the interim regions are a lower field region, then such an expanding shell would certainly "burst out" at the limits of spiral arms or into supernova galactic tunnels as suggested by Cox (1972). Even if the connected tunnel model does not exist and if the same field pervades the whole galaxy, we would expect such an expanding shell to be highly unstable to flute modes (two dimensional analogy of the fluid-fluid Taylor instability). The smallest wavelengths to grow when the cosmic ray shell of matter is slightly decelerated by the magnetic field are those wavelengths equal to a larmor radius or $\lambda_{min} \simeq 10^{12}(\Gamma-1)$ cm. This wavelength is so small compared to the radius of the shell, when slowing down is important, that a very large range of wavelength $\lambda \lesssim 1/10\, R$ will have grown to the nonlinear limit, and the particle orbits will soon become nonadiabatic due to the instability. They then can enter the magnetic field and diffuse off without further "pushing" the field as a piston.

This phenomenon of an explosion in a magnetic field was dramatically demonstrated in the "Star Fish" high altitude nuclear bomb test over Johnston Island (Colgate, 1965: D'Arcy and Colgate, 1965) in the Pacific. The experimental results and theoretical prediction were in agreement with the above hypothesis of instability growth and particle diffusion onto the field lines <u>without</u> giving up much of their individual energy. The critical parameter for this to be true was that the radius of the magnetic bubble (bottle) blown in the earth's magnetic field R_{cavity} $\simeq$400 kilometers, had to be large compared to the particle larmor orbits, $R_L \simeq 3$ km. The ionized debris then escaped at full energy along the field lines to the southern conjugate point. We expect the same thing to happen to a supernova explosion except to a very much higher degree. In this case R_{cavity} $\simeq 2 \times 10^{20}$cm and $R_L \sim 10^{12}$ cm so that fluid field instabilities are ensured and so escape along field lines becomes dominant before a sizeable energy has been converted into magnetic field compression.

IX UPPER ENERGY LIMITS

The condition for the breakout of the shock was that the dynamic friction of the pair fluid should be sufficient to dissipate the momentum of the ions in the shock in a distance equal to the residual thickness of the star. This gave a breakout thickness of roughly 1 g /cm^2 at an energy shock fluid factor $(\Gamma-1) \simeq 100$. The maximum energy of protons after expansion corresponded to $(\Gamma-1) = 2.5 \times 10^5$ or 2.5×10^{14}eV.

If this were the absolute maximum energy of the shock ejection mechanism, then we would expect a <u>priori</u> a rather drastic change in spectrum where a new mechanism might take over. Instead only a modest and debatable change in slope occurs at $E \simeq 10^{15}$eV. It would be a great deal more satisfying if the shock mechanism of acceleration was to be transformed rather than supplanted by an entirely different mechanism.

We believe such a transformation will occur due to the influence of the magnetic field.

The pre-supernova star must have a dipole magnetic field of the order of 1.5×10^8 Gauss at the pre-supernova surface of $R = 8 \times 10^7$ cm, if the conserved flux will result in a field of 10^{12} Gauss at the neutron star radius of 10^6cm. The larmor radius of protons of $(\Gamma-1) = 100$ in this field is approximately 1 cm compared to a scale height of the breakout layer of 1.2×10^3 cm. The shock strength can, therefore, increase by 10^3

before the larmor radius would exceed the breakout surface layer thickness. We next must ask if the shock continues to propagate in the combined magnetic field and stellar photospheric matter, what will be the effective scale height and what will be the limiting condition where the shock propagation law breaks down. We have already estimated an optical surface of $h_{surf} \simeq 500$ cm, and $\rho_{surf} \simeq 5 \times 10^{-6}$ g cm^{-3}. Beyond this point in the photosphere we expect the scale height to remain sensibly constant until matching onto an equivalent of a corona. The shock should follow the same power law versus mass fraction out to this surface, but thereafter the scale height should be constant so that $(\Gamma-1)$ shock $\propto F^{0.236}$ corresponding to the change from the radiative zero solution where $\rho \propto F^{3/4}$ to the constant scale height condition where $\rho \propto F$. The expected change in slope is shown in Fig. 1.

The energy factor $(\Gamma-1)$ after expansion should correspond to a flatter cosmic ray energy spectrum.

$$N(>E) \propto E^{-1.6} \qquad (9.1)$$

Since the density falls linearly with mass fraction, we see that the energy factor increases to 10^{10} (10^{19} eV/nucleon) when ρ has fallen to 5×10^{-12} g cm^{-3}. This is small compared to the mass density of the magnetic field $B^2/8\pi c^2 = 4 \times 10^{-7}$ g cm^{-3}, but we see no reason that the mass density of the magnetic field should not return all its energy back to the shocked particles. We have yet to investigate the pair formation in the shocked magnetic field, but we believe, these too will return their energy to the nucleons during the expansion phase.

A density decrease from 4×10^{-6} to 5×10^{-12} g cm^{-3} in the photosphere at near constant scale height is reasonable. Therefore, we feel that a shock acceleration up to $(\Gamma-1) \approx 10^{10}$ is probable and possibly significantly higher. Since the dynamic friction is with the field and not relative to a large photon flux, we expect the nuclei to remain bound and therefore Fe nuclei can explain the few events $>10^{20}$ eV.

X   COMPOSITION

We expect that the outer layers will be enriched in heavy nuclei. The question is: "Will these nuclei survive the shock?"

The primary criticism of the shock origin of cosmic rays has been that the heavier nuclei would be spalled to free nucleons in the passage through the shock transition (Ginzburg and Syrovatsky, 1964; Kinsey, 1969). The primary rebuttal to these objections was the recognition (Colgate, 1970) that the pair dynamic friction was so large that the nuclei would come to rest in the moving fluid having traversed far less than a nucleon-nucleon scattering mean free path. This can be seen from the calculation of the shock breakout equations (5.2) and (5.3) where even at the highest energies considered for the fluid dynamic shock $(\Gamma-1) \simeq 100$, the slowing down length of a proton becomes $h\rho \simeq 1.2$ g cm^{-2}. Since the nuclear destruction mean free path must be almost x50 longer, we can neglect destruction by the nucleon-nucleon collision cross section.

Instead, we must look to the electromagnetic interactions as the possible source of nuclear destruction because the pair and photon density are overwhelmingly the major constituents of the fluid. When the temperature of the shocked fluid is $\geq mc^2/3$ as is the case for $(\Gamma_s - 1) \leq 50$, and the fluid expanded energy $E \leq 10^5$ Mc^2, the pair density is roughly (x7/4) of the photon density. Since the photonuclear cross section is significantly larger $\simeq 10$ to 100 times that of the electronuclear cross section, the pair fluid can be neglected and only the photons will contribute to nuclear destruction in the shock transition. Since the mean photon energy in the Planck distribution is $\simeq 3$ kT, and the temperature $\simeq mc^2$, the photon energy relative to the streaming nucleons will be $\simeq 1.5$ (Γ_s-1) MeV. The photonuclear cross section, on the other hand, has a resonant peak at $\simeq 30$ MeV, and a full width at half max of $\simeq 30\%$. Therefore, the shock energy factor (Γ_s-1) must be of the order of 20 before major spallation takes place. By this point in the stellar envelope, $kT \simeq 0.4$ mc^2 and the final energy is 2×10^4 so that it is unlikely that any significant spallation will occur. To show this in greater detail, we can calculate the photon spectral distribution $N(h\nu', \Gamma)$ relative to the slowing down nucleus.

Since $\beta \to 1$, the flux of photons can be treated as a plane wave viewed from the moving particle. The spectral distribution of these photons is

$$N((h\nu'),\Gamma) = N_o \int_o^{2\pi} (\tfrac{1}{2}) \, n \, (h\nu) \sin \theta \, d\theta \qquad (10.1)$$

where $h\nu = \dfrac{h\nu'}{\Gamma(1 + \cos \theta)}$

and the Planck distribution of N_o photons cm^{-3} is given by

$$n(h\nu) = \frac{1}{1.202} \left[\frac{(h\nu)^2}{\exp(h\nu/kT)-1} \right] \qquad (10.2)$$

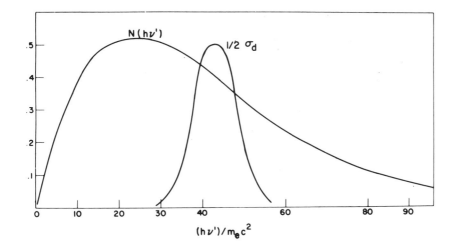

Fig. 2: Photon spectral distribution function and photonuclear cross section behaviour.

N_0 is measured in the frame of the fluid and 1/1.202 is the normalization factor. The number of photo-disintegrations per unit of path dx becomes

$$d\,n_d(\Gamma) = dx \int_0^\infty N(h\nu',\Gamma)\sigma_d(h\nu')d(h\nu') \qquad (10.3)$$

where $\sigma_d(h\nu')$ is the photonuclear cross section given by Kinsey, (1969)

$$\sigma_d(h\nu') \simeq \tfrac{1}{3} A^{4/3} \exp\left\{-\tfrac{1}{2}\left[(h\nu' - E_m)/\Gamma_n\right]^2\right\} \times 10^{-27} \text{cm}^2,$$

where $(h\nu')$ is the photon energy in the nuclear frame of reference, E_m = 22 MeV and Γ_n = 2.55 MeV, the nuclear width.

The increment of path length dx is given by equation (5.3) so that

$$dx = d\Gamma \left[n_{\pm}\, \sigma_{\pm}\, \frac{Z^2}{A}\, (\ln\Gamma + 3.4) \right]^{-1}$$

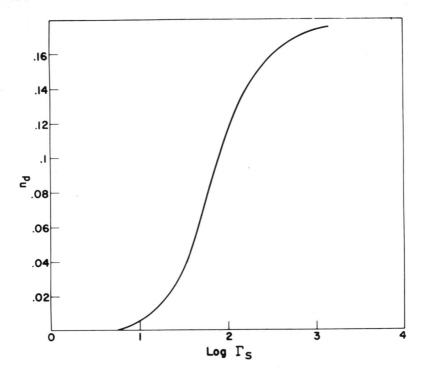

Fig. 3: Dependence of spallation, due to interactions with photons, on shock velocity.

Therefore

$$n_d(\Gamma_s) = \int_o^{\Gamma_s} d\, n_d = \frac{N_o}{n_+} \frac{\sigma_o}{\sigma_+} \frac{A^{7/3}}{Z^2} =$$

$$\int_o^{\Gamma_s} \frac{d\Gamma \int_o^\infty d(h\nu')(2\Gamma_n^2 \pi)^{-\frac{1}{2}} \exp\left\{-\frac{1}{2}\left[(h\nu'-E_m)/\Gamma_n\right]^2\right\} \int_o^{2\pi} \frac{1}{2} n(h\nu)\sin\theta\, d\theta}{\ln\Gamma + 3.4}$$

(10.5)

where σ_o is the spallation cross section 3.3×10^{-27} cm^2 and σ_+ the effective pair slowing down cross section, equation (5.3).

The integration was performed on a computer. In Figure 2, the spectral distribution function $N(h\nu',\Gamma)$ is shown for $\Gamma = 30$, $kT = mc^2$, as well as the cross section behaviour. The integral of $N(h\nu',\Gamma) \, d(h\nu')$ is normalised to unity. The spallation n_d for $A^{7/3}/Z^2 = 1$ is shown in Figure 3, where at $kT = mc^2$, $N_o/n_+ = 1.57$, $\sigma_o/\sigma_+ = 1.28$.

If we use $A^{7/3}/Z^2 = 9.2$ for carbon, 50% of the nuclei will undergo one photonuclear reaction when $(\Gamma_s - 1) kT/mc^2 = 20$. From equations (4.1) and (4.4) we obtain $(\Gamma_s-1) kT \leq 0.89 F^{-0.076}$ which is always less than 20 even at breakout. Therefore, we do not feel that significant spallation will take place.

XI SUMMARY

We have discussed the supernova shock origin theory of cosmic rays and shown that a pre-supernova model of a compact star $\rho_{central} \simeq 2 \times 10^{10}$ gm cm^{-3} gives rise to an ejected matter distribution that is in reasonable agreement in both total energy, energy spectra, and enriched composition with that observed in cosmic rays.

REFERENCES

Arnett, W.D., and Schramm, D.N., Ap. J., 184, L47, 1973.

Barkat, Z., Wheeler, J.C., and Buchler, J.R.,
    Ap. J., 171, 651, 1972.

Bruenn, S.W., Ap. J., 177, 459, 1972.

Chandrasekhar, S., "An Introduction to the Study of Stellar
    Structure", Chicago, Ill. University of Chicago Press, 1939.

Colgate, S.A., J. Geophys. Res., 70, 3161, 1965.

Colgate, S.A., and White, R.H., Ap. J., 143, 626, 1966.

Colgate, S.A., "Acta. Physica Academiae Scientiarum Hungaricae",
    29, Suppl. 1, pp. 353-359, 1970.

Colgate, S.A., Ap. J., 163, 221, 1971.

Colgate, S.A., McKee, C.R., and Blevins, B., Ap. J. 173, L87,
    1972.

Colgate, S.A., and McKee, C.R., Ap. J., 181, 903, 1973.

Colgate, S.A., Ap. J., 187, 321, 1974.

Couch, R.G., and Arnett, W.D., Ap. J., 180, L65, 1973.

Cox, D.P., Ap. J., 178, 159, 1972.

D'Arcy, R., and Colgate, S.A., J. Geophys. Res., 70, 3147, 1965.

Eltgroth, P.G., Phys. Fluids, 14, No. 12, 2631, 1971.

Ezer, D., Cameron, A.G.W., and Icarus, I., Los Alamos Rept. No. 5-6, 1963.

Freedman, D.Z., National Accelerator Laboratory Pub. 73/76-THY, Batavia, Ill., 1973.

Ginzburg, V.L., and Syrovatskii, S.I., "The Origin of Cosmic Rays", New York, N.Y: Pergamon Press, 1964.

Johnson, M.H., and McKee, C.F., Phys. Rev. D., 3, No. 4, 858, 1971.

Kinsey, J.A., Ap. J., 158, 295, 1969.

Noerdlinger, P.D., Ap. J., 165, 509, 1971.

Ono, Y., Sakashita, S., and Ohyama, N., Progr. Theoret. Phys. Suppl. No. 20, p. 85, 1961.

Ramaty, R., Boldt, E.A., Colgate, S.A., and Silk, J., Ap. J., 169, 87, 1971.

Schwarzschild, M., "Structure and Evolution of the Stars", Princeton, N.J: Princeton University Press, 1958.

Weinberg, S., Phys. Rev. Letters, 27, 1688, 1971.

Wilson, J.R., Ap. J., 163, 209, 1971.

Wilson, J.R., UCRL Preprint 75266, Lawrence Livermore Laboratory (to be published in Phys. Rev. Letters),1974.